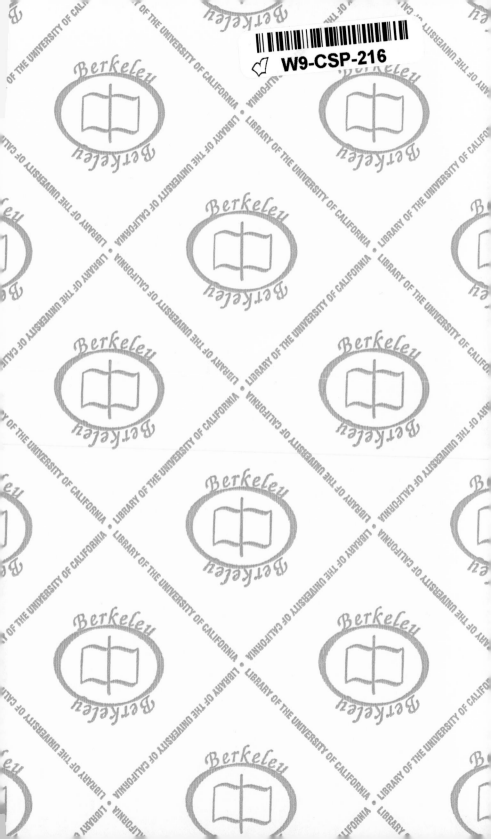
W9-CSP-216

Polymeric Delivery Systems

ACS SYMPOSIUM SERIES **520**

Polymeric Delivery Systems

Properties and Applications

Magda A. El-Nokaly, EDITOR
The Procter & Gamble Company

David M. Piatt, EDITOR
The Procter & Gamble Company

Bonnie A. Charpentier, EDITOR
Syntex Research

Developed from a symposium sponsored
by the Division of Cellulose, Paper and Textile Chemistry
and the Biotechnology Secretariat
at the 203rd National Meeting
of the American Chemical Society,
San Francisco, California,
April 5–10, 1992

American Chemical Society, Washington, DC 1993

CHEM
Sep Lae

Library of Congress Cataloging-in-Publication Data

Polymeric delivery systems: properties and applications/ Magda A. El-Nokaly, David M. Piatt, Bonnie A. Charpentier [editors].

p.   cm.—(ACS Symposium Series, 0097–6156; 520).

"Developed from a symposium sponsored by the Cellulose, Paper and Textile Chemistry Division at the 203rd National Meeting of the American Chemical Society, San Francisco, California, April 5–10, 1992."
Includes bibliographical references and index.
ISBN 0–8412–2624–5
1. Polymeric drug delivery sytems—Congresses.   I. El-Nokaly, Magda A., 1945–   . II. Piatt, David M., 1954–   . III. Charpentier, Bonnie A., 1952–   . IV. American Chemical Society. Cellulose, Paper and Textile Chemistry Division.   V. American Chemical Society. Meeting (203rd: 1992: San Francisco, Calif.).   VI. Series.

RS201.P65P64   1993
615'.19—dc20   92–40099
                                          CIP

The paper used in this publication meets the minimum requirements of American National Standard for Information Sciences—Permanence of Paper for Printed Library Materials, ANSI Z39.48–1984.   ∞

Copyright © 1993

American Chemical Society

All Rights Reserved. The appearance of the code at the bottom of the first page of each chapter in this volume indicates the copyright owner's consent that reprographic copies of the chapter may be made for personal or internal use or for the personal or internal use of specific clients. This consent is given on the condition, however, that the copier pay the stated per-copy fee through the Copyright Clearance Center, Inc., 27 Congress Street, Salem, MA 01970, for copying beyond that permitted by Sections 107 or 108 of the U.S. Copyright Law. This consent does not extend to copying or transmission by any means—graphic or electronic—for any other purpose, such as for general distribution, for advertising or promotional purposes, for creating a new collective work, for resale, or for information storage and retrieval systems. The copying fee for each chapter is indicated in the code at the bottom of the first page of the chapter.

The citation of trade names and/or names of manufacturers in this publication is not to be construed as an endorsement or as approval by ACS of the commercial products or services referenced herein; nor should the mere reference herein to any drawing, specification, chemical process, or other data be regarded as a license or as a conveyance of any right or permission to the holder, reader, or any other person or corporation, to manufacture, reproduce, use, or sell any patented invention or copyrighted work that may in any way be related thereto. Registered names, trademarks, etc., used in this publication, even without specific indication thereof, are not to be considered unprotected by law.

PRINTED IN THE UNITED STATES OF AMERICA

1993 Advisory Board

ACS Symposium Series

M. Joan Comstock, *Series Editor*

V. Dean Adams
Tennessee Technological
University

Robert J. Alaimo
Procter & Gamble
Pharmaceuticals, Inc.

Mark Arnold
University of Iowa

David Baker
University of Tennessee

Arindam Bose
Pfizer Central Research

Robert F. Brady, Jr.
Naval Research Laboratory

Margaret A. Cavanaugh
National Science Foundation

Dennis W. Hess
Lehigh University

Hiroshi Ito
IBM Almaden Research Center

Madeleine M. Joullie
University of Pennsylvania

Gretchen S. Kohl
Dow-Corning Corporation

Bonnie Lawlor
Institute for Scientific Information

Douglas R. Lloyd
The University of Texas at Austin

Robert McGorrin
Kraft General Foods

Julius J. Menn
Plant Sciences Institute,
    U.S. Department of Agriculture

Vincent Pecoraro
University of Michigan

Marshall Phillips
Delmont Laboratories

George W. Roberts
North Carolina State University

A. Truman Schwartz
Macalaster College

John R. Shapley
University of Illinois
    at Urbana–Champaign

L. Somasundaram
E. I. du Pont de Nemours and Company

Peter Willett
University of Sheffield (England)

Foreword

THE ACS SYMPOSIUM SERIES was first published in 1974 to provide a mechanism for publishing symposia quickly in book form. The purpose of this series is to publish comprehensive books developed from symposia, which are usually "snapshots in time" of the current research being done on a topic, plus some review material on the topic. For this reason, it is necessary that the papers be published as quickly as possible.

Before a symposium-based book is put under contract, the proposed table of contents is reviewed for appropriateness to the topic and for comprehensiveness of the collection. Some papers are excluded at this point, and others are added to round out the scope of the volume. In addition, a draft of each paper is peer-reviewed prior to final acceptance or rejection. This anonymous review process is supervised by the organizer(s) of the symposium, who become the editor(s) of the book. The authors then revise their papers according to the recommendations of both the reviewers and the editors, prepare camera-ready copy, and submit the final papers to the editors, who check that all necessary revisions have been made.

As a rule, only original research papers and original review papers are included in the volumes. Verbatim reproductions of previously published papers are not accepted.

M. Joan Comstock
Series Editor

Contents

INDEXES

Preface

POLYMERIC DRUG DELIVERY SYSTEMS are generating great interest in the scientific community. The thorough study of these systems is leading to increasingly frequent discoveries. However, equivalent systems for foods, cosmetics, and herbicide delivery are very hard to find. They have received a much smaller share of the fundamental research, especially from academia, and are still very much an applied art in the hands of the excellent, but unpublished, formulators of industry. Information on such delivery systems is lost as those formulators retire, their work never to be published or documented. Exceptions are found in the burgeoning work with polymeric delivery systems applied to skin. These research efforts started in pharmacology and are slowly moving into the cosmetic field.

The purpose of *Polymeric Delivery Systems* is to pull together the current work in the field as it applies to drugs, cosmetics, food, and herbicides. Although the editors have worked very hard to present a balanced view of the various disciplines, the resulting collection reflects the fact that most published work in the field is still on polymeric drug delivery systems—in itself a strong argument for such a book.

This volume is intended to inform and teach industrial technologists and academic researchers who are concerned with delivery systems for drugs, herbicides, and cosmetics. The advances described here are meant to stimulate new ideas for applying drug delivery technology to systems in cosmetics, food, pharmaceuticals, and pesticides. International authors representing academia, industry, and governmental research centers have provided a balanced perspective in their presentations on advances, basic research, and practical considerations of application techniques.

We put much thought into arranging the flow of the chapters so that they may help build the reader's knowledge in an orderly fashion. This book was not meant to be just a collection of papers for reference purposes, but a stimulus to the thinking process, designed to promote ideas and to present new developments in those areas lacking in the previously mentioned resource material.

The opening chapter outlines the scope of polymeric delivery systems: preparations, properties, and applications. Other topics include controlled release in the food and cosmetics industries, the general properties of various polymers used as delivery systems, and the processing of delivery systems. The contents continue on to applications in drugs, pesticides, and foods, starting with simple systems that can be widely applied (e.g., cyclodextrin—used to deliver water-insoluble drugs—has also been reported to carry cosmetics, herbicides, and flavors for beverages). Following discussion of the kinetics of release (vital to any polymeric delivery system), the book concludes with the more specialized polymeric drug delivery systems applied in vivo, i.e., polysaccharides such as biodegradable chitin, glycosylated dextran, and polylactides. *Polymeric Delivery Systems* presents varied aspects of the latest research on polymeric delivery systems in the hope that it will inspire more exchange, dialogue, and learning among the disciplines.

Acknowledgments

We thank the American Chemical Society's Division of Cellulose, Paper and Textile Chemistry and the Biotechnology Secretariat for sponsoring the symposium and Syntex Research and the Procter & Gamble Company for their partial financial support. Special thanks are due to Procter & Gamble managers Ted Logan, Ph.D. Recruiting Office; Ken Smith and Dave Bruno, Food and Beverage Technology Division; Robert Boggs and Ray Martodam, Health Care Technology Division; and Lynda Sanders and Boyd Poulsen of Syntex Research.

We thank each of the contributing authors for their cooperation, without which there would not have been a book. Many thanks to the editorial staff of the ACS Books Department, especially Anne Wilson, for cheerful and professional support. We also acknowledge with thanks the secretarial support of Peggy Sehlhorst.

MAGDA A. EL-NOKALY
The Procter & Gamble Company
Cincinnati, OH 45239–8707

DAVID M. PIATT
The Procter & Gamble Company
Cincinnati, OH 45232

BONNIE A. CHARPENTIER
Syntex Research
Palo Alto, CA 94303

October 19, 1992

Chapter 1

Polymer Delivery Systems Concepts

I. C. Jacobs and N. S. Mason

Department of Chemical Engineering, Washington University,
One Brookings Drive, St. Louis, MO 63130

Examples of successful applications of controlled release are mentioned and commonly used concepts are defined. The chapter describes how polymers are used in devices and how their properties affect device performance. Processes of wide applicability including those covered in succeeding chapters are summarized.

The task of controlled release is deceptively simple: get the right amount of the active agent at the right time to the right place. Controlled release is a term that represents an increasing number of techniques by which active chemicals are made available to a specified target at a rate and duration designed to accomplish an intended effect. In their most elegant implementation, these systems can mimic processes of living cells such as secretion of hormones or enzymes. Many other terms besides "controlled" have been used to describe somewhat different delivery system concepts from "continuous" release to "timed" release. Some of the terms are defined more precisely by Ballard (*1*). Only a few examples will be cited in the paragraphs following.

Controlled Release of Drugs. Controlled release is often used to extend the time the effective therapeutic dose is present at the target from a single administration, and to avoid or minimize concentrations that exceed therapeutic requirements. It also can decrease the needed dose of an expensive active ingredient. Protecting certain tissues is often desirable, e.g. the stomach from irritation. Targeting tissues may be desirable, avoiding toxic effects. This can make the administration of the drug less invasive. Taste-masking of a drug can improve patient compliance. Incompatibilities between ingredients can be mitigated. A few commercial examples may be mentioned: Occusert, a device for releasing pilocarpine, a drug to treat glaucoma, to the eye (*2,3*), an implantable osmotic pump capable of delivering a nearly constant 20 mg/hr of solution to the body for almost a day (*4*) and Norplant, a long-term contraceptive implant (*5*).

Controlled Release Pesticides. Controlled release can increase the effectiveness of an agent, and its specificity. It can decrease the possibility of damage to the environment.

0097–6156/93/0520–0001$06.00/0
© 1993 American Chemical Society

Penncap M and E, represent examples of microencapsulated forms of methyl and ethyl Parathion (6). Systems have been developed for releasing pheromones to confuse insects (7), collars to keep fleas away from pets (8), even keep mollusks from the hulls of ships (8).

Controlled Release Fertilizers. Controlled release can decrease the number of applications. It can make one application last longer. A reduction in run-off into rivers or aquifers may be another benefit.

Special Control Release Applications. "Carbonless carbon paper" was the first product involving microencapsulation. A latent dye, crystal violet lactone is contained in 5 to 25 micron microcapsules that are coated on the back of a page. The action of a sharp pencil, typewriter, or impact printer ruptures the microcapsules, releasing the lactone to react with an acid clay present on the copy sheet, activating the color (9).

Side effects in the treatment of diabetes would be decreased, if encapsulated implants of islets of Langerhans from the pancreas could be used. The islets could originate from a different species. Encapsulation would keep the islets from being destroyed by the immune system of the host, allowing the cells to release insulin in response to the concentration of glucose (10).

Improvements in oil well stimulation with hydraulic fracturing fluids have been realized with the use of controlled release viscosity breakers (11). Increasing the likelihood of germination of seeds under adverse conditions is another application of controlled release. Because controlled release is a rapidly changing field, new ideas are being generated and applications are being evaluated every day.

Functions of Polymers in Controlled Release.

Polymers are uniquely suited as materials of construction for delivery systems because their permeability can be modified and controlled. They can be shaped and applied relatively easily by a large variety of methods. Active ingredients and property modifiers can be incorporated either physically or chemically. In general, polymers have little or no toxicity. Despite the diversity of applications, they principally serve as membranes or envelopes, as matrices in which the active ingredient is dispersed or dissolved, or as carriers which are chemically attached to the active ingredient. Not all controlled release devices use polymers explicitly. For example liposomes do not.

Delivery Systems Terms.

Devices can range in size from as small as one molecule to coated tablets and boluses used in cattle.

Availability, or bioavailability is an important property of most drug-containing devices. Tests in glass-ware (*in vitro*) or in the living system (*in vivo*) must be conducted to ascertain that the active ingredient is available under specified conditions.

Biodegradable refers to polymers that under certain conditions undergo a decrease in molecular weight eventually disintegrating or dissolving in the medium. It is most often applied to polymers and copolymers of lactic, glycolic, hydroxybutyric acids as well as poly(ε-caprolactone). For these substances the degradation is hydrolytic and no enzymes are involved. Kinetics generally follow first order (*12*, *13*) and are often insensitive to pH. Natural polymers such as starch and cellulose are also biodegradable and enzymes are involved in these cases. Polyphosphazenes, polypeptides, and proteins have also been proposed for drug delivery. Other polymers mentioned as biodegradable were, poly(dihydropyrans), poly(acetals), poly(anhydrides) (*14*), polyurethanes, and poly(dioxinones).

Biocompatible devices are devices that can be applied without causing undesirable effects in living systems.

Bioabsorbable devices are those which are degraded by a living system and can be utilized or metabolized by it.

Erodible systems are designed to control the release of the active substance by erosion of the matrix.

Zero order release means the active substance is released at a constant rate. By analogy to reaction kinetics, the rate of release, $-dc/dt = k$. It can be approximated by a reservoir device containing a saturated solution of the active ingredient surrounded by a rate-limiting membrane. This requires a supply of undissolved active ingredient in the device and a non-changing or zero sink condition (*15*).

First order or pseudo-first order means the release rate is proportional to the concentration remaining in the device i.e., with a rate, $-dc/dt = kc$. This implies a decrease in rate with time. It would be approximated by a device in which the concentration decreased as the amount remaining in the device decreased.

Burst effect refers to the tendency of some devices to release the active substance more rapidly during some period, usually the initial test period, than during the steady-state, or a later period.

Time lag is observed when the rate limiting membrane must first establish a concentration gradient before releasing at the designed rate. Therefore some time elapses before the release rate reaches its designed value. Mathematical expressions of this and the above phenomenon are given in (*15*).

Reservoir devices consist of a drug or other active agent enclosed within an inert controlling membrane. Examples would be a tube filled with the active substance, where the wall of the tube would serve as the limiting membrane, a sphere of the active substance coated with a film controlling the diffusion of the active substance, or a slab of the active substance closed off from the medium by a film which controls the diffusion.

Matrix devices in which the active drug is dispersed throughout the polymer are called monolithic devices or monoliths by a number of authors. One can distinguish between monolithic devices in which the active ingredient is dissolved, dispersed, located in connected pores, or granular. The equations describing the time behavior of such devices predict that initially the fraction released varies as time$^{1/2}$ and the rate of release is proportional to time $^{-1/2}$. Baker and Lonsdale (15) in their classic paper, treat each of the cases separately including the different geometries such as slab, sphere and cylinders. As time increases, the release rate is better approximated by an exponential decay (Late time approximation). A more recent review of this material by Roseman may be found in (16).

Targeting of drugs is a concept that is as old as the dream of the magic bullet of Paul Ehrlich (17). Thies (18) distinguishes between active and passive targeting. Passive targeting takes advantage of an existing body process such as the rapid concentration of particles smaller than 1 micron in the liver or spleen after intravenous injection. In active targeting, the natural tendency of the body to distribute the active substance is altered. The drug or carrier, because it has a special affinity e.g., by certain molecular interactions such as a lock and key mechanism, interacts with specific cells. If this could be implemented, it could improve cancer chemotherapy considerably.

Enteric coatings, coatings which are less soluble in acidic aqueous solutions than neutral ones, provide a means for a drug to by-pass the stomach and only become available in the intestine. This is a method to direct rather than target a drug. Many enteric coatings are based on polymers containing phthalic acid residues attached to cellulosic or vinyl polymer chains. Reverse enterics, on the other hand, do not dissolve in neutral media such as the mouth, but dissolve very rapidly in an acid medium such as the stomach.

Polymer Properties that Affect the Release of Active Substances.

Consider a polymer membrane and a diffusing active substance in solution. The amount of the substance which diffuses per unit time across the membrane at steady-state is proportional to the diffusion coefficient and to the concentration difference across the membrane measured on each side of the membrane just inside the membrane. (It is also proportional to the area and inversely proportional to the thickness of the membrane). Because the concentration inside the membrane is not known and is often much different from the concentration in solution, the concentration difference normally measured in the solution must be multiplied by the distribution coefficient. The distribution coefficient is the ratio of the solubility of the agent in the membrane to its solubility in the external medium. The permeability is the product of the diffusion coefficient and the solubility coefficient. Its dimensions are the same as the diffusion coefficient, $(length)^2/time$.

It must be pointed out, as Baker and Lonsdale (*15*) have done, that the relatively low diffusivity and solubility of large molecules in polymers often constrain the release rates attainable with delivery systems which employ unplasticized dense membranes. Even silicone rubbers which have high diffusion coefficients and solubilities for many organic active molecules of interest, rarely permit maximum daily fluxes greater than 2.5 mg/cm^2 through a 0.1 mm thick membrane. For less permeable polymers, fluxes two orders of magnitude lower are typical. This is the reason why applications have favored silicone rubbers and only very potent drugs such as steroids. Some values of these parameters are listed in Table I. It illustrates the range of values encountered. The diffusion coefficient and the solubility (i.e. distribution coefficients) have very different dependencies in the polymer phase. This will be discussed below.

Table I. Selected Values of Diffusion Coefficients
and Partition Coefficients in Polymers

Solute	Polymer	Diffusion Coefficient cm^2/sec	Partition Coefficient (polymer/water)
Acetophenone	Polyethylene	3.55×10^{-8}	3.16
Chlormadione Acetate	Silicone Rubber	3.03×10^{-7}	82
Estriol	Polyurethane Ether	2×10^{-9}	133
Fluphenazine	Polymethyl-methacrylate	1.74×10^{-17}	
Hydrocortisone	Polycaprolactone	1.58×10^{-10}	
17α-hydroxy-progesterone	Silicone Rubber	5.65×10^{-7}	0.89
Progesterone	Silicone Rubber	5.78×10^{-7}	45 to 60
Salicylic Acid	Polyvinyl Acetate	4.37×10^{-11}	

SOURCE: Adapted from ref. 16 with permission. Copyright 1980 CRC Press.

Hydrophilic polymers, or polymers that can swell in water or other solvents, are not subject to the limitations of the dense polymers in the above table and will be covered in a later section. Porous polymers also can have higher fluxes.

Diffusion Coefficient. Diffusion coefficients are much more sensitive to molecular weight of the diffusing substance in polymers than they are in low viscosity liquids, Figure 1. Diffusion coefficients depend on the mobility of segments of the polymer chain. Table II summarizes the trend of several factors on the diffusion coefficient. Diffusion of a molecule in a polymer requires the cooperative movement of several polymer chain segments. This is why substances have higher diffusion coefficients in polymers that have lower interchain forces such as silicone rubber and natural rubber in comparison to polystyrene. The lower the molecular weight, or more accurately, the size of the diffusing substance, the smaller is the required segmental motion. Therefore it is more likely for the smaller molecule to find sufficient space for diffusion. For the same reasons, an increase in crystallinity of the polymer will decrease the diffusion coefficient of the substance as will an increase in the degree of cross-linking.

Plasticizers or solvents that increase the segmental mobility lead to increases in diffusion coefficients, whereas fillers which decrease the segmental mobility lead to decreases in the diffusion coefficient. The diffusing substance itself, if it plasticizes the polymer, will increase its own diffusion coefficient. Diffusion coefficients are always higher above the glass transition, T_g, of the polymer than below. (The glass transition temperature is the temperature at which the polymer changes from a glassy to a rubbery state). The activation energy for diffusion decreases above the T_g as well. Whether above the T_g or below, the diffusion coefficient increases with temperature. The diffusion coefficient can be increased by copolymerization, e.g. by random insertion of groups especially if the groups increase the flexibility or decrease the crystallinity.

Table II - Factors which Change the Diffusion Coefficient

Increases in Factor Listed Below	Effect on Diffusion Coefficient
Interchain Forces	-
Segmental Mobility	+
Permeant Molecular Weight	-
Polymer Crystallinity	-
Plasticizer	+
Copolymerization	+
Temperature	+
Glass Transition	-

Solubility. The solubility of substances in polymers is very sensitive to small changes in the molecule such as the addition of a substituent. The most widely used attempts to predict solubilities in polymers have been the use of solubility parameters (*19*). The closeness of the solubility parameter between the active substance and the polymer

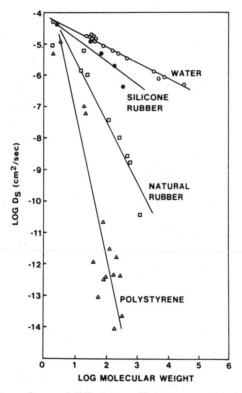

Figure 1. Dependence of diffusion coefficients on molecular weight. (Reproduced with permission from ref. 16. Copyright 1980 CRC Press, Inc.)

would favor compatibility and solubility. However, the degree of hydrogen bonding and polarity effects are also important (20).

Structural Considerations.

Hydrophobic Polymers. When active agents are dispersed in polymers the geometry may play a role regarding availability and release rate. The first models were tortuosity models (21). Pores were assumed to be of uniform cross-section but tortuous. Therefore, a molecule must travel a greater distance than if the pores were straight. Constriction or corrugation models assume that there are places where the pores become narrower. This geometry may more closely approximate porous polymers. The arrangement of pores can also affect the release, i.e. whether the pores are isolated or connected. As the porosity increases more of the pores connect to each other and availability of the agent increases rapidly. The percolation threshold, which is the value of the porosity at which the availability increases rapidly, ranges from 50 to 70 % for two dimensional matrices and from 20 to 43 % for three dimensional ones (21). This depends to some degree on the pore shape.

Hydrophilic Polymers and gels are used extensively in controlled release devices. Gelatin is a versatile hydrophilic polymer. It is soluble in water above 30 to 35 C but is a gel below this temperature. It can also be cross-linked so that it is no longer soluble in water at any temperature. Other substances which gel on temperature decrease include agar, agarose, and iota-carrageenan. (22). Hydrophilic polymers that can be cross-linked ionically include alginates, and low methoxyl pectins with calcium ions, and kappa-carrageenan with potassium or calcium ions. All these substances have been used in microcapsules as well as other controlled release devices.

Hydrogels, e.g., cross-linked copolymers of hydroxyethyl methacrylate and ethylene dimethacrylate, have been studied extensively for medical applications because of their biocompatibility (23). Hydrophilicity can be controlled by copolymerization with methyl methacrylate or other less water soluble monomers.

Hydrophilic polymer properties can be used to engineer devices which can respond to their environment and thus, potentially could be used as elements in feed-back control loops. Cross-linked gels that contain ionic groups can show substantial changes in swelling with small changes in pH (24). Such a gel can respond to the concentration of glucose if it also incorporates glucose oxidase. Glucose oxidase catalyzes the oxidation of glucose to gluconic acid changing the pH. Okano (25) constructed interpenetrating gels from N-acryloylpyrrolidine and polyoxyethylene which showed a change in equilibrium swelling in water from 150 % to 500 % when the temperature was cycled from 10 to 30°C. The rate of release of a drug more than doubled in response to the same step change in temperature. Controlled release devices can be fabricated also from polymers which are normally water soluble but dissolve more slowly than the active ingredient. Such polymers can also be cross-linked by a variety of methods. Diffusion coefficients of water swollen polymers are much higher than those in non-swollen polymers. In fact they approach diffusion coefficients in solution. A simplified equation from (26) may be useful to correlate data on diffusion of small molecules in gels.

$$\ln(D/D_w) = (kA_s)(1-H)/H$$

where D and D_w are the diffusion coefficients in the gel and in water respectively, A_s is the effective cross-section of the solute molecule, H is the volume fraction of water in the gel, and k is a constant at constant temperature.

Processes for Achieving Controlled Release.

Methods for the manufacture of polymer delivery systems may be divided into physical and chemical processes depending upon whether the active ingredient becomes covalently attached during the process. Physical methods are most often used to place the active ingredient into a reservoir or matrix form. Of the many processes described in the literature, the following physical processes are used most often:

Spray coating
  Pan Coating
  Air Suspension coating

Wall deposition from solution
  Complex coacervation
  Polymer-Polymer Induced Coacervation

Interfacial polymerization
  Solid-Gas
  Liquid-Liquid
  Interfacial Polycondensation

Matrix Solidification
  Spray Drying
  Spray Chilling
  Solvent Evaporation from Emulsions
  Solvent Evaporation in Castings
  Castings or Particle Formation using Polymerization
  Flavor and Fragrance Trapping

Centrifugal Methods
  Annular Jet Encapsulation
  Suspension Separation

Spray Coating. Particles to be coated are put in motion either by rotation as in a drum, or coating pan, or by a carrier gas as in a fluidized bed (also called air suspension coating). The coating as a solution, latex, or low-viscosity melt is sprayed onto or into the bed of particles. The motion of the particles needs to be sufficient to prevent particle agglomeration while the solvent is drying. Coatings are built up in layers on the surface of the particles and the quality of the barrier will be affected by the nature of the coating formulation as well as process conditions. Spraying rates, coating droplet size, bed temperature, humidity, particle motion, and surface wettabilities are all important factors. When latexes are used as alternatives to organic solutions of polymers, post-heat treatment above the glass transition temperature is often used to improve film formation. Plasticizers which lower the T_g are also used for the same purpose. For melts, one is limited to low viscosities to obtain sufficiently fine droplets.

Pan coating is among the most widely used procedures for controlled release, especially pharmaceuticals. The current pan coating designs evolved from confectionary pans invented approximately 140 years ago (27). Newer designs improve mixing efficiency with baffles and improved drying air flow across or through the particle bed with the most well known variants among what are known as side vented coating pans. Lack of reproducibility and operator error plague the process.

Air suspension or fluidized bed coating may place the spray nozzle on top of the bed, on the side, or near the bottom. The Wurster process (28), which places the nozzle in a draft tube near the bottom of the vessel, can often be fully automated (Figure 2). Coating uniformity is excellent, and irregular small particles can be used with a variety of coating materials. Particles down to approximately 100 microns can be coated.

Wall Deposition from Solution.
Complex Coacervation-phase separation was developed by NCR in 1954 (9). Gelatin and gelatin-gum acacia were described for the wall of microcapsules containing the dye used in the manufacture of carbonless carbon paper.

The technique has been adapted for the controlled release of many materials. For example, an organic phase is dispersed in a dilute aqueous gelatin solution and a second water soluble polymer is added that can associate with the gelatin, e.g., a polycarboxylic compound, or possibly a polysulfonic acid. The pH of the solution is adjusted below the isoelectric point of the normally amphoteric gelatin making it cationic and thus causing an association of the polymers to occur and the appearance of a polymer-rich phase or "coacervate". The coacervate droplets must wet the surface of the core particles before coalescing into a continuous coating. Poor wetting of particles and agglomeration of the capsules during formation and drying are common problems. Coacervation also can be induced by conditions or materials that influence polymer hydration such as temperature, water soluble solvents, or by addition of large amounts

TOP SPRAY COATER

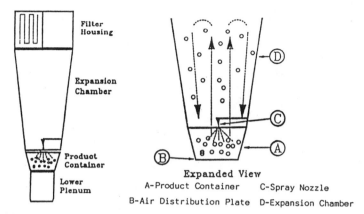

Filter Housing

Expansion Chamber

Product Container

Lower Plenum

Expanded View

A-Product Container   C-Spray Nozzle

B-Air Distribution Plate   D-Expansion Chamber

WURSTER —BOTTOM SPRAY— COATER

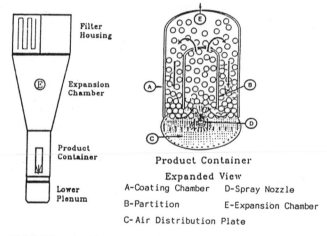

Filter Housing

Expansion Chamber

Product Container

Lower Plenum

Product Container

Expanded View

A-Coating Chamber   D-Spray Nozzle

B-Partition   E-Expansion Chamber

C- Air Distribution Plate

ROTOR —TANGENTIAL SPRAY— COATER

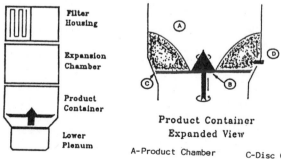

Filter Housing

Expansion Chamber

Product Container

Lower Plenum

Product Container
Expanded View

A-Product Chamber   C-Disc Gap or Slit

B-Variable Speed Disc   D-Spray Nozzle

Figure 2. Spray Coating Apparatus. (Reproduced by permission of Glatt Air Techniques)

of salts such as sodium sulfate. The resulting gelatin coatings can be cross-linked with formaldehyde or glutaraldehyde.

Polymer-Polymer and Non-Solvent Induced Coacervation. Analogous processes exist for the encapsulation of water soluble solids using organic phase separation (*29*). In these processes, the core particles are suspended in an organic solution of a polymer such as ethyl cellulose or cellulose acetate butyrate that is induced to separate as a coacervate either through cooling, slow addition of non-solvents for the polymer, or the addition of a more soluble polymer, called a phase inducer.

Interfacial Polymerization. Barriers can be formed around liquid droplets or on solid surfaces by causing a chemical reaction to occur at the phase boundary be it a liquid-liquid, solid-liquid, or solid-gas.

Solid-Gas. Free radical condensation of gas phase para-xylylene was developed as a coating process by Union Carbide Corp. during the 1960's (*30*). Solid para-xylylene is sublimed at 500°C under vacuum to produce free radicals that then undergo polymerization on cold surfaces. Uniform walls of linear poly (p-xylene) have been deposited on the surfaces of a variety of particulate solids.

Liquid-Liquid encapsulation may be applied to any water insoluble liquid, solid or gas. Because the conditions are so mild it has been used for the encapsulation of living cells. The method involves the formation of droplets of a sodium alginate solution containing the cells that are then placed into a solution of calcium chloride converting the drops into gel beads. Encapsulation of viable insulin producing islet cells for transplantation into diabetics was first disclosed by Lim and Sun (*10*). The washed gel beads containing the islets are then placed in a solution of polylysine for sufficient time to form the association polymer of lysine and alginate on the gel bead surface. The capsules are then placed in a sodium citrate solution to extract the calcium from the bead interior and reliquefy the droplet.

Interfacial polycondensation is widely used for the formation of barriers around organic liquid droplets. Uniform walls of polyureas, polyamides, and mixed systems of the two are widely known. The process involves a polyfunctional electrophile such as a diisocyanate or diacyl chloride dissolved in an organic phase that is then dispersed in an aqueous phase containing the active substance. A water soluble polyfunctional nucleophile is added and the reaction occurs at the boundary of the phases. One of the two reactants needs to be at least tri-functional for cross-linking of the walls. These barriers can have good strength and integrity.

Matrix Solidification. Polymer delivery systems may not necessarily consist of small particles but also may be slabs or cylinders protecting an active ingredient. Likewise, the processes that involve the formation of small particles protecting an active ingredient may not be in the form of a single core surrounded by a barrier polymer.

Products which consist of a dispersed phase, either solid or liquid, contained or imbedded in a particle can be termed matrix particles. These are often less costly to manufacture. Matrix devices can be formed by casting, extrusion or otherwise forming monoliths that contain a dispersed phase.

Spray Drying. This process is widely used to produce free flowing granules for the food, pharmaceutical, and agrichemical industries. The material to be dried is suspended or dissolved in a liquid that is atomized by either nozzles or spinning discs at the top of a chamber (Fig. 3). In most spray dryers hot air flows concurrently with the particles as they fall through the tower. The liquid is evaporated rapidly and the overall drying time is short with production rates varying from a few kilos per hour to tons per hour. Water soluble polymers such as starches, cellulose derivatives, and gums are commonly used. The use of organic solvents is being replaced with water-based dispersions (latexes) because of environmental health and safety concerns. Much work has been done to improve the stability of the oil in water emulsions prepared for drying with the use of modified polymers (*31*). Latex compositions and plasticizers have been investigated to improve product performance (*32*).

Spray Chilling uses similar equipment as spray-drying but does not require evaporation. The liquid or solid active ingredient is dispersed into a molten continuous phase that is then atomized and cooled to form solid particles.

Solvent Evaporation from Emulsions. Here, the active ingredient is dissolved in a polymer solution that is then dispersed in a second immiscible phase. Conditions of pressure and temperature are adjusted to slowly evaporate the dispersed solvent phase thus converting the mixture into a dispersion of solid matrix particles containing the active ingredient (*33*). Control of the dispersion step is critical to getting the desired particle size. Particle agglomeration during the time-consuming evaporation step is often a problem as the particles can go through a sticky stage in the process.

Solvent Evaporation in Castings. Polymer films and slabs can be cast routinely by dissolving the polymer of interest in a suitable solvent along with the active ingredient and then forming the film or slab using a Gardner knife or an appropriate mold. The solvent is allowed to evaporate slowly. Some polymers, which swell in water, are loaded with water soluble ingredients by soaking the device in a concentrated solution of the active agent, removing the swollen polymer from the solution, and surface washing of the device. This latter technique allows only low loadings of active agent, often under 20 %.

Castings or Particle Formation using Polymerization Reactions. Monomer mixtures or pre-polymers can be mixed with an active ingredient and poured into molds or otherwise formed into droplets. The polymerization reactions are initiated with heat, ionizing radiation, or catalysts and when completely cured become an effective matrix

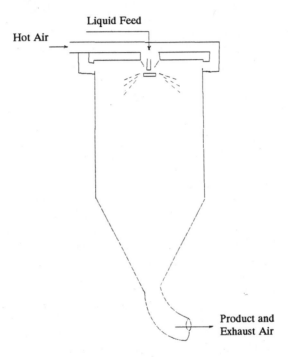

Figure 3. Conventional Cocurrent Spray-Dryer.

for controlled release. The active ingredient generally must be inert to the polymerization process.

Flavor, Fragrance or Pesticide Trapping. Processes have been developed (*34*) that allow trapping of hydrophobic liquid droplets in a starch matrix. The starches are chemically derivatized, then mixed with the oil droplets and finally dried and ground. In other processes flavor or fragrance/starch mixtures are extruded, subdivided into particles and dried.

Centrifugal Processes.
   Annular Jet Encapsulation. Southwest Research Institute, in San Antonio, Texas pioneered a technique for microencapsulating droplets from 300 to 1500 microns (*35*). In this method, the core liquid is pumped through an orifice to form a jet. The coating material is forced through a concentric annular space around the core jet with the velocity matching the velocity of the internal jet. As the concentric jet moves away from the orifice, the stream breaks up into droplets. The droplets are hardened by cooling or solvent removal, or by complexation with calcium in the case of alginate walls.

   Suspension/Separation Process. In this process, developed at Washington University in St. Louis, Missouri, the core particles are first suspended in a coating liquid that is either a melt or a polymer solution (*36*). The suspension is delivered to a rotating disk of specific design that separates the now coated particles from the excess coating. The rotational speed of the disk is adjusted so that the coating material forms a film on the edge of the disk that is thinner than the diameter of the cores. When the coating film atomizes, it forms droplets smaller than the coated particles that are then cooled or dried to give the product. The smaller particles of excess coating are recycled after separation. The method is useful for particles in the range of 30 micron to 2 mm. Since the process is rapid and continuous, the throughputs can be very high with low operating costs. The same plant also may be used to produce granules up to 600 micron.

Chemical Processes.

The attachment of biologically active molecules to polymer backbones for controlled release has been extensively studied since the early 1960's (*37*). A typical synthesis involves an active ingredient that contains a reactive functional group that can form a bond at sites along the polymer. The bonds are often ester, amide, or anhydride. Hydroxy or carboxy containing natural polymers such as cellulose, chitin, lignin and hyaluronic acid have been studied. Synthetic polymers such as, polyvinyl alcohol, polyethylenimine, and acrylic or methacrylic copolymers have also been used. These processes also can offer the intriguing possibility that both drug and targeting functionalities can be attached to the polymer chain.

Literature Cited

1. Ballard, B. E. In Sustained and Controlled Release Drug Delivery Systems; Robinson J. R. Ed.; Drugs and the Pharmaceutical Sciences Marcel Dekker: New York 1978 Vol 6; pp. 3-6
2. Chandrasekaran, S. K.; Benson, H.; Urquhart, J. In Sustained and Controlled Release Drug Delivery Systems; Robinson, J. R., Ed.; Marcel Dekker: New York 1978, pp 569-72.
3. Yates, F. E.; Benson, H.; Buckles, R.; Urquhart, J.; Zaffaroni, A. In Advances in Biomedical Engineering, Brown, J. H.; Dickson, J. F., III, Eds; Academic Press, New York, 1975   Vol 5, p 15.
4. Theeuwes, F. In Controlled Release Technologies: Methods, Theory and Applications; Kydonieus, A. F., Ed.;   CRC Press New York 1980 Vol. 2; pp 195-205.
5. Woutersz T. B. Int. J. Fertil. 1991 36 Suppl 3 pp. 51-56
6. Koestler R. C. In Controlled Release Technologies: Methods, Theory, and Applications; Kydonieus A. F., Ed.; CRC Press: Boca Raton, FL, 1980, Vol 2; p 130.
7. Brooks, T. W. In Controlled Release Technologies: Methods, Theory, and Applications; Kydonieus A. F., Ed.; CRC Press: Boca Raton, FL, 1980, Vol 2; p 165.
8. Cardarelli, N. F. In Controlled Release Technologies: Methods,Theory, and Applications; Kydonieus, A. F., Ed.; CRC Press: Boca Raton, FL, 1980 Vol 1; pp 58-59.
9. Green, B. K.; Schleicher L. S. U.S. Pat. 2,800,457(July 23, 1957) to NCR Co.
10. Lim, F.; Sun A. M. Science 1980, 210, 908.
11. Walles, W. E.; Williamson T. D.; Tomkinson D. L.   U.S. Pat. 4,741,401.
12. Schindler A.; et al. Biodegradable Polymers for Sustained Drug Delivery, Continuing Topics in Polymer Science; Plenum Press 1977 Vol 2.
13. Mason, N. S.; Miles C. S.; Sparks R. E. In Biomedical and Dental Applications of Polymers; Gebelein, C. G.; Koblitz, F. F., Eds.; Polymer Science and Technology, Plenum: New York 1981 vol 14; pp 279-291.
14. Laurencin, C.; Domb A.; Langer R. In Proceedings of the 17th International Symposium on Controlled Release of Bioactive Materials; Lee, V. H. L., Ed.; Controlled Release Society: Lincolnshire, IL 1990 pp 158-159.
15. Baker R. W.; Lonsdale H. K. In Controlled Release of Biologically Active Agents; Tanquary A. C.; Lacey R. E. Eds.; Advances in Experimental Medicine and Biology, Plenum: New York, 1974 Vol 47; pp 15 -71.
16. Roseman, T. J. In Controlled Release Technologies: Methods, Theory, and Applications; Kydonieus A. F., Ed.; CRC Press: Boca Raton, FL 1980 pp 21-54.
17. Encyclopedia Britannica; 15th ed, Encyclopedia Britannica Inc. Chicago IL, 1991 Vol 4 p 395.

18. Thies, C. In Controlled Release of Drugs: Polymers and Aggregate Systems; Rosoff, M., Ed.; VCH: New York, 1988 pp 97-123.
19. Hildebrandt, J. H.; Scott, R. L. The Solubilities of Nonelectrolytes; Reinhold: New York 1950.
20. Burrell, H.; Offic. Diag. Feder. Paint Technol 1957 29 pp 1069-1076 (1957).
21. Siegel, R. A. In Controlled Release of Drugs: Polymers and Aggregate Systems; Rosoff, M. Ed.; VCH: New York, 1988 pp 1 - 51.
22. Handbook of Water-Soluble Gums and Resins; Davidson, R. L., Ed.; McGraw-Hill, New York 1980.
23. Hydrogels for Medical and Related Applications; Andrade, J. D., Ed.; ACS Symposium Series No. 31, ACS: Washington, DC, 1976.
24. Park, T. G.; Hoffman, A. S. In Proceedings of the 17th International Symposium on Controlled Release of Bioactive Materials; Lee V. H. L., Ed.; Controlled Release Society: Lincolnshire, IL, 1990 pp 112-113.
25. Okano, T. In Proceedings of the 17th International Symposium on Controlled Release of Bioactive Materials; Lee V. H. L., Ed.; Controlled Release Society: Lincolnshire, IL 1990   pp 19-20.
26. Yasuda, H.; Peterlin, A.; Colton, C. K.; Smith, K. A.; Merrill, E. W. Makromol. Chem. 1969, 126 pp 177-186.
27. Colorcon, Inc, A Seminar on Film-Coating Technology; Colorcon, Inc, West Point PA 19486
28. Wurster, D. E. U.S. Pat. 2,648,609 and 2,799,241 To Wisconsin Alumni Research Foundation.
29. Sparks, R. E. In Encyclopedia of Chemical Processing and Design; Mc Ketta J. J.; Cunningham W. A. Eds.;   Marcel Dekker: New York, 1989, Vol 30 pp 162-180.
30. Gorham, W. F. J. Poly. Sci. Part A-1 1966, 4, 3027.
31. Marotta, N.G; Boettger, R.M.; Nappen, B.H.; Szymanski, C.D. U.S. Patent 3,455,838.
32. Aquacoat - Aqueous Polymeric Dispersion; FMC.
33. Redding, T.W.; Schally, A.V.; Tice, T.R.; Myers, W.E. Proc.Natl.Acad.Sci. 1984, 81, 5845.
34. Shasha, B.S. In Controlled Release Technologies: Methods,Theory, and Applications; Kydoneius, A.F., Ed.; CRC Press: Boca Raton, Florida, 1980, Vol.2; pp.207-224.
35. Somerville, G.R. U.S. Pat. 3,015,128; 3,310,612; 3,389,194.
36. Sparks, R.E.; Mason, N.S. U.S. Pat. 4,675,140; 5,100,592.
37. McCormick,C.L. et al In Controlled Release of Pesticides and Pharmaceuticals; Lewis, D.H., Ed.; Plenum Press: New York, NY 1981; pp 147-158.

RECEIVED October 5, 1992

Chapter 2

Polymeric Drug Delivery Systems
An Overview

Patrick Sinko[1] and Joachim Kohn[2,3]

[1]Department of Pharmaceutics, College of Pharmacy, and [2]Department
of Chemistry, Rutgers University, New Brunswick, NJ 08903

Although already in 1971 Yolles et al. discussed the design of polymeric devices for the delivery of drugs (1) and a patent was filed by
Boswell and Scribner on the use of polylactic acid in drug delivery
systems (2), two widely cited papers by Yolles et al. from the year
1973 (3,4) are often regarded as the starting point of the
development of polymeric drug delivery systems. From the modest
beginnings in 1973 when drugs were simply mixed into a polymeric
matrix, an avalanche of theoretical and practical advances led to the
rapid development of a large number of distinct system configurations
and device designs. Today, drug delivery devices can be conveniently
divided into two large categories relating to "controlled drug release
systems" and "targeted drug delivery systems."

The definition of controlled release versus targeted drug delivery
is based on the relationship between the site of drug release and the
site of drug action. Controlled release systems deliver drug into the
systemic circulation at a predetermined rate. Thus, the site of drug
release and the site of drug action are not the same. Targeted delivery systems, on the other hand, release medications at or near the
site of action. An advantage of targeted drug delivery is that high local
concentrations of drug can be achieved, since the drug is delivered
predominantly to the site of action rather than being distributed
throughout the whole body.

It is noteworthy that there is surprisingly little practical overlap
between "controlled release" and "targeted delivery". For the successful development of a controlled release device, a water insoluble,
inert polymer is usually needed. Device formulation is predominantly
an engineering problem requiring detailed knowledge of physico-
chemical phenomena such as diffusion, swelling, erosion and/or

[3]Corresponding author: Department of Chemistry, Rutgers University, P.O. Box 939,
Piscataway, NJ 08855–0939

0097–6156/93/0520–0018$07.00/0
© 1993 American Chemical Society

degradation. On the other hand, for the successful development of a targeted (pendent chain type) delivery system a water soluble, easily functionalized polymer is usually needed and an intimate knowledge of the biological and/or physiological interactions that lead to the desired targeting effect is indispensable.

As a scientific field, the science of drug delivery is highly inter-disciplinary. Almost all of the major disciplines related to chemistry, biology, pharmaceutics, pharmacology, and medicine have a bearing on drug delivery (Figure 1). This diversity makes it almost impossible to collect a set of research papers that is representative of the field as a whole and, *at the same time,* internally cohesive. This point is clearly illustrated by the selection of manuscripts included in this volume which has been designed to provide a representative overview of the wide range of applications of polymeric delivery systems.

Particularly noteworthy is the central role occupied by polymer chemistry within the field of "drug delivery". The rapid advances made in the design and development of new drug delivery systems were fueled predominantly by advances made in polymer chemistry. Implantable, or insertable controlled release devices, targeted drug carriers, and transdermal systems all contain polymeric materials that were often specifically formulated for these applications. On the other hand, the obvious need to find new materials for the growing research effort in drug delivery provided the impetus (and often the financial support) for the development of a wide range of new, degradable polymers. A more detailed overview of some polymers used for drug and peptide delivery is provided in a separate chapter.

Over the last 10 to 20 years, the field of drug delivery has grown and progressed at a truly remarkable rate. This growth was fueled not only by a convincing need but also by powerful commercial interests. The ability to maintain market shares and prolong the effective period of patent protection for a given drug by reformulation of the drug as a "controlled release" or "targeted delivery" system has been recognized as an important reason for the interest of the pharmaceutical industry in advanced drug delivery devices. Thus, significant contributions to the development of advanced drug de-livery were made by researchers associated with industrial research programs and close collaborative efforts between scientists in aca-demia and industry have always been a characteristic of the field. This is clearly demonstrated by the large proportion of industrial contributions included in this volume.

Controlled Drug Release

The objective in the design of a controlled drug release system is to release a pharmacologically active agent in a predetermined, pre-dictable, and reproducible fashion. Originally, the underlying ration-ale of controlled release formulations was that a drug is more effective and exhibits less side effects when the drug concentration in circulation is kept constant at some optimum level for prolonged periods of time. This rationale can be illustrated by a graph showing the expected drug concentration profiles for a number of different methods of drug administration (Figure 2). Briefly, the drug concentration in blood reaches a maximum very rapidly after administration of a standard dosage form and then decreases to a minimum, at which point repeated administration becomes necessary. Often the initial maximum concentration is above the therapeutically desirable level, increasing the risk of side effects. On the other hand, the minimum concentration may be below the therapeutically effective level. In this way, standard dosage forms can result in a drug regimen in which the patient oscillates between alternating periods of drug overdose and drug inefficacy. Controlled release systems, ideally, smooth the peaks and valleys in the drug concentration in blood providing a more effective drug regimen.

Researchers soon realized that constant drug concentrations are not necessarily the best treatment regimen. During the 1980s, diabetes, where widely fluctuating levels of insulin are required to mimic the natural biofeedback mechanisms, became a widely investigated "test case" for the application of a modulated release system. Thus, the term "controlled release" was expanded to include also those systems that were intentionally designed to provide nonlinear release characteristics. In this context, external stimuli such as temperature changes (5), pH changes (6), magnetic (7) and electric fields (8), ultrasound (9), microwave irradiation (10), and visible light (11) were used to produce changes in the rate of drug release. In many cases, these studies explored whether external stimuli can imitate natural biofeedback mechanisms. This challenging line of research may lead ultimately to controlled drug release systems that release drugs in response to a specific biological process.

Possible Sites of Action for Controlled Release Systems. Advanced controlled release systems offer a significant degree of freedom in the choice of their site of action. Whereas most "traditional" formulations have to be either injected or ingested, polymeric controlled release systems can be placed into virtually any one of the available body cavities, can be implanted, or can be attached externally to the skin. Thus a wide choice of new routes of drug administration has become available.

Nose drops, eye drops, lozenges, and skin ointments have been available for decades, however, these formulations have been limited in the past to the local administration of drugs intended to act in the nose, eye, mouth/throat, or skin respectively. In recent years, a number of academic and industrial laboratories have investigated the *systemic* administration of drugs via the nasal membranes (nasal route), the mucous membranes of the mouth (buccal route), the eye (ophthalmic route), or the skin (transdermal delivery). These routes have been explored in particular for peptide drugs (such as insulin) which cannot currently be administered orally. In this context, transdermal delivery and buccal delivery have been recognized as particularly promising and are widely investigated.

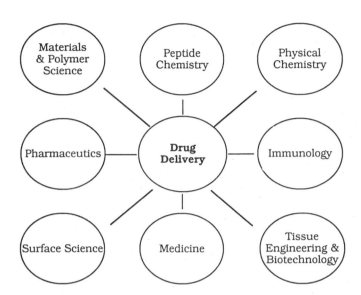

Figure 1:   Scientific disciplines that relate to drug delivery. Often, advances made in the peripheral disciplines contributed significantly to advances made in drug delivery. A particularly strong relationship exists between polymer chemistry and drug delivery, since most drug delivery systems depend on polymeric materials.

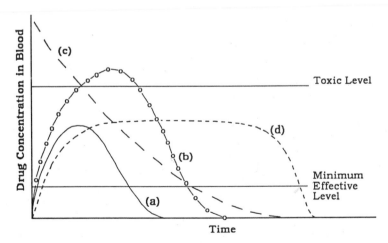

Figure 2:   Theoretical plasma concentrations after administration of various dosage forms: (a) Standard oral dose; (b) Oral overdose; (c) I.V. injection; and (d) Controlled release system.

Transdermal Delivery Systems. Transdermal delivery, where a controlled release system is attached externally to the skin, has emerged as the commercially most successful new route of drug administration. Several transdermal delivery systems are already on the market (Table I).

The commercial products listed in Table I were made possible in part by advances in the design and fabrication of polymeric membranes and films. The use of advanced polymer technologies in the formulation of transdermal drug delivery systems provides thus a good illustration for the dependency of progress in drug delivery on progress in the material sciences.

Table I: Commercially Available Transdermal Drug Delivery
Systems

Trade Name	Drug Delivered	System Type
Transderm-Scop	Scopolamine	Reservoir with RLM[1]
Duragesic Transdermal System	Fentanyl	Reservoir with RLM[1]
Catapres TTS	Clonidine	Reservoir with RLM[1]
Estraderm	Estradiol	Reservoir with RLM[1]
Transderm-Nitro	Nitroglycerine	Reservoir with RLM[1]
Nitro-Dur II	Nitroglycerine	Matrix dispersion in adhesive polymer
Nitro-Dur	Nitroglycerine	Matrix dispersion
Nitrodisc	Nitroglycerine	Hybrid reservoir and matrix dispersion
Habitrol	Nicotine	Matrix with adhesive layer
Nicotrol	Nicotine	Dispersion of matrix and drug
Prostep	Nicotine	Matrix dispersion

[1]   RLM: Rate Limiting Membrane

Commonly recognized advantages of transdermal drug delivery are the ability to interrupt the flow of drug by simply removing the transdermal patch from the skin and the circumvention of first-pass metabolism in the liver - one of the most important disadvantages of the oral route of drug administration. Transdermal systems consist usually of a drug reservoir, a diffusion-limiting membrane, and an adhesive layer for attachment to the skin. Penetration enhancers are often added to the formulation to increase the rate of drug diffusion through the skin. A major limitation of transdermal systems is that only minute quantities of drug can be administered in this way since the rate of drug diffusion through intact skin is generally low. Therefore, transdermal systems are applicable only for highly potent drugs. A further limitation relates to the observation that some degree of skin irritation can apparently not be avoided when the skin is exposed to relatively high local concentrations of drug and/or various penetration enhancers, and is covered for prolonged periods by the transdermal patch.

Buccal Delivery. In addition to transdermal delivery, buccal delivery is now recognized as particularly promising and is widely investigated in industry. Interestingly, academic laboratories have so far shown less interest in this research effort. Buccal delivery offers excellent accessibility so that drug delivery systems can be easily attached and removed. Drugs absorbed by the buccal route will avoid first pass hepatic metabolism, an advantage for the delivery of peptides and small proteins. Buccal drug delivery devices are effective because they increase the retention time in the oral mucosa by a bioadhesion mechanism.

Buccal drug delivery has been used for both local and systemic administration of drugs. The challenge of controlled buccal drug delivery is to develop systems that increase buccal residence time by mucosal adhesion. Mucosal adhesion is achieved using a polymer or combination of polymers such as hydroxypropylcellulose, ethylcellulose, polymethylmethacrylate and sodium polyacrylate that exhibit adhesive properties when in contact with saliva. The three commonly used delivery systems include adhesive tablets (12), adhesive gels (13) and adhesive patches (14). An example of a commercial buccal tablet is Susadrin. Susadrin delivers nitroglycerine in the treatment of angina pectoris for 5 h compared to about 5 min for conventional sublingual tablets (15). Susadrin uses an adhesive comprised of hydroxypropylcellulose and ethylcellulose (12).

Classification of Drug Release Systems. A commonly used classification scheme is based on the physical design of the drug release device and leads to the differentiation between two fundamentally different types of devices, the *matrix system* and the *reservoir system* (16). A matrix system consists of a drug uniformly dispersed within a polymer, while a reservoir system consists of a separate drug phase (e.g., drug particles or droplets) physically dispersed within a surrounding, rate-limiting polymeric phase. Both matrix systems and reservoir systems can be formulated in a wide range of shapes and sizes ranging from microparticles to large disks or slabs.

Occasionally, additional device-based classifications are used. For example, controlled release devices may be further classified as hydrogels, transdermal systems, osmotic systems etc. Those classifications, however, tend to be less useful than the fundamental differentiation between matrix systems and reservoir systems.

Three different mechanisms of drug release can be identified and can be referred to as "solvent controlled", "diffusion controlled", and "chemically controlled" release. These classifications represent theoretical situations where the rate of drug release is controlled predominantly by solvent interactions such as swelling of the polymer, by the diffusion of drug through a polymeric matrix or a membrane, or by chemical processes such as polymer degradation, erosion, or the cleavage of a drug from a polymeric carrier. Obviously, in many practical devices, the rate of drug release may be affected by various combinations of the above mechanisms. These mechanisms are also being used as a basis for the classification of the release devices themselves and thus one may distinguish between solvent controlled, diffusion controlled or chemically controlled systems.

Solvent Controlled Systems. The two primary solvent-controlled mechanisms of drug release include polymer swelling and osmosis. Most swelling controlled systems are hydrogels, e.g., water soluble polymers that have been rendered insoluble by crosslinking. In these systems, the rate of swelling (and thus the rate of drug release) depends on the hydrophilic/hydrophobic balance of the polymeric matrix and the degree of crosslinking (17). For example, Korsmeyer and Peppas (18) investigated the release of theophylline from hydrogels made of highly crosslinked poly(vinyl alcohol). Other examples of swellable systems were reported by Conte et al. (19) who used diclofenac sodium and cimetidine to investigate the mechanism of drug release from dosage forms based on poly(vinyl alcohol), hydroxypropylmethylcellulose, and carboxymethylcellulose. In these systems, release was controlled by the simultaneous swelling and erosion of the polymeric matrix. Brondsted and Kopecek (17) have reported the use of pH sensitive hydrogels that swell in the distal intestine and colon. Once swelling begins, the crosslinks become accessible to specific colonic enzymes (azoreductases) leading to the degradation of the hydrogel and the concomitant release of drug into the colon. This is an example for the use of both solvent control and chemical control within the same system.

Osmosis represents the second mechanism of solvent control. Osmotic systems are usually composed of a drug reservoir enclosed by a water-selective polymeric membrane. The membrane allows for transport of water but does not allow for the passage of drug. The polymeric membrane has a small opening through which drug is released as a result of the built-up of hydrostatic pressure within the device. An example of this technology is the implantable Alzet minipump and the oral version, Osmet, for controlled oral drug delivery. The widely known Oros tablet system is an extension of the Osmet technology. An example of the commercial success of osmotic delivery systems is Procardia XL. Procardia XL is a once-a-day product using the Oros technology. Commercially, the system is intended to prolong the nifedipine product line. One of the advantages of this system is that the dosage frequency could be reduced from three times daily for the traditional dosage form to once daily.

Diffusion Controlled Systems. Reservoir and matrix systems represent fundamentally different system designs for the diffusion-controlled release of drugs. As defined above, a typical reservoir system consists of a nondegradable, rate-limiting polymeric membrane, separating a core of drug from the biological environment. Typically, reservoir systems have been formulated as capsules, microcapsules, hollow fibers, or tubes with sealed ends (Figure 3). Several nondegradable polymers with FDA approval history (such as polysiloxanes used in the formulation of the Norplant system) are available. In reservoir systems consisting of nondegradable polymers, the rate of drug release is strictly controlled by the rate of drug diffusion through the polymeric membrane. Thus, the formulation of the drug core and the fabrication of functional systems are relatively simple tasks.

Two different types of controlling membranes, homogeneous or microporous, are being used in the formulation of reservoir systems. Microporous membranes have the advantage that drug diffuses through pores that are filled with the same medium as the reservoir. Diffusion control in homogeneous membranes, on the other hand, depends on membrane-drug partitioning. Transdermal drug delivery systems utilize both of these mechanisms (Table I). For example, the Transderm-Nitro system uses a homogeneous EVA copolymer membrane while the Transderm-Scop system is based on a microporous rate-controlling polypropylene membrane (20).

One of the potential disadvantages of implantable reservoir systems (such as the Norplant contraceptive device) is the danger of "dose dumping". If the surrounding polymeric membrane should become leaky due to cracks or sudden rupture, the entire drug core could be released into circulation within a very short time. Therefore, special consideration has to be given to the use of degradable polymers in the design of reservoir systems. Here, the need to keep the polymeric membrane mechanically intact throughout the period of drug release requires a very careful optimization of the polymer properties so that the process of drug release is essentially completed by the time the polymeric membrane looses its mechanical strength due to degradation. As a rule of thumb, one can expect that the formulation of a device with an active release time of X will require a polymer whose complete bioabsorption from the implant site will occur over a period of 3X. As a consequence, upon repeated administration of the device, a "steady-state" number of active and partially degraded, empty "shells" within the body of the patient is established. Polycaprolactone, for example, has been used in the formulation of a reservoir-type implantable contraceptive device (the Capronor system) that is being tested clinically in phase II trials in many countries (21). In that system, drug release is diffusion controlled over a period of about one year, with the polycaprolactone membrane completely degrading after about 3 years from the date of implantation. Thus, upon long-term, repeated use of the device, one active and two "empty" devices would accumulate in the patient. Whether the Capronor system has significant advantages over the Norplant system (a nondegradable reservoir-type contraceptive based on a polysiloxane membrane) depends to a large extend on the circumstances of the patient, the ease at which the nondegradable system can be removed, and the general medical environment. In developing countries where trained surgeons may not be available for implant removal or where patients may not return to the clinic to have their implant removed, the degradable system may indeed be the preferred system configuration.

Although short-duration reservoir systems can be formulated with fast-degrading polymeric membranes, as a general rule, slowly degrading polymers appear to be particularly useful for the design of degradable reservoir systems. Fortunately, several such polymers with promising tissue compatibility and suitable physicomechanical properties are commercially available (Table II).

Table II: Some Slow-Degrading Polymers that can be Considered for the Design of Long-Acting, Degradable Reservoir Systems

Polymer	Comments and References
Poly(lactic acid)	High molecular weight preparations preferred, medical grades readily available from commercial suppliers (e.g., Medisorb) (22).
Polycaprolactone	Readily available from commercial suppliers (e.g., Polysciences) (21).
Polyhydroxybutyrate and co-polymers with valeric acid	Prepared by biosynthesis, different grades with valeric acid contents from about 7 to 30% are available from ICI under the trade name Biopol (23,24).
Poly(N-palmitoyl-trans-4-hydroxy-L-proline ester)	An amino acid derived polyester, belonging to a new class of potential implant materials, the pseudo-poly(amino acids). Available from Sigma Chemical Company (25,26).
Poly(DTH carbonate)	A polycarbonate made of derivatives of the natural amino acid L-tyrosine. Available from Sigma Chemical Company (16,27-29).

In a matrix system, the drug is uniformly distributed within the polymeric phase. Like reservoir systems, matrix systems can be fabricated in a variety of shapes, including microspheres that would be suitable for injection. Matrix systems are safer than reservoir systems, since no potentially lethal "drug spill" can occur even if the device would break into several pieces. If the matrix system is fabricated from a degradable polymer, the release mechanism is often a combination of diffusional release and chemically controlled release. This case will be discussed in more detail in the the section on chemically controlled systems.

As drug is being released from a nondegradable matrix system, the rate of diffusion tends to decrease. For this reason, it can be quite difficult to formulate matrix systems that will exhibit constant drug release rates in a reproducible fashion over extended periods of time (30). A possible solution to this intrinsic problem was suggested by Langer et al. who used matrix systems with special geometries (coated hemispheric devices) to compensate for the decrease in the rate of diffusion with time (30). In spite of these difficulties, the matrix system is a widely studied system design that is suitable for various drug release applications. For example, the polyanhydride based release system for BCNU (currently in phase 3 clinical trials) is a matrix system (31). Likewise, among the transdermal systems, Nitro-Dur and Nitro-Dur 11 are based on a nondegradable matrix design.

Chemically Controlled Systems. Systems in which the rate of drug release is predominantly controlled by the rate of polymer degradation, the rate of the physical erosion of the polymer, or the rate at which a drug is cleaved from the polymer backbone are considered chemically controlled release systems. The two main types of chemically controlled systems are matrix systems based on degradable polymers and pendent chain systems (Figure 4).

Within the context of this chapter, we limit our discussion to the case of a solid, polymeric matrix-type release system. The transformation of such an implant into water soluble material(s) is best described by the term "bioerosion". This process is associated with macroscopic changes in the appearance of the device, changes in the physicomechanical properties of the polymeric material, physical processes such as swelling, deformation or structural disintegration, weight loss, and the eventual loss of function.

All of these phenomena represent distinct and often interconnected aspects of the complex bioerosion behavior of a degradable matrix device. It is important to note that the bioerosion of a solid device is not necessarily due to the chemical cleavage of the polymer backbone, or the chemical cleavage of crosslinks or side chains. Rather, simple solubilization of the intact polymer, for instance, due to changes in pH, may also lead to the erosion of a solid device.

Two distinct modes of bioerosion have been described in the literature (32). In "bulk erosion", the rate of water penetration into the solid device exceeds the rate at which the polymer is

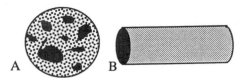

Figure 3: Schematic representations of a typical reservoir system: (A) Cross-sectional view showing drug particles (black) dispersed within the drug core; (B) Reservoir system formulated in the shape of a tube. The drug core is filled into the tube whose ends are then sealed. Drug release is controlled by the diffusion of the drug through the surrounding, rate-limiting polymeric membrane.

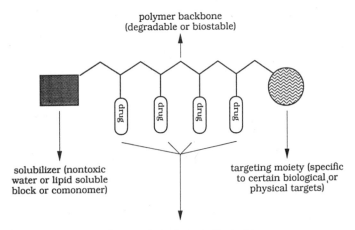

Figure 4: Schematic representation of a "pendent chain system", used predominantly in the design of soluble, macromolecular drug conjugates. Various modifications of this figure have been published in a large number papers to illustrate the concept of a pendent chain system. Such a system consists usually of a degradable or nondegradable polymeric backbone to which a pharmacologically active agent is covalently attached through reactive pendent chains. A spacer may be placed between the drug and the polymer backbone. Targeting moieties (such as monoclonal antibodies) and solubilizing elements (such as polyethylene glycol units) can be incorporated into the system design to improve the overall performance of the drug carrier. Numerous polymers such as dextran, polylysine, poly(glutamic acid), or functionalized derivatives of polyacrylamide have been used as backbones.

transformed into water soluble material(s). Consequently, the uptake of water is followed by an erosion process that occurs throughout the entire volume of the solid device. Due to the rapid penetration of water into the matrix of hydrophilic polymers, most of the currently available polymers will give rise to bulk eroding devices. In a typical "bulk erosion" process, cracks and crevices will form throughout the device which may rapidly crumble into pieces. A good illustration for a typical bulk erosion process is the disintegration of a sugar cube that has been placed into water. Depending on the specific application, the often uncontrollable tendency of bulk eroding devices to crumble into little pieces can be a disadvantage. It is easy to see that the rate of drug release cannot be adequately controlled while the release device disintegrates into random fragments.

Alternatively, in "surface erosion" the rate at which water penetrates into the polymeric device is slower than the rate of transformation of the polymer into water soluble material(s). In this case, the transformation of the polymer into water soluble material(s) is limited to the outer surface of the solid device. The device will therefore become thinner with time, while maintaining its structural integrity throughout much of the erosion process. In order to observe surface erosion, the polymer must be hydrophobic enough to impede the rapid imbibition of water into the interior of the device. In addition, the rate at which the polymer is transformed into water soluble material(s) has to be reasonably fast. Under these conditions, scanning electron microscopic evaluation of surface eroding devices has sometimes shown a sharp border between the eroding surface layer and the intact polymer in the core of the device (*33*). So far, true surface erosion has been observed only in a small number of polymers. Currently, polyanhydrides (*34-37*) and poly(ortho esters) (*38-41*) are the best known examples of polymers that can be fabricated into surface eroding devices.

While in theory surface eroding matrix systems appear to be preferable over bulk eroding systems, ideal surface erosion can only be achieved in "unloaded" devices or when hydrophobic drugs are incorporated into the polymer matrix. In practice, hydrophilic drugs (that tend to cause the imbibition of water into the device) tend to lead to complex drug release profiles that are only partially controlled by polymer erosion. On the other hand, by formulating bulk eroding matrix systems in such a way that most drug is released before the device physically disintegrates, bulk eroding matrix systems can provide constant drug release rates. Contrary to a widely held misconception, surface erosion is therefore not a necessary requirement for the formulation of degradable matrix systems that show constant release.

Factors Influencing the Degradation of a Polymeric Device. In order to successfully design chemically controlled drug release systems, it is necessary to understand the main factors that determine the overall rate of the erosion process. In the case of a solid matrix system, the main parameters are the chemical stability of the polymer backbone, the hydrophobicity of the polymeric matrix, the morphology of the polymer, the initial molecular weight of the polymer, the degree of swelling of the drug-loaded polymeric matrix, the fabrication process, the presence of catalysts, additives or plasticizers, and the geometry of the implanted device.

The susceptibility of the polymeric backbone toward hydrolytic cleavage is probably the most fundamental parameter. Generally speaking, anhydrides tend to hydrolyze faster than ester bonds which in turn hydrolyze faster than amide bonds. Thus, polyanhydrides will tend to degrade faster than polyesters which in turn will have a higher tendency to bioerode than polyamides. However, solely based on the known susceptibility of the polymer backbone structure toward hydrolysis it is not possible to predict the rate at which any given polymeric device will undergo bioerosion.

The observed erosion rate of a drug-loaded matrix system is strongly dependant on the ability of water molecules to penetrate into the polymeric matrix. The hydrophobicity of the polymer which is a function of the structure of the monomeric starting materials as well as the nature and loading of the drug can therefore have an over-whelming influence on the observed bioerosion rate. For instance, the erosion rate of polyanhydrides can be slowed by about three orders of magnitude when the hydrophilic sebacic acid is replaced by hydrophobic bis(carboxy phenoxy)propane as the monomeric starting material (37). Likewise, devices made of poly(glycolic acid) erode faster than identical devices made of the more hydrophobic poly-(lactic acid), although the ester bonds have about the same chemical reactivity toward water in both polymers.

The observed bioerosion rate is further influenced by the mor-phology of the polymer. Within the framework of this discussion, three distinct morphological states (semicrystalline, amorphous-glassy, and amorphous-rubbery) have to be considered.

In a semicrystalline polymer, the crystalline regions are most densely packed and offer the highest resistance to the penetration of water. Consequently, the rate of backbone hydrolysis tends to be higher in the amorphous regions of a semicrystalline polymer than in the crystalline regions. A good illustration of the influence of the polymer morphology on the rate of bioerosion is provided by a comparison of poly(L-lactic acid) and poly(D,L-lactic acid): although these two polymers have chemically identical backbone structures and an identical degree of hydrophobicity, devices made of poly(L-lactic acid) tend to degrade much slower than identical devices made of poly(D,L-lactic acid). The slower rate of bioerosion of poly(L-lactic acid) is due to the fact that this stereoregular polymer is semi-crystalline, while the racemic poly(D,L-lactic acid) is amorphous.

Likewise, a polymer in its glassy state is less permeable to water than the same polymer when it is in its rubbery state. This observa-tion could be of importance in cases where an amorphous polymer has a glass transition temperature (Tg) that is not far above body tem-perature (37 °C). In this situation, the incorporation of a drug into the polymeric matrix could lower the Tg of the "loaded" device below 37 °C, resulting in unexpected and abrupt changes in the bioerosion rate.

The manufacturing process may also have a significant effect on the erosion profile. For example, Mathiowitz and coworkers (33) showed that polyanhydride microspheres produced by melt encapsulation were very dense and eroded slowly, whereas the same microspheres, formed by solvent evaporation, were more porous (and therefore more water permeable) and eroded more rapidly.

The above examples illustrate an important technological principle in the design of degradable matrix systems: The bioerosion rate of a given polymer is not an unchangeable property, but depends to a very large degree on readily controllable factors such as the presence of plasticizers or additives, the manufacturing process, the nature of the drug and the level of drug loading, the initial molecular weight of the polymer, and the geometry of the device.

From an applied perspective, the most commonly investigated degradable polymers for the formulation of matrix-type drug release systems are poly(lactic acid) and lactic/glycolic acid copolymers (22). However, at this point, no implantable, degradable release system has been approved in the USA. One of the most advanced systems is a degradable matrix system, developed by Langer's group at MIT. This system is used to release an antineoplastic agent from a polyanhydride matrix (31) and is currently undergoing phase III clinical trials in several medical centers in the USA.

Targeted Drug Delivery

The basic premise of "targeted drug delivery" is the assumption that the therapeutic index of a drug can be improved when the drug accumulates selectively in specific tissues, organs, or cell types. Drugs can be targeted by a wide range of mechanisms and therefore numerous approaches have been developed, some of which will be reviewed in the following sections. On a very basic level, one can distinguish between *insoluble, particulate drug carriers*, and *soluble, macromolecular drug conjugates*. These two designs represent fundamentally different approaches to the targeting of drugs. Furthermore, one has to distinguish between actively and passively targeted systems. *Passively targeted systems* utilize existing body mechanisms to reach their destination. For example, due to the fact that many particulate drug carriers are rapidly taken up by the reticuloendothelial system (RES), the tendency of microparticulates to accumulate in the liver is well documented. Thus, the liver uptake of microparticulates represents probably the best studied example for passive targeting. In contrast, *actively targeted systems* utilize a specific biological interaction to actively seek their target. The use of monoclonal antibodies as "targeting moieties" against specific cancer cells represents a good example for active targeting. Other specific biological interactions, such as the interaction between an enzyme and its substrate, or the interaction between a hormone and its receptor can also be exploited for the design of actively targeted systems (42).

Insoluble Particulate Carriers. A drug carrier, as implied by the name, is a pharmacologically inactive polymer whose task is to carry a drug to its site of action. Microspheres, and micro- or nanoparticles belong to the group of insoluble, particulate carriers. Liposomes are also important particulate carriers, but will not be reviewed here since they are not polymeric.

A primary consideration for the use of microparticulate carriers is biodegradability. Although crosslinked derivatives of poly(methyl methacrylate) or poly(acrylamide) have been suggested for the preparation of nano- and microparticles, these particles are not readily degradable in vivo and, when injected or implanted, tend to accumulate in the body upon repeated administration. For this reason, nondegradable microparticulate carriers are not practical candidates for most drug delivery applications. On the other hand, particles composed, for example, of poly(alkyl 2-cyanoacrylate), copolymers of acrylamide and dextran (43), derivatives of poly(glutamic acid) (44), or poly(lactic acid) (22) are biodegradable. Such carrier systems are currently being intensely investigated.

Microparticulate carriers are usually *passively targeted systems*. As mentioned above, the high uptake of microparticulate carriers by Kupffer cells located in the liver provides an excellent mechanism for the passive targeting of the liver. Microparticulate carriers have also been suggested for oral applications: Nefzger et al. (45) demonstrated that a certain fraction of orally administered poly(methyl methacrylate) (PMMA) nanoparticles were systemically absorbed. This surprising result was obtained when radioactively labelled poly(methyl-(1-^{14}C)-methacrylate) nanoparticles were orally administered to rats and about 10-15% of the total radioactivity was recovered in the urine and the bile. Since PMMA is biostable, the authors suggested that the nanoparticles were indeed absorbed intact from the GI tract.

Macromolecular conjugates. Macromolecular conjugates are usually *actively targeted systems*. A schematic of a macromolecular conjugate is shown in Figure 4. A comprehensive review of the literature relating to macromolecular conjugates revealed several hundred major publications which can obviously not be covered within the context of this overview. We will therefore present only some fundamental considerations and a few selected samples. For more information on soluble, macromolecular carriers, the reader is referred to a recent review by Drobnik (46).

The successful design of a clinically useful, macromolecular carrier is an exceedingly difficult task, as evidenced by the fact that a truly heroic international research effort has so far not resulted in any significant clinical application of such systems. An often cited rationale for the design and synthesis of a macromolecular carrier is the expectation that the macromolecular carrier will prolong the drug residence time in the desired body compartment. In addition, the carrier is often claimed to reduce the immunogenicity of the drug and protect it from inactivation. Finally, the design of a polymeric drug conjugate offers the opportunity to attach specific targeting moieties which should provide some control over the biodistribution of the drug.

A major problem in the design of macromolecular carriers is the timing of drug release. The carrier is usually expected to be stable in circulation for at least some time to reach its target site. Once at its target site, however, the attached drug must usually be released from the carrier in order to be active. In spite of some ingenious work, attempting to tailor the release of drug to specific enzymatic activities or environments (such as the acidic environment found in lysosomes) (47), no generally applicable approach for the exact timing of drug cleavage from the carrier has so far been identified. Another serious problem in the evaluation of macromolecular carriers is the almost complete absence of useful correlations between in vitro and in vivo testing. Obviously, cell cultures, with their lack of compartmental barriers are a generally poor model system for the evaluation of macromolecular carriers whose activity is shaped to a large extent by their ability (or inability) to cross compartmental barriers in vivo.

Over the years, a wide range of polymeric backbones, both degradable and nondegradable, were explored. Noteworthy are the early use of dextrans and other polysaccharides (48), various poly(amino acids) such as poly-L-lysine (49) and poly(glutamic acid) (50), and derivatives of polymethacrylamide (51) as polymeric backbones. In recent years, the use of nondegradable polymeric backbones such as copolymers of maleic anhydride and divinyl ether, poly(vinyl alcohol), or derivatives of polymethacrylamide has somewhat decreased in favor of degradable polymers of which the poly(amino acids) are often regarded as most promising (46).

The use of poly(ethylene oxide) (also referred to as poly(ethylene glycol) or PEG) as a copolymer component was widely investigated as a "solubilizer" for drug carriers that included marginally soluble polymers or particularly hydrophobic drugs. For example, an AB type block copolymer of poly(aspartic acid) with PEG has been studied as a water soluble macromolecular prodrug for the sparingly soluble anthracycline antibiotic adriamycin (52). The major advantage of that copolymer over simple poly(aspartic acid) was the retention of water solubility of the conjugate despite the introduction of a large number of hydrophobic adriamycin residues. The polymer had a micellar structure in aqueous buffer with a hydrophilic outer shell comprised of PEG and a hydrophobic inner core of poly(aspartic acid). After uptake by target cells, the hydrolysis of the poly(aspartic acid) backbone resulted in the release of adriamycin from the conjugate.

Although a wide range of biological interactions could be utilized for drug targeting, polyclonal or monoclonal antibodies represent the by far most commonly considered targeting moieties. These systems have usually been far more active in vitro than in vivo. Numerous examples of antibody-drug, antibody-polymer, or antibody-polymer-drug conjugates are available in the literature. For example, monoclonal antibody conjugates of adriamycin with a poly(PEG-aspartic acid) copolymer have been investigated for targeted drug delivery (53,54). In many ways, this particular macromolecular carrier encompasses many of the "preferred" structural elements: PEG as a solubilizer, a biocompatible, degradable backbone structure, and an important antineoplastic agent with significant toxicity that could be alleviated by a targeted dosage form.

Examples of Specific Targeted Delivery Systems. In the following section, some examples of specific targeted delivery systems will be presented in more detail. In arranging this section, we grouped the delivery systems based on their mode of targeting. This common classification scheme leads to the distinction between (a) systems that are mechanically placed into a specific site, (b) passively targeted systems, (c) actively targeted systems, and (d) systems that utilize physical means for drug targeting.

Mechanically Placed Delivery Systems. In the simplest approach to targeting, a controlled release system is *mechanically placed* into a specific site. The targeting effect of such a system is based on the confinement of drug release to a certain volume around the delivery system.

Currently, implantable pumps and biostable or bioerodible polymeric devices are being used in this manner. An example of a mechanically placed device is the Progestasert Intrauterine Device (IUD), a contraceptive system that delivers 65 µg of progesterone for over 400 days. The Progestasert IUD utilizes a drug reservoir with a rate limiting ethylene-vinyl acetate copolymer membrane. Another example is the Ocusert pilocarpine ophthalmic insert for the treatment of glaucoma. The Ocusert insert is placed in the cul-de-sac of the lower eyelid. Zero order release of pilocarpine through a rate limiting ethylene-vinyl acetate copolymer membrane at rates of either 20 µg/h (Ocusert Pilo-20) or 40 µg/h (Ocusert Pilo-40) is achieved for 7 days. Although the Ocusert device is inserted only once weekly, it has not become popular with glaucoma patients who prefer to instill timolol eye drops twice daily.

The mouth is also a useful site for mechanically placed devices since they can be easily inserted and removed and patient acceptance of the oral route is high. An example of an oral insertable system is the use of tetracycline for the treatment of periodontal disease. (55). Tetracycline has been incorporated into an ethylene-vinyl acetate copolymer matrix for the long-term delivery of low levels of tetracycline. Although the device is effective in preventing periodontal disease, comfortable tooth anchoring devices are needed to make this regimen a success.

Implantable systems have found utility in the treatment of certain brain tumors. The intracranial delivery of mitomycin, adriamycin, BCNU and 5-fluorouracil in 10% polymethylmethacrylate (PMMA) composites were shown to increase the survival of patients with glioblastoma and anaplastic astrocytoma (56). Another example is the use of PMMA pellets containing methotrexate to treat brain tumors. Pellets that were implanted in rat brain tumors were shown to significantly reduce tumor volume and increase the survival of the rats (57).

Passively Targeted Delivery Systems. Passive targeting involves the random movement of the delivery system through the body to its site of release. Passively targeted systems use natural flow in areas such as the blood stream or gastrointestinal tract and, at a specific physiological site of uptake or reconversion, the therapeutic moiety is released.

Colonic drug delivery is an example of an area where passively targeted drug delivery devices may be effective. Once swallowed, a colonic drug delivery device will traverse the small intestine in about 1.5 to 4.5 h (58). At this point, it enters the colon. Colonic drug delivery strategies usually include the use of polymers or polymer-prodrugs that are substrates for enzymes found only in the colon; therefore, the device is protected from digestive enzymes until it gets to the colon. Colon specific strategies have included the delivery of low molecular weight prodrugs (59) and polymeric prodrugs (60). These systems release drug only in the presence of glycosidases or azoreductases which are only present in the colon (61). Brondsted and Kopecek (17) have studied hydrogels that are susceptible to azoreductases in the colon. These authors synthesized hydrogels containing both acidic comonomers and enzymatically degradable azoaromatic crosslinks. The gels have a pH sensitive swelling mechanism with maximal swelling at the increased pH of the distal small intestine. In the colon the gels reach a degree of swelling that makes the cross-links accessible to azoreductases. The gel is then degraded and the drug is released. Saffran et al. have used a hydrophobic azopolymer coating susceptible to azo reduction to deliver vasopressin and insulin orally in rats (10) and dogs (62). In the dog study, bovine insulin was mixed with 5-methoxysalicylate (an absorption enhancer) and placed into a gelatin capsule. The capsule was coated with a terpolymer of styrene, hydroxyethylmethacrylate and N,N'-bis-(β-styrylsulphonyl)-4,4'-diaminoazobenzene as a cross linking agent. After multiple oral doses in dogs, profound decreases in hepatic glucose production and plasma glucagon-like activity were observed. Glycosidic polysaccharides have also been used for targeting glycosidic enzymes of the colon such as β-glucosidase, β-glucuronidase, and β-xylodase as reported by Lancaster et al. (63).

Actively Targeted Delivery Systems. During the 1970s and 1980s unrealistic expectations led to predictions that actively targeted drug delivery systems would, within a short time, revolutionize the treatment of cancer and other diseases. During that time, terms like "magic bullet", "guided missile", "warhead carrier" were used to describe actively targeted delivery systems. These terms are not only an oversimplification, they are wrong given the biological characteristics of antibody-antigen interactions as compared to the behavior of modern weaponry. The early research efforts on actively targeted delivery systems were uniformly unsuccessful and consequently not a single antibody-based cancer therapeutic is currently on the market (there are, however, some antibody-based diagnostics available).

The purpose of actively targeted drug delivery systems is to alter the natural distribution pattern to direct drugs to specific organs, tissues or cells. In principle, a large variety of biological interactions can be used for targeting purposes. In practice, however, the specific interactions between antibodies and their antigens are most commonly employed.

The research effort on actively targeted drug delivery systems is fueled by powerful business interests, since cancer chemotherapeutics currently constitute a $4 billion/year market. Over 30 North American and European companies are developing antibody based therapeutics. While the development of clinically useful targeted drug delivery systems was technologically not feasible during the 1970s and 1980s, significantly improved antibodies and superior chemical methods for the assembly of macromolecular conjugates have now become available. Due to these advances, several products are currently in clinical trials and will become available in the near future. These products will be used in conjunction with conventional therapies, rather than replacing them.

A major technological problem of actively targeted drug delivery systems relates to the source of the antibody. Since for ethical reasons humans cannot be used as hosts in the production of monoclonal antibodies (MAb), human monoclonal antibodies against human cancers are not readily available. Most of the early work on actively targeted drug delivery systems was therefore done with murine antibodies. When such antibodies are injected into humans, the human immune system recognizes the murine antibody as foreign. The subsequent "human antimouse antibody" (HAMA) response neutralizes the murine antibody and thus eliminates its targeting effect. In addition, murine antibodies have a very short half-live in circulation which is acceptable for diagnostic applications but a severe disadvantage for therapeutic applications. Currently, chimeric antibodies, humanized antibodies, and antibody fragments are being explored in an attempt to overcome the limitations of non-human antibodies.

Another serious problem is posed by the antibody targets themselves. Although monoclonal antibodies are highly specific in their interaction with antigens, the effectiveness of their action as targeting moieties depends on the proper choice of the target. Potential targets include, for example, differentiation antigens, oncogene products, and growth factors. The need to carefully choose the target is illustrated by the behavior of many tumor-associated differentiation antigens which are normally expressed by fetal cells and are often present in low levels on normal cells. Furthermore, tumor-associated differentiation antigens are not always bound to the cell membrane but can be found in circulation. Circulating antigens may react with the antibody of the drug delivery system - thereby preventing the system to reach its intended target.

The actively targeted drug delivery systems that are now becoming available for clinical trials are simple antibody-drug conjugates that do not contain the polymeric backbone shown in Figure 4. Although it would usually be advantageous to attach a drug-loaded polymer to the antibody (rather than attaching multiple drug units to the antibody directly), the polymeric carrier can be a serious source of complications.

First, most anticancer drugs have to be internalized by the target cell in order to be effective. One way to achieve this goal is to use "internalizing" antibodies, e.g., antibodies that are naturally taken up by the cell. The attachment of a large polymer to the antibody may interfere with the process of internalization and thus render the system ineffective. Second, most polymeric drug conjugates have a tendency to be taken up by elements of the reticuloendothelial system (RES). This tendency can lead to the predominant accumulation of the polymeric conjugate in the liver, preventing the system to reach its intended target. Further research is therefore needed toward the identification of polymeric carriers that, by themselves, show as little nonspecific uptake as possible. One such carrier is a recently synthesized, highly water soluble poly(ether urethane) derived from PEG and the natural amino acid L-lysine (64-66). In preliminary biodistribution studies, this carrier was shown to remain in the blood compartment without organ-specific uptake (67).

Physically Controlled Targeted Delivery Systems. Using physical controls such as localized magnetic fields, pH differences or temperature gradients, a physically controlled delivery system can be directed to a specific site or its contents can be released at a certain site. One of the promising developments in physically controlled delivery systems is the use of localized magnetic fields. Widder and coworkers (68) were the first to report the use of magnetically responsive microspheres for targeted delivery. More recently, Gupta and Hung (69) studied the effects of magnetic albumin microspheres in the delivery of doxorubicin, an anticancer agent, in rats. After an intravenous injection of doxorubicin microspheres, a magnetic field was applied to the tail for 30 min. The magnetic field resulted in a significantly higher doxorubicin concentration in the tail as compared to the concentration measured in appropriate control studies. In addition, the delivery of doxorubicin to all non-target tissues, including the liver and heart, was substantially reduced. Although reports of the use of polymeric magnetic microspheres in the literature are lacking, the studies of Gupta et al. (69) may lead to the development of magnetically targeted polymeric systems.

Summary

After an initial period of rapid advances in the fundamental understanding of polymeric drug delivery systems and a period of excessive expectations, the field is now showing signs of scientific maturity. The often significant research efforts during the 1970s and 1980s (which were almost always unsuccessful from a commercial point of view), are now starting to "pay off": A wave of transdermal delivery systems has been introduced into the market over the last few years and several additional systems are in advanced clinical trials. The final approval of the contraceptive device Norplant in 1990 introduced the first polymeric, implantable device into the US market. The polyanhydrides used as chemotherapeutic delivery system for brain tumors may become the first FDA approved implantable polymeric matrix system based on a degradable polymer. Finally, a number of targeted anticancer systems using antibody-drug (but not yet antibody-drug-polymer) conjugates will become available in the near future.

The experience of the last 20 years has shown that there is a strong connection between advances in the material sciences (particularly polymer chemistry) and advances in the design and implementation of new drug delivery systems. In this context, the slow pace in which new, degradable polymers are being adopted for drug delivery applications is a concern. In an attempt to reduce the cost of the FDA approval process for new drug delivery systems, industrial development efforts have strongly favored the well-established polymers of lactic and glycolic acid, almost to the exclusion of promising alternatives such as poly(ortho esters), poly(hydroxybutyrates), pseudo-poly(amino acids), polyphosphazenes, and a number of natural polymers such as derivatives of cellulose, albumin, collagen, or gelatin. The successful development of a wide range of degradable delivery systems will almost certainly depend on the availability of a wider choice of degradable polymers.

Currently, one of the most important challenges in the development of controlled release systems is the design of systems that are capable of modulated release profiles, in response to an external stimulus or in response to a biofeedback mechanism. In the area of targeted delivery systems, one of the most significant challenges is to better understand how the biological properties of polymeric drug conjugates can be improved in order to optimize the in vivo performance of these systems.

Acknowledgments

Joachim Kohn acknowledges the support of a NIH Research Career Development Award (GM00550).

Literature Cited

1. Yolles, S.; Eldridge, J. E.; Woodland, J. H. R. *Polym. News* **1971**, *1 (4&5)*, 9-15.
2. Boswell, G. A.; Scribner, R. M., German Patent 2051580, filed May 6, 1971, assigned to duPont de Nemours and Co.
3. Woodland, J. H. R.; Yolles, S.; Blake, D. A.; Helrich, M.; Meyer, F. J. *J. Med. Chem.* **1973**, *16 (8)*, 897-901.
4. Yolles, S.; Eldridge, J. E.; Leafe, T. D.; Woodland, J. H. R.; Blake, D. A.; Meyer, F. J. *Adv. Exp. Med. Biol.* **1973**, *47*, 177.
5. Hoffman, A. S.; Afrassiabi, A.; Dong, L. C. *J. Control. Rel.* **1986**, *4*, 213-222.
6. Kuhn, W.; Hargitay, B.; Katchalsky, A.; Eisenberg, H. *Nature* **1950**, *165*, 514-516.
7. Hsieh, D. S. T.; Langer, R.; Folkman, J. *Proc. Natl. Acad. Sci. (USA)* **1981**, *78 (3)*, 1863-1867.
8. Eisenberg, S. R. *J. Membr. Sci.* **1984**, *19 (2)*, 173-194.
9. Miyazaki, S.; Hou, W. M.; Takada, M. *Chem. Pharm. Bull.* **1985**, *33 (1)*, 428-431.
10. Miyazaki, S.; Yokouchi, C.; Takada, M. *Chem. Pharm. Bull* **1989**, *37 (1)*, 208-210.
11. Suzuki, A.; Tanaka, T. *Nature* **1990**, *346*, 345-347.
12. Bremecker, K. D.; Strempel, H.; Klein, G. *J. Pharm. Sci.* **1984**, *73 (4)*, 548-552.
13. Brown, J. P.; McGarraugh, G. V.; Parkinson, T. M.; Wingard, R. E.; Onderdonk, A. B. *J. Med. Chem.* **1983**, *26*, 1300-1307.
14. Baker, R. W.; Lonsdale, H. K. *Chemtechnology* **1975**, *5 (11)*, 668-674.
15. Schor, J. M.; Davis, S. S.; Nigalaye, A.; Bolton, S. *Drug Develop. Ind. Pharm.* **1983**, *9 (7)*, 1359-1377.
16. Kohn, J. *Drug News and Perspectives* **1991**, *4 (5)*, 289-294.
17. Brondsted, H.; Kopecek, J. *Biomaterials* **1991**, *12 (6)*, 584-592.
18. Korsmeyer, R. W.; Peppas, N. A. *J. Membr. Sci.* **1981**, *9*, 22-39.
19. Conte, U.; Colombo, P.; Gazzaniga, A.; Sangalli, M. E.; LaManna, A. *Biomaterials* **1988**, *9*, 489-493.
20. Yum, S. I.; Wright, R. M. In *Controlled Drug Delivery*; Bruck, S. D., Ed.; Vol. 2, CRC Press: Boca Raton, Fl, 1983; pp 65-88.
21. Pitt, C. G. In *Biodegradable Polymers as Drug Delivery Systems*; Chasin, M. and Langer, R., Ed., Marcel Dekker Inc.: New York, NY, 1990; pp 71-120.
22. Lewis, D. H. In *Biodegradable Polymers as Drug Delivery Systems*; Chasin, M. and Langer, R., Ed., Marcel Dekker Inc.: New York, NY, 1990; pp 1-41.
23. Barham, P. J.; Keller, A.; Otun, E. L.; Holmes, P. A. *J. Mater. Sci.* **1984**, *19*, 2781-2794.
24. Miller, N. D.; Williams, D. F. *Biomaterials* **1987**, *8*, 129-137.
25. Kohn, J.; Langer, R. *J. Am. Chem. Soc.* **1987**, *109*, 817-820.

26. Kohn, J. In *Biodegradable polymers in drug delivery systems*; Chasin, M. and Langer, R., Ed., Marcel Dekker: New York, NY, 1990; pp 195-229.
27. Silver, F. H.; Marks, M.; Kato, Y. P.; Li, C.; Pulapura, S.; Kohn, J. *J. Long-Term Effects Med. Implants* **1992**, *1 (4)*, 329-346.
28. Kohn, J. In *Polymeric Drugs and Drug Delivery Systems*; Dunn, R. L. and Ottenbrite, R. M., Ed.; ACS Symposium Series; Vol. 469, American Chemical Society: Washington, DC, 1991; pp 155-169.
29. Pulapura, S.; Kohn, J. *Biopolymers* **1992**, *32*, 411-417.
30. Rhine, W. D.; Hsieh, D. S. T.; Langer, R. *J. Pharm. Sci.* **1980**, *69 (3)*, 265-270.
31. Chasin, M.; Domb, A.; Ron, E.; Mathiowitz, E.; Langer, R.; Leong, K.; Laurencin, C.; Brem, H.; Grossman, S. In *Biodegradable Polymers as Drug Delivery Systems*; Chasin, M. and Langer, R., Ed., Marcel Dekker Inc.: New York, NY, 1990; pp 43-70.
32. Heller, J. In *Controlled Drug Delivery, Fundamentals and Applications, 2nd edition*; Robinson, J. R. and Lee, V. H. L., Ed., Marcel Dekker: New York, NY, 1987; pp 180-210.
33. Mathiowitz, E.; Kline, D.; Langer, R. *J. Scanning Micros.* **1990**, *4 (2)*, 329-340.
34. Leong, K. W.; Brott, B. C.; Langer, R. *J. Biomed. Mater. Res.* **1985**, *19*, 941-955.
35. Leong, K. W.; D'Amore, P. D.; Marletta, M.; Langer, R. *J. Biomed. Mater. Res.* **1986**, *20*, 51-64.
36. Leong, K. W.; Simonte, V.; Langer, R. *Macromolecules* **1987**, *20 (4)*, 705-712.
37. Langer, R. *Chemistry in Britain* **1990**, *26 (3)*, 232-236.
38. Heller, J. *J. Control. Rel.* **1985**, *2*, 167-177.
39. Heller, J.; Ng, S. Y.; Penhale, D. W. H.; Fritzinger, B. K.; Sanders, L. M.; Bruns, R. A.; Gaynon, M. G.; Bhosale, S. S. *J. Control. Rel.* **1987**, *6*, 217-224.
40. Heller, J. *J. Bioact. Compat. Polym.* **1988**, *3 (2)*, 97-105.
41. Heller, J.; Sparer, R. V.; Zentner, G. M. In *Biodegradable Polymers as Drug Delivery Systems*; Chasin, M. and Langer, R., Ed., Marcel Dekker Inc.: New York, NY, 1990; pp 121-162.
42. Goldberg, E. P. *Targeted Drugs;* Polymers in Biology and Medicine: A Series of Monographs; Vol. 2, Wiley Interscience: New York, NY, 1983.
43. Edman, P.; Ekman, B.; Sjoholm, I. *J. Pharm. Sci.* **1980**, *69 (7)*, 838-842.
44. Li, X.; Bennett, D. B.; Adams, N. W.; Kim, S. W. In *Polymeric Drug and Drug Delivery Systems*; Dunn, R. L. and Ottenbrite, R. M., Ed.; ACS Symposium Series; Vol. 469, American Chemical Society: Washington, DC, 1991; pp 101-116.
45. Nefzger, M.; Kreuter, J.; Voges, R.; Liehl, E.; Czok, R. *J. Pharm. Sci.* **1984**, *73 (9)*, 1309-1311.
46. Drobnik, J. *Adv. Drug Del. Rev.* **1989**, *3*, 229-245.
47. Shen, W. C.; Ryser, H. J. P. *Biochem. Biophys. Res. Commun.* **1981**, *102 (3)*, 1048-1054.

48. Arnon, R.; Hurwitz, E. In *Targeted Drugs*; Goldberg, E. P., Ed., Wiley Interscience: New York, NY, 1983; pp 23-56.
49. Arnold, L. J.; Dugan, A.; Kaplan, N. O. In *Targeted Drugs*; Goldberg, E. P., Ed., Wiley Interscience: New York, NY, 1983; pp 81-112.
50. van Heeswijk, W. A. R.; Hoes, C. J. T.; Stoffer, T.; Eenink, M. J. D.; Potman, W.; Feijen, J. *J. Control. Rel.* **1985**, *1* , 301-315.
51. Kopecek, J. In *Recent Advances in Drug Delivery Systems*; Anderson, J. M. and Kim, S. W., Ed., Plenum Press: New York, NY, 1984; pp 41-62.
52. Yokoyama, M.; Miyauchi, M.; Yamada, N.; Okano, T.; Sakurai, Y.; Kataoka, K.; Inoue, S. *Cancer Res.* **1990**, *50 (6)*, 1693-1700.
53. Yokoyama, M.; Inoue, S.; Kataoka, K.; Yui, N.; Okano, T.; Sakurai, Y. *Makromol. Chem.* **1989**, *190 (9)*, 2041-2054.
54. Yokoyama, M.; Okano, T.; Sakurai, Y.; Kataoka, K.; Inoue, S. *Biochem. Biophys. Res. Commun.* **1989**, *164 (3)*, 1234-1239.
55. Goodson, J. M.; Holborow, D.; Dunn, R. L.; Hogan, P.; Dunham, S. *J. Periodontol.* **1983**, *54 (10)*, 575-579.
56. Domb, A.; Maniar, M.; Bogdansky, S.; Chasin, M. *Crit. Rev. Ther. Drug Carrier Sys.* **1991**, *8 (1)*, 1-17.
57. Rama, B.; Mandel, T.; Jansen, J.; Dingeldein, E.; Mennel, H. D. *Acta Neurochir. Wien.* **1987**, *87* , 70-.
58. Davis, S. S.; Hardy, J. G.; Fara, J. W. *Gut* **1986**, *27* , 886-892.
59. Ch'ng, H. S.; Park, H.; Kelly, P.; Robinson, J. R. *J. Pharm. Sci.* **1985**, *74 (4)*, 399-405.
60. Martinez-Manautou, J. *J. Steriod Biochem.* **1975**, *6* , 889-894.
61. Peppercorn, M. A.; Goldman, P. *J. Pharm. Exp. Ther.* **1972**, *181 (3)*, 555-562.
62. Saffran, M.; Field, J. B.; Pena, J.; Jones, R. H.; Okuda, Y. *J. Endocrin.* **1991**, *131* , 267-278.
63. Lancaster, C. M.; Wheatley, M. A. *Polym. Prepr.* **1989**, *30* , 480-481.
64. Nathan, A.; Zalipsky, S.; Kohn, J. *Polym. Prepr.* **1990**, *31 (2)*, 213-214.
65. Ertel, S. I.; Nathan, A.; Zalipsky, S.; Agathos, S. N.; Kohn, J. In *Polymeric Materials, Science and Engineering*; Vol. 66, American Chemical Society: Washington, DC, 1992; pp 486-487.
66. Nathan, A.; Bolikal, D.; Vyavahare, N.; Zalipsky, S.; Kohn, J. *Macromol.* **1992**, *25* , 4476-4484.
67. Nathan, A.; Zalipsky, S.; Ertel, S. I.; Agathos, S. N.; Yarmush, M. L.; Kohn, J. *Bioconj. Chem.* **1993**, in press.
68. Widder, K. J.; Senyei, A. E.; Scarpelli, D. G. *Proc. Soc. Exp. Biol. Med.* **1978**, *58* , 141-146.
69. Gupta, P. K.; Hung, C. T. *J. Microencap.* **1990**, *7 (1)*, 85-94.

RECEIVED October 9, 1992

Chapter 3

Controlled Release in the Food and Cosmetics Industries

Lisa Brannon-Peppas

Biogel Technology, P.O. Box 681513, Indianapolis, IN 46278

Encapsulation techniques are used in the food and cosmetic industries both to control the delivery of encapsulated agents as well as to protect those agents from environmental degradation. In foods, the most important applications of encapsulation include: (i) encapsulation of flavors, (ii) shielding of oxygen- or water-sensitive components such as vitamins, and (iii) isolation of reactive ingredients until their release is desired *(1)*. Along similar lines, the cosmetics industry utilizes encapsulation to: (i) trap fragrances for controlled or sustained release, (ii) protect particularly volatile components, and (iii) provide release at a delayed time. The preparation methods for these microcapsules may be similar, whether they contain flavors, fragrances, or drugs. This paper will present an overview of the uses of microcapsules in the cosmetics and food industries, with an emphasis on the materials used and the goals and uses of the final products.

Controlled release systems have been developed in the past thirty years for a variety of "traditional" applications, i.e. applications in pharmaceutical and agricultural applications. However, it may not be realized that applications of microencapsulation in the "nontraditional" areas of cosmetics, food, fragrance, paper technology, and other fields have been prominent for far longer. Encapsulation has been used in the food industry for nearly sixty years. Encapsulation by spray drying was first used for flavor enhancement and protection in the 1930's *(2)*. Concentric nozzle techniques were developed in the 1930's and 1940's to protect vitamins. In fact, several recent marketing surveys *(3)* indicate that the market for these products is significantly higher than for pharmaceutical and agricultural products.

This review will present encapsulation techniques and uses of flavors, fragrances, food additives and agents for cosmetic applications. Most of the techniques will not be described in extreme detail, since they are presented elsewhere in this volume. However, any differences between these techniques and those usually found in pharmaceutical applications will be emphasized.

0097–6156/93/0520–0042$06.00/0
© 1993 American Chemical Society

Controlled Release in "Nontraditional" Applications

Controlled release devices and systems used in the so-called "nontraditional" areas are used in the following applications:

(1) fragrance and flavor release systems,
(2) flavor and active ingredient release in food applications,
(3) active ingredient release for cosmetic applications,
(4) bioactive agent and fragrance release for pet products,
(5) active ingredient release of bleaching and fragrance products in cleansing products,
(6) release of fragrances and related compounds for air fresheners, space release systems, botanicals, potpourri, and household devices, and
(7) release systems related to paper technology including pressure sensitive flavor-releasing systems and carbonless paper.

Depending upon the formulation that is used for the release system, the systems may be described as:

(1) encapsulated (shell or core) systems,
(2) microparticulate matrix systems,
(3) cylindrical and planar matrix devices, and
(4) osmotic and other membrane systems.

Of the above mentioned systems, microencapsulated devices (category 1) continue to be the most attractive systems in the consumer products field. Indeed, about 85% of products in the markets in those fields are encapsulated products *(3)*. The reason for this preference is that microencapsulation has been a well-understood and relatively easily applied technique that can be used to prepare release systems for a variety of volatile and non-volatile, stable and sensitive, biological and non-biological products. Since 1956, when the first patents of B. Green for NCR on the development of carbonless paper for microencapsulation of inks appeared *(4,5)*, the field has grown in sophistication and variety of products.

Mechanisms of Release. A very important classification of the controlled release devices in this field is based upon the mechanism of release or method of triggering the release behavior. From this point of view, controlled release devices in the consumer products field can be divided into the following ten categories (as shown in Table I):

(1) <u>Diffusion controlled release devices</u>: Controlled release devices based on diffusion control are systems where the release of the volatile or non-volatile active ingredient is controlled by the diffusional process of this ingredient through the polymer carrier. Such diffusion may be through the polymer proper (carrier of the device) or through pores preexisting in the polymer. If the component is a volatile material (eugenol release from a base polymer of a chewing gum) the release can be augmented by the evaporation characteristics of the active ingredient. Such diffusional mechanisms can be found in a variety of consumer products.

For example, the release of various fragrances from strips of polymers in air fresheners or the transport of flavorings in chewing gums are classical diffusional mechanisms. It must be noted that this diffusional process does not have to occur through a polymer in order for the release process to be controlled by the carrier. In fact, many food applications are based upon the utilization of the food itself as the carrier for the release. In this field, where only a small number of carriers have been approved by FDA for human consumption, the food itself is modified by a number of processes (e.g. freeze drying, other drying techniques) in order to impart porosity or other characteristics which would lead to controlled delivery of incorporated flavors. Thus, Reineccius and his associates have reported on various ingenious techniques of delaying the release of orange oil from food products by either modification of the food itself or by incorporation of small amounts of porous silica or other materials (6). Orange oil is the least stable flavor and is often used as model systems to ascertain the viability of a controlled delivery system.

Table I.   Mechanisms of Release from Controlled Delivery Systems in Consumer Products

---

Diffusion Controlled Release

Membrane Controlled Release

Pressure-Activated Release

Tearing- or Peeling-Activated Release

Solvent-Activated Release

Osmotically-Controlled Release

pH Sensitive Release

Temperature-Sensitive Release

Melting-Activated Release

Hybrid Systems

---

(2) Membrane controlled systems: Encapsulated controlled release devices are in principle membrane (reservoir) systems. As most of these systems are microspherical, containing a core of active ingredient, and surrounded by an extremely thin membrane of the encapsulant polymer, one would expect that the release behavior would be controlled by the same principles as for the classical controlled release systems (7). In theory, this would means that if all microcapsules were of the same external diameter and the same wall thickness, and if they were all placed in a fluid and released under perfect sink conditions, they should give zero-order release behavior. In reality, this is almost impossible as the microcapsules are of a wide size range and may have different wall thicknesses (8). Still, the rate of release is controlled by the concentration difference across the wall of the microcapsules, the thickness of the wall, permeability of the active ingredient through the wall and, of course, the diffusion coefficient of the active ingredient with respect to the surrounding environment. All of the microencapsulation systems that will

release their contents by that classical release process may be considered to be membrane-controlled release systems.

(3) Pressure-activated controlled release systems: A number of controlled release systems, including many of those used in paper technology, are triggered by the application of pressure on the walls. Such systems are usually microencapsulated devices but their walls are rather dense and the incorporated active ingredients are relatively non-volatile. Thus, neither diffusional release due to concentration gradients nor evaporation of active ingredient through any pores should be expected to contribute to the release mechanism. Instead, the prominent mechanism of release is rupture of the wall due to application of external pressure. Such is, for example, the case with microencapsulated inks adhering to the back of a paper in carbonless paper. The pressure applied by a pen on the front of the paper ruptures the microcapsules, immediately releasing their contents. Similar systems are those based upon Scratch 'n Sniff technology where, again, a shear action of a fingernail on a piece of paper would rupture the microcapsules.

(4) Tearing or peeling activated systems: Tearing or peeling activated release systems are an unusual type of controlled release devices which have been finding more and more application in the field of fragrance release or in applications related to print media advertising. In such systems, microcapsules are usually placed, with an adhesive, between two layers of paper or polymer film. The microencapsulated systems may contain fragrance oils, inks, colorants, and other related agents. The user usually applies a peeling action by pulling apart the two layers of paper (or polymer film) thus rupturing the microcapsules and releasing their contents. Such systems have found wide use in various systems including the technologies of Microfragrance, Colorburst, and Fragrance Burst.

(5) Solvent-activated systems: A number of controlled release products and devices in the consumer field are based upon diffusion of the surrounding fluid through the surrounding polymer carrier. Typically, water or a related aqueous solution penetrates the polymer carrier, swells it, and leads to a faster release of the incorporated active ingredient. In some cases, the polymer (or other encapsulating material) will completely dissolve. Thus, the swelling process controls the overall release of the incorporated active ingredient. A number of consumer products reported in the patent literature utilize this system, especially in the field of fragrances and flavor release. This is a particularly favored mechanism in a large number of Japanese patents (*9-11*).

(6) Osmotically-controlled devices: Certain types of microencapsulated controlled release devices are triggered by a large osmotic pressure that is created in the interior of the capsule due to the presence of an osmotically active agent. Indeed, if a chemical compound encapsulated in a particular device has a very high solubility in water, a large osmotic pressure can be created. Narkis and Narkis (*12*) and Theeuwes (*13*) have calculated the typical values of osmotic pressure that can be created by some of these compounds. If the osmotic pressure created is higher than the mechanical integrity of the microcapsule walls, then the microcapsules will burst, releasing their contents. A typical device triggered by such an osmotic phenomena would be a microencapsulated device containing bleaches that would not release during the early stage of the washing cycle, but would release only after a certain time due to osmotic rupture.

(7) pH sensitive systems: In the field of cosmetic products, there has been a recent explosion with respect to intelligent devices that respond to changes in pH of

the user. For example, systems have been reported in the patented literature *(14)* that are sensitive to changes of the pH of the skin. As a result, of that change, the polymer carrier collapses, thus "squeezing out" its active contents. This process is, in principle, reversible. Systems like these are still at their infancy in the cosmetic area *(15)*, but there is a strong indication that they will be more acceptable in the future because the cosmetic market is one where the mark-up is extremely high, thus allowing for a higher cost of production (due to the relatively high cost of pH-sensitive materials) if the final product is to exhibit unusual properties such as pH sensitivity.

(8) <u>Temperature-sensitive systems</u>: Certain types of controlled release devices are being researched which are based on polymeric carriers that are temperature sensitive. In recent presentations, both Miles and Peppas *(15,16)* have noted that certain polymeric materials are sensitive to small changes to temperature, around room temperature, leading to either expansion or collapse of their structure. Such systems, could therefore exhibit dimensional reduction leading to exudation of their contents upon a small change of temperature. As in the previous category, such systems could be beneficial in the cosmetic and personal care fields. It must be noted that the temperature sensitivity is centered around the critical temperature of the polymeric carrier and does not involve melting.

(9) <u>Melting-activated release systems</u>: A number of controlled release systems for consumer applications are based upon a phenomena of melting of either the polymer carrier itself or the active ingredient. The most useful case, is that of melting of the encapsulant walls, leading to an abrupt release of the content of the microcapsules. Again, systems like this have been proposed for release in cleansing products, especially in situations where one would expect a particular ingredient to be release only during a drying cycle or during ironing. In addition, such encapsulated products are also used in baking mixes and are designed to be released only during baking.

(10) <u>Hybrid systems</u>: Numerous controlled release devices may exhibit a combination of several triggering events and release mechanisms. For example, a novel deodorant stick device may contain microencapsulated fragrance oils which would release by a combination of temperature increase, sweat production, and pH change. Such systems are presently under development and are even commercially available.

Encapsulation in the Food Industry

The effectiveness of encapsulation of flavors depends strongly on the concentration of the flavor as well as the method of encapsulation. The most successful encapsulant must be molecularly impenetrable to the flavor within it. In addition, the wall surrounding the flavor must have outstanding physical integrity which allows for no undesired breakage. The encapsulant should also be impenetrable to oxygen to increase the oxidation stability of the flavor.

Methods of Encapsulation of Essential Oils. Encapsulated flavors are produced by a variety of techniques, depending upon the desired final form of the controlled release system as well as the nature of the flavor (usually an essential oil) to be encapsulated and the encapsulating material. In addition to controlled release and flavor protection, microencapsulation of flavor may provide an additional advantage in that it converts liquid material into a solid which is significantly easier to handle*(17)*. Approximately 30 million pounds of encapsulated flavoring are used in the United States each year. These products may be formed by spray drying, extrusion, molecular inclusion, or coacervation. A summary of the advantages and disadvantages of each method are given in Table II. The methods used to encapsulate fragrances are similar, if not identical, to the methods used to encapsulate flavors. The essential oils, which are the active agents in both applications, are closely related and sometimes the same oil may be used for applications in both both flavor and fragrance controlled release.

Table II.   Encapsulation Processes for Flavors

Microencapsulation Technique	Advantages	Disadvantages
Spray Drying	Low cost Availability of equipment Good protection Variety of wall materials	Small particles (dispersibility) Not for extremely low boilers Limited shelf life (oxidation)
Extrusion	Excellent shelf-life Visible "pieces" of flavor	Low flavor load (8-10 wt%) Slowly soluble (variable) High temperature process
Molecular Inclusion	Low process costs Very stable	Low flavor load (6-15 wt%) Low solubility Nonuniform component binding Cost
Coacervation	True encapsulation High load (25-97 wt%) Low cost Good shelf-life	Large particle size (20μm-2mm) Material and wall must be immiscible Boiling point limitations

The characteristics of spray dried products are highly dependent upon the processing conditions of the encapsulation procedure. The oxidation of the essential oil in the final microcapsule (and therefore its shelf life) is directly related to the nature of the encapsulating material (usually a wax), the storage time, and the emulsion size prior to the spray drying. The factors which most directly influence the shelf life of these products are the presence of surface oil, microcapsule porosity, the presence of trace pro-oxidants or anti-oxidants, and the quality of the oil. Spray dried products are readily producible at low cost, with available equipment, but tend to have a shorter shelf life than other encapsulated systems due to their lack of oxidation protection for the essential oil.

Extruded products are usually formed by passing a mixture of the essential oil, a sugar-starch hydrolysis mixture as an encapsulant, and possibly an antioxidant, through an extrusion die into a bath of non-solvent where the extruded pieces are broken into particles by impact breakage. This preparation procedure yields products with an excellent shelf-life (1500 days or more at 25°C), but with a low flavor load (8-10 wt%).

Molecular inclusion is accomplished by trapping an essential oil within the center of molecules of cyclodextrins. This process involves solubilizing the cyclodextrin to be used, adding the essential oil, mixing, filtering, and drying. The process is very simple, giving a very stable product which, unfortunately, has a low flavor load (6-15 wt%), is expensive, and exhibits nonuniform binding.

Coacervation may produce microcapsules with walls of waxes, fats, proteins, carbohydrates, or gelatins. These procedures are usually more complex than those previously mentioned, with a multitude of ingredients necessary: essential oil, antioxidant, surfactant, hydrophilic colloid (wall material), plasticizer, oxygen scavenger, and sequestrant. However, the flavor loading may reach 25-97 wt%, unlike other methods. The procedures utilize available equipment and well-developed processes, so they are comparatively low cost. In addition to essential oils, hydrophilic and hydrophobic solids as well as suspensions may be encapsulated using coacervation.

Uses of Encapsulated Flavors. There are several hundred types of microcapsules being used as food additives in the U.S. today. Direct, intentional addition of artificial flavors, natural flavors, spices, nutrients, preservatives and other additives to enhance or alter the appearance, composition, and quality of food is widespread *(18)*. Encapsulation is used to protect flavors and to retard their release so that the flavor of a food product will last for a longer time than with unencapsulated flavor. Some food products which may use encapsulated flavors are dry beverage mixes, food mixes (cakes, biscuits, pastries), and instant or ground coffee. In addition, chewing gums almost always contain encapsulated flavors, and may well contain more than one flavor. Aspartame, an artificial sweetener, is usually encapsulated because it easily decomposes due to the presence of water, elevated temperatures, and may react unfavorably with other flavoring compounds.

The most common release mechanisms for release of flavors (and other food additives) are:

(1)   Compressive force which breaks the capsule open by mechanical means (i.e. pressing or chewing),

(2)   Shear force which breaks open the capsule (i.e. as in a blender),

(3)   Dissolution of capsule wall (i.e. dissolution of dry beverage mix in water),

(4)   Melting of capsule wall (i.e. during baking), and

(5)   Core diffusion through the wall at a slow rate due to presence of external fluid (i.e. water) or elevated temperature.

These and other methods of release for consumer products are summarized in Table I and have been fully described earlier in this chapter.

Encapsulation of Other Food Ingredients and Additives. Other encapsulated food additives may serve as preservatives, antioxidants, redox agents, acidity buffers, colors, nutrients, and enzymes. The methods already described for encapsulation of flavors (see Table I) may be used as well for entrapment within sugar crystals of compounds which are strongly absorbed physically or chemically *(19)*. The triggers to release these additives are usually temperature or moisture levels (for hydrophilic encapsulants) or temperature level alone (for fat-based encapsulants). Other, less common, release triggers are pH, enzymatic release, ultrasonics, grinding and photo-induced release.

Cheese Ripening and Flavor Enhancement.   A slightly different approach to the encapsulation and controlled release of enzymes in food applications is the use of liposomes to immobilize enzymes. These enzyme-loaded lipid vessicles may assist in improved flavor as well, through controlled modification of enzymatic processes. One of the most useful applications of this technology is the use of these encapsulated enzymes in cheese ripening *(20)*. Both milkfat-coated microcapsules and liposomes have been used to successfully encapsulate cell-free extracts, viable cells, purified enzymes, and spores. Different techniques have been used to achieve the goals of acceleration of cheese ripening and improvement of cheese flavor. Cell-free extracts of *Streptococcus lactis* subsp. *diacetylactis* were encapsulated with substrate and cofactors for generation of diacetyl and acetoin, two of the flavors found in cheeses. It was found that cheese containing the intact microcapsules contained up to eight times more acetyl than control cheese *(21)*. Other cell-free extracts have been studied which can enhance the production of other cheese flavors such as 3-methyl butanal and 3-methyl butanol *(22)*. Therefore, by including appropriately encapsulated cell extracts, the flavors of cheeses may be enhanced and modified by creating a cheese with different ratios of flavoring compounds than would be found in the control cheese.

A number of methods for reducing the ripening time for cheese have also been investigated *(23-25)*. One method *(23)* is to encapsulate enzymes such as neutrase to protect them during curd formation, but to encourage release during ripening and therefore to lead to an increased hydrolysis of ß-casein and faster ripening. These capsules are well retained in the cheeses (90%) and, if used appropriately, can cut the ripening times in half. Encapsulated enzymes have proven to be significantly more useful in cheesemaking than unencapsulated enzymes *(19)*. Unencapsulated enzymes, added to the milk prior to cheesemaking, produce a cheese with poor textural quality, high levels of whey contamination and a low yield. Unencapsulated enzymes, added to the curd, produce a better cheese yield, but the distribution of the enzyme is poor and the cheese is crumbly. Encapsulated enzymes may be added to the milk before cheese-making and do release the enzymes appropriately to increase cheese yield as will as to decrease ripening time. Unfortunately, the encapsulation efficiency of these techniques must be increased (currently < 35%) before encapsulated enzymes will be widely used in cheese manufacture. In addition, more research must be conducted to evaluate the findings that these capsules migrate to areas between the fat globules and casein matrix and whether or not that is the optimal location for enzyme release during cheese production.

Meat Acidulant.   Especially in meat products such as sausages, it is desirable to acidify the meat after processing to stabilize the meat and inhibit undesirable bacterial formation *(26)*. In the past, this has been accomplished by inoculating the sausage with lactic-acid producing bacteria, direct acidification, addition of glucono delta lactone, and coating of granular acids with fat or starch before incorporation. All of these methods have their disadvantages, with the two coated systems either releasing their acidulant too quickly (starch coating) or too late if at all (fat coating melting at 100-150 °F, which only occurs during <u>some</u> cooking processes). An improved encapsulation method uses a fluidized bed to coat granular acid with a 40:60 mixture of hydrogenated vegetable oil and water-soluble glycerides. These microcapsules have been shown to keep the pH of sausage in the acidic range (4.4-8.85) for up to 24 hours at 80 °F and 48 hours at 32 °F.

Encapsulation in the Cosmetics Industry

In many fragrance and cosmetic applications described in this chapter, it is interesting to note that one product may contain more than one encapsulated product, prepared by different technologies. For example, a liquid makeup foundation may very well contain encapsulated fragrance, sunscreen, humectant, and a sebum oil absorber. The most emphasis will be placed on the release of fragrances, since encapsulated fragrances are used significantly more often in cosmetics products than any other encapsulated materials.

Encapsulation of Fragrances. Fragrance release is of importance not just in perfumes, colognes, and related products. In fact, a report on cosmetics raw materials states that the fragrance ingredients are the crucial materials for all cosmetics, with the influence of scent extending far beyond that of the classic perfume blends (27). Controlled release of fragrance is far more pervasive than most realize. A list of some products which <u>currently</u> use encapsulated fragrance in marketed products is shown in Table III. This list, even though it is lengthy, probably does not include all of the currently marketed applications of controlled fragrance release and it certainly does not even touch on potential markets (28). What is most interesting to note is that, almost unseen, controlled release has become a major part of everyday life in the use of these products. The methods by which fragrances can be released are similar to that of flavor release and have been summarized in Table I and already have been described in detail.

Table III. Products Currently Using Controlled Release of Fragrance

---

Laundry Care	Paper Products
Specialty Cleaners	Air Fresheners
Pet Products	Hair Care Products
Bath Products	Body Powders
Soaps	Lipsticks
Skin Care Products	Antiperspirants
Deodorants	Athlete's Foot Powders
Potpourri	Perfume
Cologne	Fragrance Samplers

SOURCE: Adapted from ref. 28.

Two major advantages in utilizing controlled release of fragrances, which are not as important in flavor release, are fragrance marketing and cost savings. Fragrance samples, as included in magazines, give consumers an opportunity to try a fragrance (29). The consumer is then more likely to purchase that perfume, if they like the perfume, than if they had not had the opportunity to sample the product before purchase. For most fragrances, and especially the finer fragrances which can cost more that $6000 per kilogram, encapsulation will stabilize the fragrance and controlled release will prolong the lifetime of one application. Therefore, an encapsulated product will have a longer shelf life (of benefit to manufacturer and consumer) as well as a longer utility per application (of benefit to consumer).

Encapsulation of Other Agents in Cosmetic Applications. Other than fragrances, humectants and "anti-aging" compound have been most often formulated in controlled release systems in cosmetics *(30)*. Typical coatings for encapsulated products are shown in Table IV. Mixtures of these materials may also be used as coatings. The release from these formulations is usually triggered by shear force (rubbing), temperature, volatility, solubility and diffusivity *(31)*. The "anti-aging" products are also often formulated in liposomes and have sales of $100 million per year in the United States alone. It is in the release of these agents such as retinoic acid that the distinction between cosmetics and pharmaceuticals narrows *(27)*. These agents actually increase the turnover rate of skin cells, leaving skin in better condition when the use is stopped.

Table IV. Typical Coatings for Encapsulated Products in Cosmetics

Acrylates	Glycolide Copolymers
Aminoplasts	Gum Arabic
Cellulose Acetate Phthalate	Shellac
Ethylcellulose	Starch/Dextrin
Gelatin	Polystyrene
Gelatin Complexes	Waxes

SOURCE: Adapted from ref. 30.

Conclusions

Controlled release, especially in the form of microencapsulated products, has been used for fragrance and flavor release for decades. In fact, controlled release is significantly more widespread in such applications in consumer products than in pharmaceutical and agricultural applications. Microencapsulated products have enhanced the flavor, stability, and ease of use of products as various as baking mixes, drink mixes, sausage, magazine perfume samples and sunscreens. The challenges of producing large amounts of encapsulated product at very low costs has developed technology which can be and has already been transferred to many pharmaceutical encapsulation processes, to the benefit of all concerned.

Literature Cited

1.   Langer, R.; Karel, M. *Polymer News* **1981**, 7, pp. 250-258.
2.   Thies, C. Second Workshop on *Controlled Delivery in Consumer Products: Encapsulation Processes for Controlled Delivery Applications*, Controlled Release Society, May 13-15, 1992.
3.   Second Workshop on *Controlled Delivery in Consumer Products: Encapsulation Processes for Controlled Delivery Applications*, Controlled Release Society, May 13-15, 1992.

4. Green, B. U.S. Patent 2,730,456.
5. Green, B. U.S. Patent 2,730,457.
6. Bolton, T.A.; Reineccius, G.A. *Perfumer and Flavorist* **1992**, *17(2)*, pp. 17-22.
7. Langer, R.; Peppas, N.A. *J. Macromol. Sci.-Rev. Macromol. Chem. Phys.* **1983**, *C23*, pp. 61-126.
8. Sparks, R. *Microencapsulation, Book of a Short Course*, Center for Professional Advancement, 1983.
9. Yamazaki, S. Japanese Patent 79 70,373, June 6, 1979.
10. Miyashita, T. Japanese Patent 62 67,016, March 26, 1987.
11. Kumagai, S. Japanese Patent 61 60,605, March 26, 1986.
12. Narkis, N.; Narkis, M.; *J. Appl. Polym. Sci.* **1976**, *20*, pp. 3431-3436.
13. Theeuwes, F. *J. Pharm Sci.* **1975**, *64(12)*, pp. 1987-1991.
14. Manning, A.J. Mathur, K.K., Berman Patent 2,551,891, July 8, 1976.
15. Miles, J. Second Workshop on *Controlled Delivery in Consumer Products: Encapsulation and Delivery of Perfumes and Fragrances*, Controlled Release Society, May 13-15, 1992.
16. Peppas, N.A.; Brannon-Peppas, L. *J. Membr. Sci.*, **1990**,*48*, pp. 281-290.
17. Reineccius, G. First Workshop on *Controlled Delivery in Consumer Products: Flavor Encapsulation,* Controlled Release Society, October 18-19, 1990.
18. Versic, R.J. In *Flavor Encapsulation*; Risch, S.J., Reineccius, G.A., Eds.; American Chemical Society, Washington, D.C., 1991; pp. 1-6.
19. Karel, M.; Langer, R. In *Flavor Encapsulation*; Risch, S.J., Reineccius, G.A., Eds.; American Chemical Society, Washington, D.C., 1991; pp. 177-191.
20. El Soda, M.; Pannell, L.; Olson, N. *J. Microencapsulation* **1989**, *6(3)*, pp. 319-326.
21. Magee, E.L.; Olson, N.F. *J. Dairy Science* **1981**, *64(4)*, pp. 616-621.
22. Braun, S.D.; Olson, N.F. Lindsay, R.C.; *J. Food Biochemistry* **1983**, *7*, pp.23-41.
23. Law, B.A., King, J.S. *J. Dairy Research* **1985**, *52*, pp. 183-188.
24. Piaro, J.C.; El Soda, M.; Alkahalaf, W.; Rousseau, M.; Desmazeaud, M.; Vassal, L.; Gripon, J.C. *Biotech. Letters* **1986**, *8(4)*, pp. 241-246.
25. Kirby, C.J.; Brooker, B.E.; Law, B.A. *Int. J. Food Sci. Tech.* **1987**, *22*, pp. 355-375.
26. Reineccius, G. First Workshop on *Controlled Delivery in Consumer Products: Controlled Delivery Applications in the Food Industry*, Controlled Release Society, October 18-19, 1990.
27. Layman, P.L. *C & E News* **1988**, April 4, pp. 12-19.
28. Miles, J. First Workshop on *Controlled Delivery in Consumer Products: Encapsulation of Fragrances*, Controlled Release Society, October 18-19, 1990.
29. Fraser, A. First Workshop on *Controlled Delivery in Consumer Products: Controlled Release Technology in Print Media Advertising* Controlled Release Society, October 18-19, 1990.
30. Magill, M.C. First Workshop on *Controlled Delivery in Consumer Products: Controlled Delivery in Cosmetic and Personal Care Products*, Controlled Release Society, October 18-19, 1990.
31. Abrutyn, E. Second Workshop on *Controlled Delivery in Consumer Products: Polymeric Transport Systems for Personal Care Applications*, Controlled Release Society, May 13-15, 1992.

RECEIVED October 5, 1992

Chapter 4

Biodegradable Polyesters for Drug and Polypeptide Delivery

Patrick P. DeLuca, Rahul C. Mehta, Angie G. Hausberger, and
B. C. Thanoo

College of Pharmacy, University of Kentucky, Lexington, KY 40536-0082

A thorough knowledge of the physico-chemical properties of biodegradable polymers of lactide/glycolides and their delivery systems is a much needed but usually missing asset in formulating high efficiency delivery systems for drugs and polypeptides. Polymer properties including chemical composition, sterioisomeric form and molecular weights affect the rate and extent of drug delivery and rate of polymer degradation following parenteral administration. Properties of microsphere delivery systems such as size, surface area, porosity, drug content and moisture content affect the route of administration, distribution and deposition as well as ultimate clearance of drugs and delivery systems. The importance and application of these physico-chemical properties are presented with examples of three drugs with different intended therapeutic applications.

Biodegradable materials which are also biocompatible are being utilized in medical applications as drug carriers, surgical sutures and even various prostheses that, once administered to a patient, do not require surgical removal. These biodegradable materials can be made from natural substances, such as albumin, gelatin, collagen and polysaccharides or from synthetic polymers. The advantage of the latter are that they can be chemically classified, are non-immunogenic and can be prepared economically in large quantities. These attributes are among those which are essential for a drug carrier which is to be administered by the parenteral route. Additional requirements for an acceptable biodegradable polymer for drug release systems include degradation into non-toxic metabolites, absence of

0097-6156/93/0520-0053$07.75/0
© 1993 American Chemical Society

impurities such as significant amounts of catalyst, stabilizer or emulsifier residues, and ease of processing.

Synthetic Biodegradable Polymers. Yolles, et. al., were the first to report the use of a synthetic biodegradable polymer for parenteral administration of a drug, specifically cyclazocine from a poly(L-lactic acid) implant (*1*). A drawback of this device was the lengthy <u>in vivo</u> degradation time of several years for the polymer device.   Frazza and Schmitt developed an absorbable suture composed of poly(glycolic acid) which possessed a more acceptable period of 60 days for virtually complete <u>in vivo</u> degradation (*2*). Following these early studies, many different drugs, including anticancer agents, narcotic antagonists and steroids, have been incorporated into microparticles, films and similar devices fabricated from various biodegradable polyester copolymer systems resulting in a wide range of degradation and release behavior (*3-8*). Because of the amorphous nature of the copolymers and the reduced particle size of the carriers, degradation occurred in a matter of weeks instead of months.

Upon review of the literature it becomes apparent that there is a need to understand the terminology for describing polymer behavior in the living organism. Drug carriers for parenteral applications are considered <u>biomaterials</u>, a term which encompasses all materials introduced into body tissues for specific therapeutic, diagnostic or preventive purposes.   These materials must be <u>biocompatible</u> or histocompatible, meaning they must not cause any significant adverse response to the physiologic environment or damage to the biomaterial. Upon interaction with body tissue and fluids they must undergo <u>biodegradation</u> into non-toxic components, either chemically or biologically, or a combination of both.   Other interchangeable terms for biodegradation include <u>bioerosion</u> and <u>bioabsorption</u>.

Application in Drug Delivery. As drug delivery devices, it is desirable to have polymers which degrade by a hydrolytic process. In some cases, release of the incorporated agent will depend on the rate of degradation of the matrix material. The nature of degradation, heterogeneous or homogeneous, will determine the release kinetics, i.e., zero or higher order. <u>Heterogeneous degradation</u> occurs at the surface of the polymeric carrier where it is interfaced with the physiologic environment. In this case, the degradation rate is constant, and, as illustrated in Figure 1, the undegraded carrier retains its chemical integrity during the process. Logically, carriers possessing high surface to volume ratios would undergo faster degradation than their lower ratioed counterparts. <u>Homogeneous degradation</u> involves a random cleavage throughout the bulk of the polymer matrix. While the molecular weight of the polymer steadily decreases, the carrier can remain in its original shape and retain mass until the polymer has undergone significant degradation, i.e., as much as 90%, and reaches a critical molecular weight, at which time solubilization and mass loss commences.

Classes of Biodegradable Polymers. Some of the classes of biodegradable polymers which have been reported for applications as drug carriers are listed in Figure 2.

HETEROGENOUS

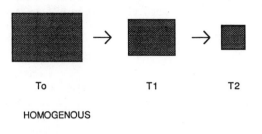

To   T1   T2

HOMOGENOUS

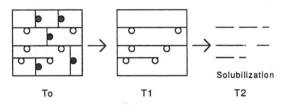

Solubilization

To   T1   T2

Figure 1. Mechanisms of polymer matrix degradation.

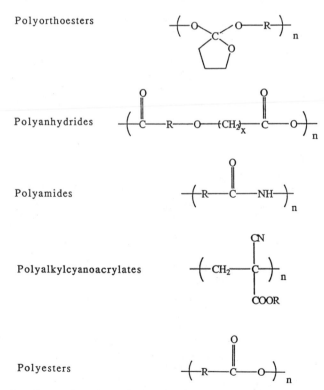

Figure 2. Structures of synthetic biodegradable polymers used for drug delivery applications.

The polyorthoesters are hydrophobic polymers which can be prepared by transesterification (9) or by simple and rapid condensation of diketene acetals with diols (10). The orthoester bond is relatively stable in alkaline conditions but undergoes hydrolysis in acidic environments. While the pure polymer exhibits homogeneous degradation, an initially heterogeneously-eroding polyorthoester has been made by incorporating a stabilizing basic salt, such as sodium carbonate, into the polymer matrix (11).

The polyanhydrides can be prepared in a range of crystallinities and hydrophobicities to enhance degradation and release. Degradation occurs by a two-step process, specifically hydrolysis of the anhydride group followed by ionization of the resulting carboxyl functional group (12). Some solubilization can also occur as the backbone is cleaved to lower molecular weight fractions. However, these polymers which are stable at acid pH degrade principally by a heterogeneous process. Erosion is pH dependent and enhanced at alkaline pH.

The synthetic polyamides, based on poly(glutamic acid) copolymers, are highly stable because of the amide group. They degrade by hydrolysis of ethyl glutamate ester groups into glutamic acid groups (10). This increases the hydrophilicity, and therefore the erosion of the polymer matrix. Degradation rates can somewhat be affected by controlling the ratio of glutamic acid units in the copolymer. However, degradation can also be influenced by the presence of non-specific amidases which cleave the amide group. This enzymatic degradation has been reported to be unreliable and non-reproducible, thus limiting the effectiveness of the polyamides as controlled drug delivery systems (10).

The poly(alkylcyanoacrylates) have received attention both as drug carriers and as surgical adhesives (9). Their spontaneous polymerization upon introduction to moist tissue makes them good candidates for the latter purpose. Erosion has been shown to be controlled by the length of the monomer chain and the pH. Heterogeneous degradation involving chain-end hydrolysis of the higher molecular weight polymers has been reported as the degradative pathway (13). While the polymers are not toxic, formaldehyde as a by-product of degradation, creates a toxicity concern. Predictably, the degree of toxicity increases with increasing degradation rate.

The polyesters are by far the most widely studied synthetic polymers for drug delivery. They are easily prepared as homo and co-polymers in a wide range of molecular weights. They undergo simple hydrolysis forming naturally occurring non-toxic metabolites. In addition, their relative strength and the long-standing safe use and acceptance of the lactide/glycolide polymers as surgical sutures has encouraged the application of the polyesters as drug carriers. The degradation pathway is by homogeneous, bulk degradation.

Biodegradable Polyesters

The synthetic biodegradable polymers which have been studied most extensively are the lactide/glycolide homo- and copolymers of the polyester class. The structures of the homopolymers are illustrated in Figure 3. These polymers may be synthesized by two methods; a) the direct polycondensation of lactic and/or glycolic acids with antimony trioxide (14) or b) the ring opening melt condensation of the cyclic diesters of lactide and/or glycolide using such catalysts as antimony, cadmium, titanium, zinc, tin, or certain amines (*14-18*). Both procedures are illustrated in Figure 4. Polycondensation produces only relatively low molecular weight polymers (< 10 kDa) which are not suitable for many applications, so the ring opening melt condensation of the cyclic dimers is the preferred method.

Terminology is also important in describing these polymers. Often polyglycolic acid or polylactic acid are used interchangeably with polyglycolide and polylactide. Polymers prepared from the cyclic dimers, the lactide or glycolide, should be designated as such while the terms poly(glycolic acid) and poly(lactic acid) should be used to describe those polymers prepared by the direct polycondensation of the monomers.

The lactide/glycolide polymers used as drug carriers are listed in Table I along with some properties. The homopolymers, poly-L-lactide and polyglycolide are crystalline in nature with crystallinities of 37% and 50% respectively as determined from x-ray diffraction and differential scanning calorimetry (DSC). The presence of these crystalline domains make the homopolymers more resistive to hydrolytic degradation.

TABLE I.   **Glass Transition and Melting Temperatures For PLA/PGA Homo- and Copolymers**

Polymer	Mol. Wt.[a]	$T_g(°C)$[b]	$T_m(°C)$[b]
poly(L-lactic acid)	2,000	40	140
poly(L-lactide)	50,000	65	175
poly(L-lactide)	100,000	60	180
poly(D,L-lactide)	-----	52	None
poly(D,L-lactide-co-glycolide)85:15	232,000	49	None
poly(D,L-lactide-co-glycolide)75:25	63,000	48	None
poly(D,L-lactide-co-glycolide)50:50	12,000	40	None
poly(D,L-lactide-co-glycolide)50:50	34,000	45	None
poly(D,L-lactide-co-glycolide)50:50	48,000	45	None
poly(D,L-lactide-co-glycolide)50:50	80,000	45	None
poly(D,L-lactide-co-glycolide)50:50	98,000	47	None
poly(glycolide)	50,000	36	210-220

a)   Information supplied by manufacturer
b)   Determined by a Perkin Elmer DSC 7

The stereoirregularity present with the lactides also will inhibit degradation. Therefore, poly L-lactide, because of its crystallinity and stereoirregularity, will take the longest of these polymers to degrade. The influence of stereoirregularity can be further emphasized by the fact that the amorphous homopolymer poly DL-lactide, ranks next in order of degradation time and ahead of the highly crystalline polyglycolide. As shown in Figure 5 for the copolymers of L-lactide and glycolide, a decrease in crystallinity (and melting point) occurred with an increase in either comonomer. Poly-L-lactide-co-gylcolide having a glycolic acid

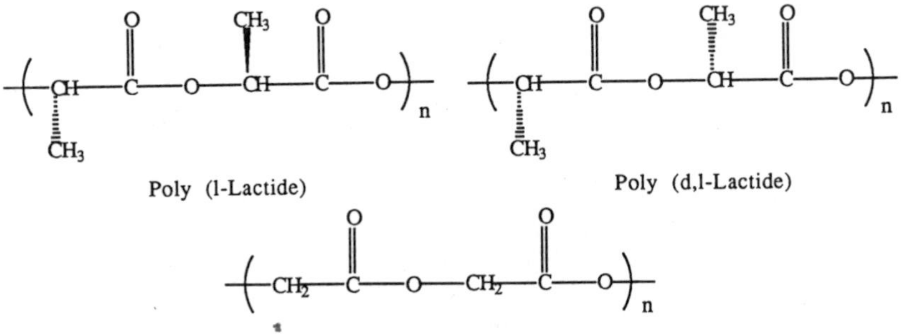

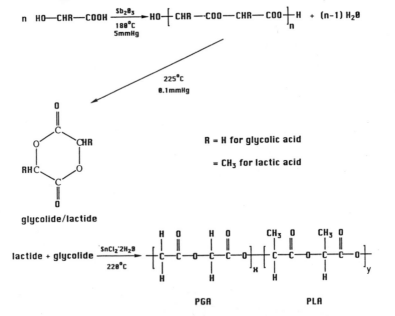

Figure 3.   Structures of poly-l-lactide, poly-d,l-lactide and polyglycolide homopolymers.

Figure 4. Schemes for the synthesis of polylactide/glycolide polymers from straight chain or cyclic monomers.

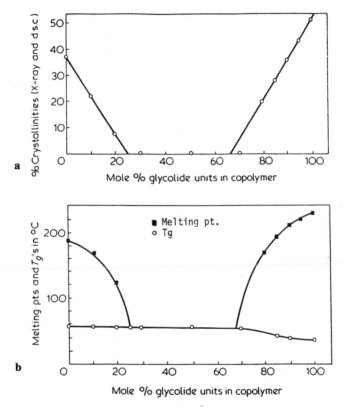

Figure 5. a) Crystallinity b) melting point and glass transition temperature of polylactide/glycolide polymers as a function of comonomer ratios. (Reproduced with permission from ref. 19. Copyright 1979 Butterworth-Heinemann Limited)

composition between 25 and 75% will be amorphous. Figure 6 shows that the co-polymers degraded faster than the homopolymers with the 50:50 composition being the least stable. Since the homopolymer of D,L-lactide is completely amorphous, the copolymer of D,L-lactide and glycolide will be amorphous from 0 to 70% glycolide. It has been established that the rate of degradation of these polymers decreases with increasing crystallinity (19,20).

Biodegradation times will vary depending on the molecular weight, the sequencing and crosslinking within the polymer backbone as well as the surface area and porosity of the matrix or carrier. The lactide/glycolide carriers degrade predominantly by a homogeneous "bulk" hydrolytic process illustrated in Figure 1. The stages involve:

1.   Hydration which disrupts hydrogen bonds and other hydrophobic forces.

2.   Initial cleavage of the covalent bonds randomly throughout the polymer.

3.   Mass loss from continued cleavage and solubilization of the low molecular weight fractions.

4.   Complete absorption.

To describe and quantitate the degradation process requires the introduction of several molecular weight terms.

Several important parameters concerning molecular weight are often mentioned when describing polymer systems. The most common of these are the number-average molecular weight (M_n) and the weight-average molecular weight (M_w), represented by equations 1 and 2, respectively.

$$M_n = \frac{\Sigma n_i M_i}{\Sigma n_i} = \frac{\Sigma w_i}{\Sigma w_i / M_i} \tag{1}$$

$$M_w = \frac{\Sigma n_i M_i^2}{\Sigma n_i M i} = \frac{\Sigma w_i M_i}{\Sigma w_i} \tag{2}$$

Here M_i, n_i and w_i are the molecular weight, number of moles, and weight of the ith monomer unit, respectively.

Both M_n and M_w are unique values for a given polymer. M_n is most affected by the lower molecular weight polymer fractions, whereas M_w is most affected by the higher fractions. Therefore M_w is always greater than M_n, except in the rare or non-existent case of a totally monodisperse polymer composed

entirely of only one molecular weight.   However, polymers are inevitably
polydisperse, meaning several or many different molecular weight fractions exist.
The extent of variability in molecular weight is determined by the polymer's
polydispersity (PD) calculated by equation 3.

$$PD = M_w/M_n \tag{3}$$

A relationship known as the Mark-Houwink equation exists between the
intrinsic viscosity, $[n]$, and the molecular weight, M, of a polymer.

$$[n] = KM^a \tag{4}$$

Both K and a are constants determined for a given polymer in a specific
solvent at a designated temperature.   Typical values for K range from 0.5 to 5 x
10^{-4}, and those for a are commonly between 0.6 and 0.8 (21).   Knowing the value
for a, one can calculate another average known as the viscosity-average molecular
weight (M_v).

$$M_v = \left[\frac{\Sigma n_i M_i^{1+a}}{\Sigma n_i M_i}\right]^{1/a} = \left[\frac{\Sigma w_i M_i^a}{\Sigma w_i}\right]^{1/a} \tag{5}$$

M_v is always greater than M_n and usually less than M_w, but can equal M_w if the
upper limit of a is reached for random coil polymers (22).
    The bulk degradation process can be characterized by both the decrease in
average molecular weight and the polymer mass as depicted in Figure 7 for 50:50
DL lactide/glycolide copolymer, 28,800 M_w (23).   The profile shows induction
periods for both M_w loss and mass loss (erosion or solubilization).   The induction
period for M_w loss was short and is believed to be due to permeation of the matrix
or the polymer backbone. M_w loss occurred abruptly and was exponential with
time.   Mass loss was much slower and occurred after about 90% of the original
polymer composition had degraded.   This suggests that the molecular weight had
to fall below a critical value (3000 in this case) before solubilization occurred.

Preparation of Microspheres for Parenteral Drug Delivery

Although primarily used in surgical procedures as suture materials,
lactide/glycolide polymers formulated as microspheres are now increasingly
employed in drug delivery applications (24-29).   There are numerous methods for
microencapsulating drugs into carriers for drug administration.   Figure 8 illustrates
the methods employed in our laboratories to prepare microspheres by a
dispersion/emulsion technique followed by solvent removal (30).
    The process involves dissolving the drug and polymer in a common solvent
and dispersing this into a continuous phase to form a microdispersion.   Solvent
removal from the microdispersion is accomplished by one of three solvent removal
techniques, the classical solvent-evaporation method which produces a relatively
non-porous matrix or by solvent-extraction or freeze-drying which produces a
product with varying degrees of porosity depending on the process conditions.

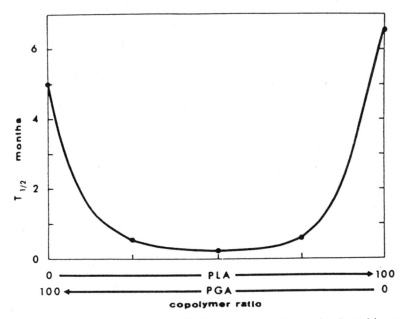

Figure 6. Half-life of PLA and PGA homo-and copolymers implanted in rat tissue. (Reproduced with permission from ref. 20. Copyright 1977 John Wiley & Sons, Inc.)

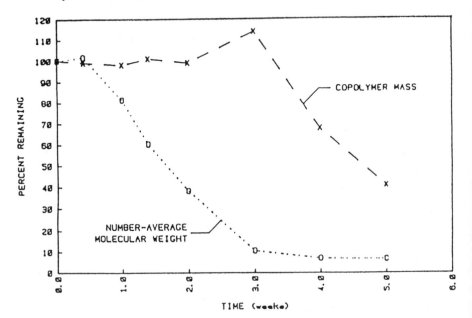

Figure 7. Degradation profile of poly(d,l-lactide-co-glycolide); 50:50; Mn 28,800. Percent mass loss and Mn as function of time in buffer. (Reproduced from ref. 23. Copyright 1987 American Chemical Society.)

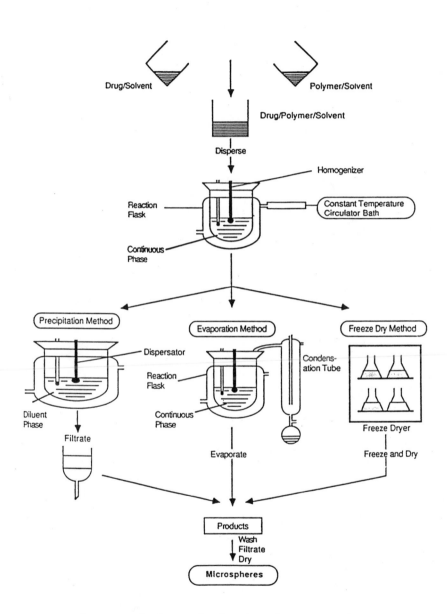

Figure 8. Schematic diagram for the microspheres preparation process.

Two critical parameters in microspheres for parenteral administration are the size and the porosity of the carrier. The size of the microsphere will be determined by the densities of the solvent phases, the mixing speeds, the surfactant type and concentration and the temperature. The porosity will be influenced by the composition of the disperse phase and the speed of solvent removal.

Other microsphere preparation processes include spray drying, rotary evaporation and press grinding (31). In the spray drying process, the polymer is dissolved in a suitable low boiling organic solvent and mixed with the drug. The solution is then transferred via a peristaltic pump into a spray dryer at an atomizing air pressure with a suitable inlet temperature and driving air at an appropriate flow rate. In the rotary evaporation process a solution of the polymer and drug (in a low boiling solvent) is emulsified in water. The resultant dispersion is then rotary-evaporated at low pressure to remove the organic solvent. In the press-grinding process, spray-dried polymer is mixed with the spray-dried drug and compressed into a pellet using a Carver press. The pellet thus formed will be ground with dry-ice in a grinder. This process produces microparticles rather than microspheres.

In order to provide safe injectable or implantable drug delivery systems free from organisms, the fabricated device should be able to withstand some type of sterilization technique with predictable assurance that the polymer will remain effective. Cobalt-60 gamma radiation offers an expedient means of terminally sterilizing polymeric drug delivery systems. The concerns are the effect of radiation on the incorporated agents and the polymer characteristics, namely, molecular weight and thermal properties which will effect polymer degradation and release of drug. A radiation dose of 1.8 to 2.6 MRads has been shown to provide an overkill of microorganisms in the sterilization of parenterals (32). In microspheres the residual moisture content will undoubtedly influence the adverse effects of radiation.

Characterization of Polymers and Microspheres

To control the manufacturing process and ensure reproducible products, rigid characterization of the polymers and the resultant carriers is essential. Several analytical procedures listed in Table II are available to determine the physical-chemical properties of the starting polymer and the formed carrier.

Gel permeation chromatography (GPC) can be used to determine both the M_w and M_n from which polydispersity of the polymer can also be calculated. Light scattering and membrane osmometry techniques can also be used to determine the molecular weight of the polymers. Dilute solution viscometry provides M_v, which, for polylactide/glycolide falls between M_n and M_w. The storage stability and the in vitro degradation can be evaluated by measurements of molecular weight distribution as a function of time. These studies give information which is useful for selection of the polymer for specific applications.

Since the rate of degradation of the polylactide/glycolide system mostly depends on the initial crystallinity, DSC can be used to determine both T_g and T_m of these polymers. The percentage crystallinity associated with the polymer can also be determined by x-ray diffraction. Since the stability of these polymers is greatly affected by moisture, the moisture content of the matrix has to be carefully controlled. Karl Fisher titration and thermogravimetric analysis (TGA) are the two commonly used methods to estimate the moisture content. The particle size of the microsphere can be determined by various techniques such as photon correlation laser technique (Laser particle size analyzer), electric resistance (Coulter counter), light blockage (HIAC-Royco), microscopy (both optical and SEM) and sieve analysis (test sieves). Evaluation of void fraction (porosity) of the microspheres is of great importance since drug dissolution and diffusion depends on this parameter. Depending on the pore size, various types of porosimetric techniques

Table II.   Techniques for Characterization of Polymer and Microspheres

1. Particle Size - Microspheres	Photon Correlation Techniques Electrical Resistance Microscopy Sieve Analysis
2. Surface Morphology - Microspheres	Scanning Electron Microscopy
3. Molecular Weight - Polymer and Microspheres	Gel Permeation Chromatography Dilute Solution Viscometry Light Scattering
4. Thermal Analysis - Polymer and Microspheres	Differential Scanning Calorimetry
5. Moisture Content - Microspheres	Karl Fisher Titration Thermogravimetry Analysis
6. Surface Area/Porosity - Microspheres	N_2 and Kr Adsorption/Desorption

such as nitrogen-krypton adsorption/desorption technique and mercury and helium porosimetry can be employed. The surface charge of the microsphere can be determine by measuring the zeta potential using a zeta meter.

Effect of Radiation Sterilization

There are very few studies reported on the effects of gamma sterilization on polymeric carriers. With polylactic acid (M_w 6400) and poly(L-lactide-co-glycolide) (M_w 183,000), chain scission was reported at doses of 2.5 MRad and higher (33). However, with the copolymer the molecular weight changes were speculated on the basis of a decrease in the T_g determined by DSC.

In our preliminary studies with poly-L-lactide (PLA) polymer and microspheres there were progressive but slight decreases in the T_g and T_m of the raw polymer with increase in radiation doses up to 2.5 MRads (Table III, Figure 9). While this change resulted in a calculated change of about 10% in the mole fraction of crystalline units, it is interesting to note that the processing of the

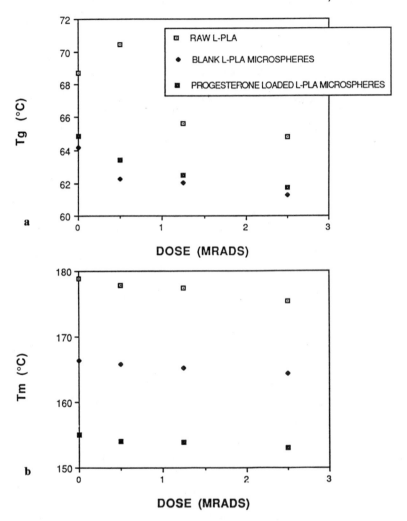

Figure 9. Effect of Co-60 gamma radiation on the a) Tg and b) Tm of raw L-PLA and L-PLA blank and progesterone loaded microspheres.

polymer into microspheres gave similar T_g values and even lower T_m. Sterilization of the blank microspheres resulted in very slight further lowering of T_g and T_m.

With progesterone loaded microspheres there were no significant changes in the T_g, T_m and recrystallization temperatures of the polymer (Table III). Additionally, the melting point of pure progesterone, 133°C, and the drug content of the microspheres were unchanged.

TABLE III   Effect of Co-60 Irradiation on Poly (L-Lactide) (M_w = 50,000)

SAMPLE	DOSE (MRads)	T_g(°C)	T_m(°C)	T_r(°C)	$(X)^2$
Raw L-PLA	0	68.68	178.9	- -	1.00
	0.50	70.42	177.9	- -	0.97
	1.25	65.62	177.3	- -	0.96
	2.50	64.82	175.4	- -	0.91
Blank L-PLA	0	64.15	166.3	84.99	- -
Microspheres	0.50	62.28	165.8	84.66	- -
	1.25	62.02	164.2	84.58	- -
	2.50	61.28	164.4	83.77	- -
Progesterone	0	64.86	155.1	83.69	- -
Loaded L-PLA	0.50	63.42	154.1	82.91	- -
Microspheres[1]	1.25	62.49	154.0	83.34	- -
	2.50	61.76	153.1	82.50	- -

1.   Melting point of progesterone is approximately 130°C.
2.   X = Mole fraction of crystalline polymer units as determined by the Flory equation *(21)*.

$$\frac{1}{T_{m,D}} - \frac{1}{T_{m,O}} = - \frac{R}{\Delta H} \ln X$$

$T_{m,D}$ = polymer melting temperature at radiation dose D.
$T_{m,O}$ = polymer leting temperature at zero radiation dose.
R = gas constant
ΔH = enthalpy of fusion per mole of crystalline units

Blank microparticles prepared from poly(D,L-lactide-co-glycolide) (PLG) 50:50 were subjected to 0, 1.5, 2.5 and 3.5 MRads of Co-60 radiation. Upon irradiation, molecular weight was reduced 7 to 8 kDa due to radiation induced random chain scission (Table IV). A trend of decreasing molecular weight with increasing radiation dose was seen for most samples.

In addition to molecular weight reduction, radiation sterilization has been found to decrease the melting temperature of PLA *(34)*. Gupta and Deshmukh discovered a linear relationship between radiation dose and the reciprocal of the PLA melting temperature. The results were verified as shown in Figure 10 *(22)*.

PLA melting temperatures decreased from 178.9°C for a non-irradiated sample to 175.4°C for a sample subjected to 2.5 MRads radiation. In recent studies with PLG microparticles, a slight lowering of the average molecular weight following 2.5 MRads was observed.

Table IV. Effect of Radiation Dose on Molecular Weight of Polymer

Dose (MRads)	M_w	M_n
0	36,919	25,269
1.5	30,622	18,648
2.5	31,321	17,778
3.5	29,308	17,540

Examples of <u>In-Vivo</u> Applications

The physico-chemical characteristics of polymers and drug delivery systems are an integral part of the ultimate in-vivo application and satisfactory performance. The required physico-chemical properties, i.e., size, surface area, porosity, drug content, release profile and biodegradation times of microspheres are governed by the type of drug and the intended in-vivo application.

Particle Size. Particle size of the microsphere is an important factor since the route of administration will determine the size of the microspheres as illustrated in Figure 11. Intravenous administration results in localization in the capillary vasculature and uptake by macrophages and phagocytes. Microspheres larger than 8 μm will lodge predominantly in the lung capillaries; those less than 8 μm will clear the lung and localize in the liver and spleen (*33*). For inhalation administration to the lung, particles in the one to 6 μm range are required. Microspheres in the size range of 10 - 100 μm can be used for subcutaneous and intramuscular administration whereas larger microspheres ($\geq$ 100 μm) have been used for embolization and also for drug delivery by implantation.

Surface Area/Porosity. Matrices of variable porosity facilitate modulation of drug release. Porous microspheres are essential to deliver high molecular weight substances which cannot diffuse from a non-porous matrix and to deliver substances which have high affinity for polymer and are not released unless the matrix is degraded. Matrix porosity is also of significance in polymeric carriers which deliver drugs, at least in part, by erosion. Polymer degradation can be controlled by altering the porosity of the matrix and hence control the rate and extent of drug release.

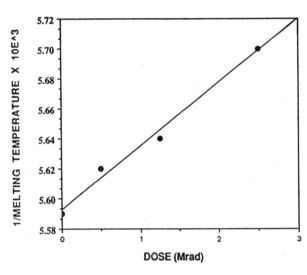

Figure 10. Effect of radiation on the reciprocal melting temperature.

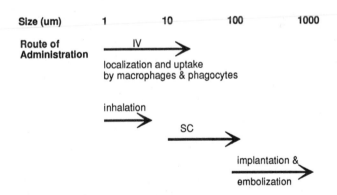

Figure 11.   Size requirements of the microspheres for various routes of administration.

Drug Content/Drug Release. Drug content and release profiles depend on the intended dose and dosing rate of the drug in a particular treatment. Drugs with low potency must be given at high dose and thus must be prepared into microspheres with high loading. Drug content also depends on the maximum amount of materials that can be delivered via the intended route of delivery. Inhalation and intra-nasal delivery will require larger drug content than other routes of delivery. Selection of the preparation media and polymers in turn depend on the intended drug content.

Biodegradation Times. The time required to completely degrade the microspheres is governed by the route of administration, and frequency of dosing. Inhalation delivery which is generally used for chronic therapy, would require rapid biodegradation to prevent systemic accumulation where as nasal delivery, where the microspheres are swallowed into the gastro-intestinal track, would not be restricted by biodegradation time. Intravenous delivery may require particles that degrade rapidly where as subcutaneous or intramuscular delivery may tolerate some accumulation of particles with slower biodegradation.

Isoproterenol-PLG Microspheres for Lung Delivery. PLG was selected for lung delivery of isoproterenol due to its amorphous nature resulting in rapid biodegradation. Designed for once a day administration, the microspheres should clear rapidly without excessive build-up in the lungs. Isoproterenol and PLG were dissolved in warm acetonitrile and added to mineral oil containing 0.1% sorbitan sesquioleate at 50°C. The dispersion was allowed to stir for 1 hour with nitrogen purge over the surface to evaporate acetonitrile from the mineral oil. The microspheres were filtered, washed and dried at room temperature.

Isoproterenol-PLG microspheres had an average drug content of 23% w/w and were spherical in shape with a smooth, non-porous surface (Figure 12). The mean diameter of the microspheres was $4.5\mu m$ with 59% of the microspheres, by weight, under $5\mu m$ diameter and 98% under $10\mu m$ diameter. The temporal changes in serotonin induced bronchoconstriction for the sustained release isoproterenol-PLG microspheres in Long-Evans rats are shown in Figure 13 (*35*). Complete protection against serotonin would result in a $Vmax_{50}$ of 100%. The challenge produced a 30 - 45% decrease in the $Vmax_{50}$ in the control group with no consistent temporal pattern, indicating a mild to moderate bronchoconstriction during the 3 to 12 hour period. In the free drug, blank-PLG and free drug with blank-PLG, the serotonin challenge produced a bronchoconstriction similar to the control group indicating that the effective period for the intratracheally instilled isoproterenol was less than three hours. This estimation of the effective period is comparable with that obtained by Choo-Kang, et. al., (*36*). Free isoproterenol at a dose of 0.1 mg/kg, was found to have a duration of action lasting only about 15 minutes. With the encapsulated drug, serotonin failed to induce any significant bronchoconstriction for at least 12 hours (*37*).

Progesterone-PLA Microspheres for Sustained Release. PLA was selected to incorporate progesterone due to the necessity of high drug content. The solvents used in the process permitted up to 50% w/w content of progesterone. PLA and

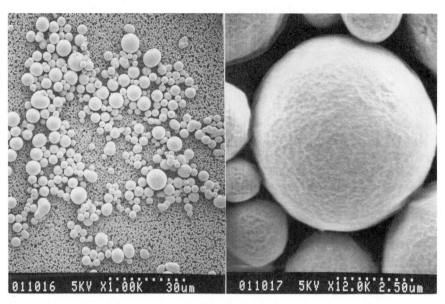

Figure 12. Scanning Electron Micrograph of Iso-PLG microspheres.

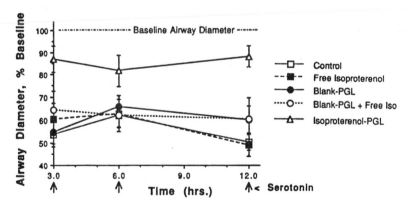

Figure 13. Bronchodilator effects of isoproterenol microspheres. (Reproduced with permission from ref. 35. Copyright 1992 Plenum Publishing Corporation)

progesterone were dissolved in methylene chloride and dispersed in glycerin at 5°C. The temperature was raised to 45°C and the dispersion was added to 5% aqueous isopropanol for extraction of methylene chloride and solidification of the microspheres. The solidified microspheres were isolated by filtration, washed and dried at room temperature overnight. This solvent-extraction-precipitation method was selected to provide porous microspheres. Non-porous microsphere would sustain the progesterone release beyond acceptable 10 - 15 days per dose for the prevention of loss of pregnancy in mares. The microspheres may be administered every two weeks. The short duration of release is necessary to terminate progesterone dosing in case the fetus is lost during pregnancy and a repeat conception is desired.

The progesterone-PLA microspheres were spherical with mean diameters of 14 to 25μm. Total drug content varied between 20% to 50% w/w. Figure 14 shows the progesterone concentrations following administration of an oil solution and microspheres (38). At a dose of 1.0g the microsphere form of progesterone resulted in a four-fold reduction in the peak serum concentration (5.3 ng/ml) when compared to the solution (19.2 ng/ml) and a two-fold reduction when compared with the suspension (9.8 ng/ml). The area under the progesterone serum concentration-time curve over 288 hrs (AUC_{0-288}) was higher with the solution (692 ng hr/ml) than the same dose in a suspension (546 ng hr/ml) or a microsphere formulation (571 ng hr/ml), probably due to the incomplete release of progesterone from microspheres or powder. The microsphere drug loading appeared to have some effect on the progesterone bioavailability relative to the oil solution (RBA). With a microsphere loading of 20% the RBA was 0.83 which increased to 0.88 for 35% loading and to 1.08 with 50% loading. This may be attributed to a relatively rapid and more complete release of progesterone due to decreased polymer content of the microsphere matrix. The RBA of the aqueous suspension of progesterone was the lowest (0.79).

In spite of the apparent drug-loading dependent changes in the bioavailability of progesterone from the microspheres, the serum progesterone levels were not significantly different and the standard deviations for a 1.4 g dose (Figure 14) are relatively low considering the in vivo nature of the experiments. A bimodal release was seen with microspheres at 1.0 g dose. This may be a result of microsphere degradation or phagocytosis by cells of the immune system.

Further analysis of fractional AUCs showed that during the first three days post administration, the solution formulation provides a higher fraction of the total area (82%) than the suspension (43%) or the microspheres (26-37%), indicating a sustained release of progesterone with the latter formulations. Further fractionation of the AUC over the therapeutic window (serum levels of 2 to 6 ng/ml; AUC_{2-6}) shows that the microsphere dose of 1.4g progesterone provides maximum AUC_{2-6} (211 ng hr/ml) followed by the microsphere dose of 1.0g (116 ng hr/ml) and solution dose (103 ng hr/ml) of 1.0g progesterone. Serum levels greater than 6 ng/ml do not provide any additional benefit and may be toxic and those below 2 ng/ml are not therapeutically effective. The results suggest that the progesterone microsphere delivery system, although not completely optimized, is a viable alternative to the daily injection of an oil solution.

Salmon Calcitonin Microspheres for Sustained Release. Polyglycolide (PGA) was initially selected for incorporation of salmon calcitonin (sCT) due to its long degradation time, providing extended release with longer dosing interval. PGA and sCT were dissolved in hexafluoroacetone sesquihydrate and dispersed into chloroform. The dispersion was then added to dioxane to extract solvents and solidify the microspheres. The microspheres were collected by filtration, freeze dried and stored at -20°C. sCT-PGA microspheres containing 0.3, 4.5 and 7.5% w/w sCT were obtained with an incorporation efficiency of 60 to 90%. The mean diameters of the microspheres were 12.0, 14.5 and 16.3μm for the three loadings, respectively. Figure 15 shows the sCT serum levels for single injections of 0.72 U/Kg of free sCT and 3.6 U/Kg of sCT-PGA microspheres. With free sCT injections, serum levels returned to baseline within 1 hour after administration but in the case of sCT-PGA, the levels were prolonged for at least 5 days after a single dose (*29*).

The release with PGA microspheres was prolonged beyond 7 days due to rapid release of sCT in-vivo, although no release could be detected in-vitro in buffer. The mechanism of release is not well understood for peptide microspheres. sCT exhibits extensive interaction with PGA, PLA and PLG in-vitro. The interaction appears to be due to hydrophobic forces and is maximum with PLG. Due to this, PLG was selected as the next polymer for sCT incorporation. A solution of PLG in methylene chloride was mixed with sCT solution in methanol and added to 0.04% aqueous sodium oleate. The solvents were extracted and evaporated from the continuous phase under nitrogen to produce porous microspheres which are collected and freeze dried. The high interaction between sCT and PLG permits a secondary loading of sCT on the surface of sCT-PLG microspheres, providing a means to achieve a pulsatile delivery system. sCT-PLG microspheres containing 5% drug load were spherical and porous. The mean particle size was 22μm (Figure 16) and incorporation efficiency ranged from 65 - 75%. Blank-PLG microspheres were also prepared by the same technique and sCT was adsorbed onto these microspheres. Up to 10% w/w sCT could be adsorbed on the PLG microspheres from an aqueous solution.

Figure 17 shows the compilation of serum concentrations resulting from a mixture of sCT entrapped and sCT adsorbed PLG microspheres. The inset shows the serum concentrations after administration of sCT as a solution. For the microspheres, a bimodal release was obtained with an initial rapid phase due to the surface adsorbed sCT followed by a phase of low but detectable release and finally sustained release of a high level of sCT, resulting from release of entrapped drug (*39*). This suggests that the interaction between sCT and PLG is significantly reduced at a definite time point resulting in an abrupt release and significant increase in serum sCT followed by maintenance at high levels for 4 - 5 days. With free sCT baseline levels were reached within 2 hours, which is expected due to the short half-life of sCT.

Summary

The physico-chemical properties of biodegradable polymers and delivery systems formulated from these polymers are very critical in successful in-vivo application

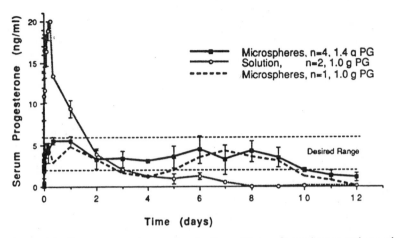

Figure 14. Serum progesterone levels from PLA microspheres and an oil solution in mares.

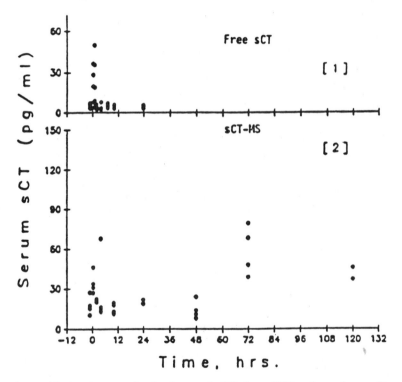

Figure 15. In vivo sustained release of sCT from PGA microspheres. Free drug - 0.72U/Kg; MS - 3.6U/Kg (0.3% load). (Reproduced with permission from ref. 29. Copyright 1991 Elsevier Science Publisher)

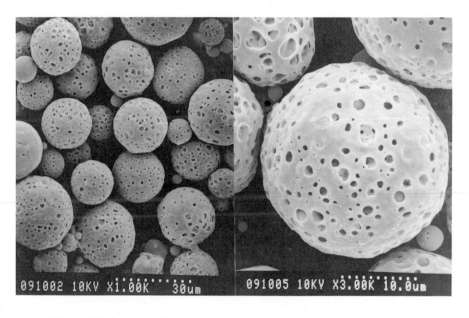

Figure 16. Scanning Electron Micrograph of sCT-PLG Microspheres.

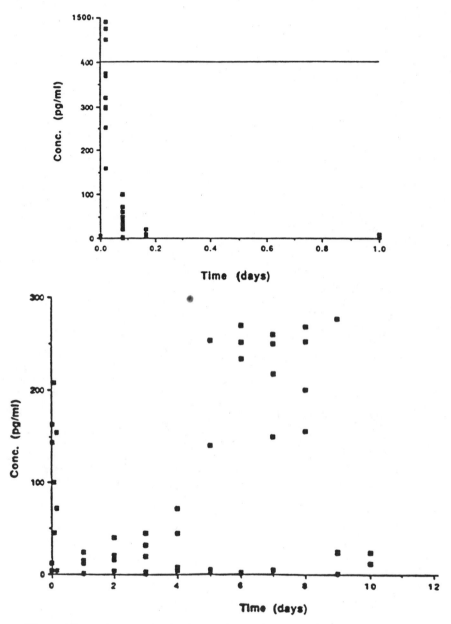

Figure 17. In vivo sustained release of sCT from PLG microspheres. Free drug - 40U/Kg; MS dose - 40U/Kg (0.3% load).

of controlled and targeted drug delivery systems. Homo- and copolymers of lactide and glycolide are the most extensively studied class of biodegradable polymers. Polymers and delivery systems provide a wide range of physico-chemical properties which can be used to tailor a delivery system for its specific in-vivo application. The chemical composition and the molecular weight of the polymer, as well as the isomeric form of the lactide monomer determine the rate of degradation of the polymer. For applications requiring long degradation times, PLA is the polymer of choice. Implants or microspheres of PLA are ideal for long term hormone delivery. The other homopolymers, Poly (D,L-lactide) and PGA have shorter degradation times than PLA. The copolymers, PLG, have shorter degradation half lives due to their amorphous nature and hence are suitable for applications requiring faster degradation. Microspheres prepared from the copolymers are well suited for inhalation and intravenous administration. Properties of microspheres, especially particle size and surface area, play an important role in the ultimate fate of the microspheres in-vivo. Particles less than 5 μm are required for efficient delivery of drugs to the lungs via inhalation, whereas particles greater than 8 μm are necessary to cause lung and liver deposition after intravenous administration. Still larger particles may be used for intramuscular and subcutaneous delivery and for embolization and implantation. Microspheres of hydrophobic drugs such as progesterone with high surface area may be used to achieve short-term sustained release of otherwise very slow releasing drug/polymer matrix. Polypeptides such as sCT may bind strongly to the hydrophobic polymers of lactide and glycolide. This interaction facilitates formulation of delivery systems with varying rates of release of polypeptide. In summary, a thorough knowledge of the physico-chemical properties of biodegradable polymers and their delivery systems along with the understanding of specific therapeutic needs significantly reduces the efforts needed to formulate the best possible delivery system for drugs and polypeptides.

Literature Cited:

1.   Yolles, S.; Eldrige, J.E.;Woodland, J.H.R. *Polym. News* **1970**, *1:9-15*.
2.   Frazza, E.J.; Schmitt, E.E. *J. Biomed. Mater. Res. Symposium* **1971**, *1:43-58*.
3.   Yolles, S; Sartori, M.F. In *Drug Delivery Systems;* Juliano, R.L. Ed.; Oxford University Press: London, **1980**, *pp 84-111*.
4.   Pitt, C.G.; Gratzi, M.M.; Jeffcoat, A.R.; Zweidinger, R.; Schindler, A. *J. Pharm. Sci.* **1979**, *68:1534-1538*.
5.   Sinclair, R.G. *Proc. 5th Inter, Sym. Contr. Rel. Bioact. Mater.* **1978**, *pp 8.2-8.17*.
6.   Spenlehauer, G; Vert, M.; Benoit, J.P.; Chabot, F.; Veillard, M. *J. Contr. Rel.* **1988**, *7:2217-2219*.
7.   Cha Y., Pitt, C.G. *J. Contr. Rel.* **1989**, *8:259-265*.
8.   Bodmeier, R., Oh, K.H., Chen, H. *Int. J. Pharm.* **1989**, *51:1-8*.
9.   Heller, J., In: *CRC Critical Reviews in Therapeutic Drug Carrier Systems.* Vol. 1, Issue 1, *pp 39-90*.

10.   Baker, R.W., *Controlled Release of Biologically Active Agents;* John Wiley
      and Sons: New York, NY, **1987**, *Chapter 4, pp 84-131.*

11.   Heller, J., Perhale, D.W.H., Helwing, R.F., Fritzinger, B.K., *Polym Eng.
      Sci.* **1981**, *21:727-731.*

12.   Chasin, M., Domb, A., Ron, E., Mathiowitz, E., Leong, K., Laurencin,
      C., Brein, H., Grossman, S., and Langer, R., in *Biodegradable Polymers
      as Drug Delivery Systems,* Chasin, M., Langer, R., Eds., Marcel Dekker:
      New York, NY, **1990**, *pp 43-70.*

13.   Cameron J.L., Woodward, S.C., Pulaski, E.J., Sleeman, H.K., Brandes
      G., Kulkarni, R.K., Leonard, F., *Surgery* **1965**, *58:424-430.*

14.   Gilding D.K., Reed, A.M., *Polymer* **1979**, *20:1459-1464.*

15.   Lewis, D.H., In: *Biodegradable Polymers as Drug Delivery Systems.* Eds.
      M. Chasin, R. Langer. Marcel Dekker Inc: New York, NY, **1990**, *pp.
      1-41.*

16.   Kulkarni, R.K., Moore E.G., Hegyelli, A.F., Leaonard, F., *J. Biomed.
      Mater. Res.* **1971**, *5:169-181.*

17.   Fukuzaki, H., Yoshida, M., Asano, M., Kumakura, M., Mashimo, T.,
      Yuasa, H., Imai, K., Yamanaka, H. *Biomaterials* **1991**, *12:433-437.*

18.   Kitchell, J.P. Wise, D.L., *Methods in Enzymology* **1985**, *112:436-448.*

19.   D.K. Gilding and A.M. Reed, *Polymer,* **1979**, *20:1459-1464.*

20.   R. Miller, J. Brady and D.E. Cutright, *J. Biomed. Mater. Res.* **1977**,
      11:711-719.

21.   Billmeyer, E.W., *Textbook of Polymer Science 3rd Ed.* John Wiley and
      Sons: New York, NY, **1984**, *p. 211.*

22.   Cooper, A.R., From Chemical Analysis Series, Ed JD Winefordner, Vol
      103. John and Wiley and Sons: New York, NY, **1989**, *pp 1-6.*

23.   Kenley, R.A., Lee, M.O., Mahoney II, T.R., Sanders, L.M.,
      Macromolecule **1987**, *20:2398-2403.*

24.   Widder, K.J., Flouret, G., Senyei, A.E., *J. Pharm. Sci.* **1979**, *68:79-82.*

25.   Sanders, L.M., Kent, J.S., McRae, G.I., Vickery, B.H., Tice, T.R., and
      Lewis, D.H., *J. Pharm. Sci.* **1984**, *73:1294-1297.*

26.   Tsai, D.C., Howard, S.A., Hogan, T.F., Malanga, C.J., Kandzari, S.J.,
      and Ma, J.K.H., *J. Microencapsulation* **1986**, *3:181-193.*

27.   Sato, T., Kanke, M., Schroeder, H.G. and DeLuca, P.P., *Pharm. Res.*,
      1988, *5:21-30.*

28.   Redmon, M.P., Hickey, A.J. and DeLuca, P.P., *J. Contr. Rel.* **1989**, *9:
      99-109.*

29.   Lee, K.C., Soltis, E.E., Newman, P.S., Burton, K.W., Mehta, R.C. and
      DeLuca, P.P., *J. Contr. Rel.* **1991**, *7:199-206.*

30.   Sato, T., Kanke, M., Schroeder, H.G. and DeLuca, P.P., *Pharm. Res.*
      1988, *5:21-30.*

31.   H.T. Wang, H. Palmer, R.J. Linhardt, D.R. Flanagan and E. Schmitt,
      Biomaterials **1990**, *11:991-1003.*

32.   Gordon, B., Agalloco, J.P., Anisfeld, M.H., Athanas, N., Chin, A.,
      Colarusso, R.J., Fogarty, M., Huggett, D.O., McGarrah, R.D., Shaw, R.,
      Shirtz, J., Wilson, J., Young, W.E., *J. Paren. Sci. & Tech.* **1988**,
      42:53-59.

33. H.G. Schroeder, G.H. Simmons and P.P. DeLuca, *J. Pharm. Sci.* **1978,** *67:504-507.*

34. Gupta, M.C.C., Deshmukh, V.G., *Polymer* **1983,** *24:827-830.*

35. Lai, Y.L., Mehta, R.C., Thacker, A.A., Yoo, S.D., McNamara, P.J. and DeLuca, P.P., *Pharm. Res.* in press.

36. Choo-Kang, Y.F.I., Simpson, W.T., Grant, I.W.B., *British Medical Journal* **1969,** *2:287-289.*

37. Lai, Y.L., Mehta, R.C., Thacker, A.A., Yoo, S.D., McNamara, P.J., DeLuca, P.P., *Pharm. Res.*, in press.

38. Gupta, P.K., Mehta, R.C., Douglas, R.H. and DeLuca, P.P., *Pharm. Res.* in press.

39. Jeyanthi, R., Tsai, T., Mehta, R.C. and DeLuca, P.P., *Pharm. Res,* **1991,** *8:s-151.*

RECEIVED October 1, 1992

Chapter 5

Plasticized Cellulose Acetate Latex as a Coating for Controlled Release

Thermal and Mechanical Properties

V. L. King and T. A. Wheatley

Pharmaceutical and Bioscience Division, FMC Corporation, Princeton, NJ 08543

Plasticizers were studied to examine their effects on the thermal and mechanical properties of films cast from a new CA latex. The latex was an aqueous dispersion of cellulose acetate (CA398-10) with a solids content of 29% and mean particle diameter of 310 nm. Plasticized films were studied by thermal mechanical analysis (TMA) and tensile testing. Plasticizers selected included glyceryl diacetate (GDA), glyceryl triacetate (GTA), triethyl citrate (TEC), acetyltriethyl citrate (ATEC), and dibutyl sebacate (DBS). They were studied at use levels of 0-160%. GDA, GTA and TEC were clearly the most effective in altering the glass transition temperature (Tg) and elastic modulus of the latex films.

Over the past decade, aqueous coatings have aroused interest in the pharmaceutical industry largely because of the need to avoid problems associated with organic solutions such as air pollution and solvent toxicity. As a result of this interest, latexes of two cellulosic derivatives (ethylcellulose and cellulose acetate phthalate) have been introduced to the market. Previously, these materials were formulated only as organic solvent-based systems.

Cellulose acetate (CA) has been used for many years in the development of osmotically controlled drug delivery systems.[1,2] Typically the CA is in the form of a membrane made by dissolving the CA and plasticizer in a non-aqueous solvent system, followed by deposition onto a solid substrate, i.e., tablet, employing typical film coating techniques. The use of a CA latex dispersion eliminates the need for solvents in the formulation and manufacture of these coated systems. There is now available a cellulose acetate latex (CA398-10) dispersion produced from direct emulsification of the polymer, a technique taught by Vanderhoff et al.[3] and Bindschaedler.[4] Bindschaedler et al.[5] have investigated the mechanical properties of films produced from cellulose acetate latexes containing various plasticizers. They concluded that "Provided that a suitable choice is made of plasticizers and film-forming conditions, cellulose acetate latexes yield strong membranes". The objective of this study was to investigate the mechanical and thermal properties of CA latex films containing other plasticizers that are widely used and accepted in the pharmaceutical industry.

0097–6156/93/0520–0080$06.00/0
© 1993 American Chemical Society

Experimental Methods

The cellulose acetate (CA) aqueous dispersion was manufactured at the FMC Coatings Plant, Newark, Delaware, USA. The latex has a solids content of 29 wt% and a median particle diameter of 310 nm as determined by centrifugal sedimentation analysis. Samples of plasticized latex films (0.50-0.70 mm thick) were studied by thermal mechanical analysis (TMA) utilizing a Perkin-Elmer TMA-7 instrument. Probe penetration into the sample was measured at 5°C/min heating with an applied force of 1 mN. The films were prepared by adding the CA latex to a dispersion of plasticizer in water, and mixing for 30 minutes to yield a final total solids content of 20%. Films were cast onto an aluminum substrate and dried for 18 hours at 60°C. The plasticizers studied were glyceryl diacetate (GDA), C.P. Hall Co.; triethyl citrate (TEC) and acetyltriethyl citrate (ATEC), Morflex; glyceryl triacetate (GTA), Eastman; and dibutyl sebacate (DBS), Union Camp.

Thin films (80-100 µm thick) for tensile testing were prepared in a similar manner, but were cast onto a glass substrate, dried at 60°C for 18 hours and stored for 5 days in dessicators prior to testing. An Instron model 1011 instrument was used for the stress/strain measurements, and was equipped with a 5-kg load transducer and adjusted to test the films at 1.0 mm/min strain rate with a 50-mm gauge length. The tested films were of rectangular shape (80 x 19.3 mm) and were cut precisely to size using a fresh razor blade and steel template. Thicknesses were measured as an average of ten micrometer measurements across the gauge length of each film.

Results and Discussion

The effectiveness of the five plasticizers for CA latex have been established under the conditions of this study. These plasticizers have GRAS status or at least some degree of approval for use in foods or food packaging as defined in their specific Title 21 CFR listings. Their chemical structures and physical properties, such as bp > 250°C, qualify them as suitable candidates as plasticizers for cellulose esters, particularly the cellulose acetate (CA398-10) used in the latex.

Thermal mechanical analysis (TMA) was used to investigate the effectiveness of the plasticizers in lowering the glass transition temperature (Tg) of CA latex films. With TMA the glass transition is often associated with the first mechanically detected softening point of a polymer. For this study the plasticizer content was varied from 0-160% of the latex solids. Plasticization plays an important role in the application of pharmaceutical latex coating formulations and is usually critical for the formation of good polymer films. Plasticizers, in general, will greatly reduce the critical film forming temperature of a polymer. They soften and swell latex spheres, which aids the particles in overcoming their resistance to deformation and facilitates film coalescence. This is best achieved by adjusting the glass transition toward reasonable coating processing temperatures. A properly plasticized film, once formed and dried, will also exhibit the desired physical strength and flexibility for its application as well as the desired permeability.

TMA data shows the effects of the five plasticizers on the onset glass transition temperature of CA latex and is presented in Table I and Figure 1. With respect to lowering the glass transition or softening temperature of the latex it is readily apparent that the rank order of plasticizer effectiveness, from best to worst is: GDA > TEC > GTA > ATEC > DBS. Glyceryl diacetate and triethyl citrate are the most similar in terms of polarity and general chemical structure to CA398-10, the polymer in CA latex. Both of these plasticizers have a hydroxyl group present along with ester functional groups. The cellulose acetate polymer has a degree of substitution of approximately 2.6 for acetate groups and 0.4 for hydroxyl groups. It is, therefore, reasonable and consistent for GDA and TEC to be the best plasticizers of the group studied in terms of lowering the glass transition temperature. Both of these plasticizers are able to lower the Tg of the latex films below room temperature as indicated by TMA. A clearly poor plasticizer is DBS, which is capable of depressing the Tg only a few degrees across the 0-160% range of plasticizer levels studied.

The three best plasticizers, which were capable of forming very good free films, were also examined for their effect on the mechanical properties of latex

Table I.   Thermal Mechanical Analysis of CA Latex Films
            Plasticizer Effect on Onset of Glass Transition,Tg

Plasticizer Level (% of Solids)	GDA Tg (°C)	TEC Tg (°C)	GTA Tg (°C)	ATEC Tg (°C)	DBS Tg (°C)
0%	164	164	164	164	164
20%	114	- -	- -	- -	- -
40%	99	107	109	131	- -
60%	65	- -	- -	- -	- -
80%	31	62	86	109	157
120%	26	37	79	103	- -
160%	<20	≤20	71	89	152

GDA (glyceryl diacetate), TEC (triethyl citrate), GTA (glyceryl triacetate)

ATEC (acetyltriethyl citrate), DBS (dibutyl sebacate)

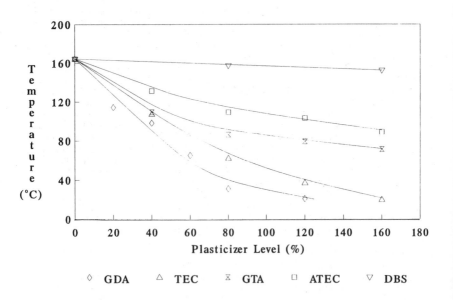

Figure 1.   CA Latex Plasticizer Content
                             vs Glass Transition Temp. (Tg)

films. Tensile testing was performed on an Instron model 1011 to determine elasticity and strength, specifically the elastic modulus and peak stress. The tensile test data showing the effect of the three plasticizers (at an 80% use level) is shown in Table II. The more efficient plasticizers, GDA and TEC, had a greater ability to increase the film's flexibility (decrease elastic modulus) and to decrease the film's peak tensile strength. Films with glyceryl triacetate (GTA) had less elasticity, but did exhibit the high tensile strength characteristic of many films from CA latexes.[5]

Table II.   Mechanical Properties of CA Latex Films
Plasticizer Effect on Elastic Modulus and Strength

Plasticizer: (80% of latex solids)	GDA	TEC	GTA
Peak Tensile Strength (kg/mm^2) ± s.d.	.47 ±.12	.44 ±.02	2.7 ±.2
Modulus of Elasticity (kg/mm^2) ± s.d.	25 ±7	27 ±2	110 ±10

GDA (glyceryl diacetate), TEC (triethyl citrate), GTA (glyceryl triacetate)

Conclusions

The effects of five different plasticizers on cellulose acetate latex films were studied by thermal mechanical analysis (TMA) and tensile testing. The CA latex tested showed properties valuable to a successful controlled release tablet coating material, and these were influenced by the type and amount of plasticizer used. It is important for this class of pharmaceutical coatings to exhibit good flexibility, elasticity and strength in order to function successfully on dosage forms with different surface curvatures and different levels of osmotic pressure that develop during drug release.

The three most effective plasticizers, glyceryl diacetate (GDA), glyceryl triacetate (GTA) and triethyl citrate (TEC) displayed very valuable but different effects on the material properties of the CA latex films. It could be seen that GDA, TEC and GTA were clearly effective in lowering the glass transition temperature (Tg) of films cast from CA latex. These three best plasticizers in terms of lowering the Tg were further studied by tensile testing thin latex films. GDA and TEC exhibited the greatest ability to modify the mechanical properties of the films as reflected in a significant change in the modulus of elasticity. By choosing the type and amount of plasticizer for the CA latex, both thermal and mechanical properties can be controlled to obtain the desired film properties.

Literature Cited

1. Theeuwes, F. *J. Pharm. Sci* **1975,** 64, 1987-1991.
2. Bechard, S.R. *Proceed. Intern. Symp. Control. Rel. Bioact. Mater.* **1990,** 17, 321-322.
3. Vanderhoff, J.W., El-Aasser, M.S., and Upgelstad, J. U.S. Patent 4 177 177, 1978.
4. Bindschaedler, et al. *Proceed. Intern. Symp. Control Rel. Bioact. Mater.* **1986,** 13, 130-131.
5. Bindschaedler, C., Gurny, R. and Doelker, E. *J. Pharm. Pharmacol.* **1987,** 39, 335-338.

RECEIVED September 11, 1992

Chapter 6

Spray Coating and Spray Drying Encapsulation
Role of Glass Transition Temperature and Latex Polymer Composition

L. Tsaur and M. P. Aronson

Unilever Research United States, 45 River Road, Edgewater, NJ 07020

Spray coating and spray drying are widely used processes to produce free flowing capsules for a diverse range of applications. Numerous studies have employed water-soluble polymers as a wall material or matrix for such capsules. However, few fundamental studies have dealt with synthetic latex polymers even though the latter can have a number of advantages in these processes. The work reported here attempts to link the performance of latex polymers in spray coating and spray drying to the material properties of the latex and ultimately to its composition. Our studies indicate that the glass transition temperature of the latex, T_g, and its relationship to the processing temperatures characteristic of spray coating and spray drying, has a major effect on the quality of the capsules formed and their yield. Routes to manipulate polymer composition and latex morphology to achieve desired material properties are illustrated by a few examples.

Encapsulation is widely used to stabilize sensitive ingredients, or control their delivery in a diverse range of applications including foods and drinks, pharmaceuticals, cosmetics, household products, and industrial applications (*1-3*). Although numerous encapsulation technologies are available (*3,4*) all have a common goal, namely, a "device", e.g., microcapsule, in which the active ingredient is either coated with or embedded in a second material. This second material, usually a polymer, is designed to isolate the active until it is triggered to release by an external stimulus. Spray coating and spray drying are widely employed physical processes used to make encapsulated particles in powder form.

Spray coating is used to coat solid powders with a polymer film and usually produces a core-shell morphology, although matrix capsules can also be produced through simultaneous coating and granulation (*5*). The process essentially involves

0097–6156/93/0520–0084$06.25/0
© 1993 American Chemical Society

spraying an appropriate polymer solution on an agitated powder (typically >~300 μM and evaporating the solvent to leave a polymer film. The process can be carried out in a pan granulator (6) or in a fluidized bed (7,8). A typical fluidized bed process is shown schematically in Figure 1. The powder to be coated is placed in chamber to the right and fluidized with heated air. The latex dispersion is fed to the nozzle spray tip and atomized with air. Various embodiments of the technology are described in Reference 9.

Spray drying is a well established process (10-12) in which active(s) and polymer are dispersed/dissolved in a solvent, atomized to form an aerosol, and the solvent flashed off with hot air. The process can be used for liquid (12,13) or solid (14) actives and most often produces matrix type capsules in the 10 to 500 μM range. A typical co-current laboratory spray dryer is diagrammed in Figure 2. An aqueous dispersion of the core material and coating polymer are pumped to the spray nozzle and atomized with air. The droplets are rapidly dried by heated air in the drying chamber and collected in the collection chamber to the right. The technology is extensively described in Reference 10.

Spray coating and spray drying encapsulation share the common feature that they involve forming a solid polymer phase by evaporation of solvent. The polymers that have been traditionally employed in these processes were soluble in the encapsulation solvent, e.g., water (12). In recent years, polymers in the form of dispersed latex have been introduced. As can be seen from the examples shown in Figure 3, the viscosity of a dispersion of a non-swelling latex is much less sensitive to polymer concentration and molecular weight than is a solution of a soluble polymer.   Thus, higher concentrations of polymer can be employed which leads to more rapid or lower temperature processing, and greater formulation flexibility.

Despite an extensive body of literature on spray coating and spray drying, relatively few studies have been reported with latex polymers (14-17). The present study was undertaken to probe some of the principles governing the use of these polymers, specially the roles of glass transition temperature, film formation and surface tack. Our goal was to understand how to tailor the latex to achieve capsules in high yield that have coherent well formed coatings and adequate storage, release and handling characteristics.

Theoretical Considerations

This section provides a simplified and qualitative physical picture of the formation and properties of films formed by latex polymers that are relevant to capsules produced by spray coating and spray drying. It is meant to be a starting point to understand how latex polymers can be tailored to efficiently produce optimal capsules by these processes. Clearly more quantitative theoretical studies will provide deeper insight into the process.

Film Formation. The preparation of capsules by evaporative processes relies on the formation of a coherent polymer film. Evaporative film formation of latex polymers differs from solution polymers in two important respects. These are the rheological

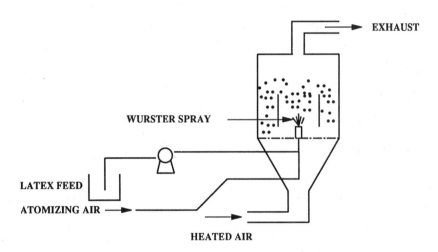

Figure 1. Schematic diagram of fluidized bed coater fitted with Wurster bottom spray.

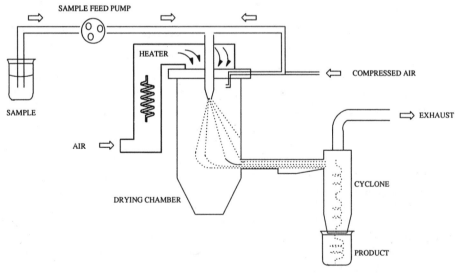

Figure 2. Schematic diagram of Yamato co-current spray drier.

behavior of the polymer solution as it dries, and the mechanism by which films are formed.

As illustrated in Figure 3, the viscosity of a polymer solution increases rapidly when the polymer concentration exceeds the overlap concentration, C^* (*18*). The overlap concentration is that polymer concentration above which polymer molecules (or polymer particles) interact in solution. To a first approximation, the value of C^* is proportional to $N^{(1-3v)}$ where N is the number of monomers per molecule (i.e., degree of polymerization) and v is the excluded volume exponent, e.g., 0.6 for a good solvent and 0.3333 for a sphere. Thus for a solution polymer, C^* is inversely proportional to molecular weight and can occur at a low polymer concentration. In contrast, C^* for a latex is independent of molecular weight and occurs much closer to the random close packing limit of 0.637 for non-interacting spheres. The net effect is that the viscosity of solution polymers increases much more rapidly during evaporation than does a latex because of chain entanglement.

Differences in the film formation mechanism between solution and emulsion polymers can be appreciated by considering the diagram in Figure 4. For solution polymers, chains become entangled relatively early in the evaporation process and this is the principal mechanism of film formation (Figure 4A). If the polymer has appreciable solubility in the solvent, it will form a continuous film over a broad molecular weight range. More importantly, the glass transition temperature, T_g, of the *dry polymer* will not have a dramatic effect on the film formation process although it will clearly affect some of the properties of the film.

In contrast, a latex which is a colloidal dispersion of polymer particles, forms films through a coalescence mechanism (Figure 4B). In the initial stages of film formation, the latex particles are subject to Brownian forces and are in constant motion. Water evaporates at the same rate from the dispersion as pure water or a dilute emulsifier solution, until a dense packed layer of latex particles forms (*19*). After the particles come into irreversible contact, some latexes form tough, transparent continuous films. This is due to deformation of the particles and interdiffusion of the polymer chains of adjacent particles. Other latexes form friable, opaque, discontinuous films with the particles maintaining their spherical shape. The properties of the film depend upon the composition and structure of the latex and the drying conditions.

The coalescence and chain interdiffusion processes discussed above require particle deformation and polymer chain mobility in the contact zone under the action of capillary forces (*20, 21*). To a first approximation, both of these processes are governed by the capillary forces and more importantly by the temperature of the particles relative to the "effective " T_g in the contact zone which we designate as T_g^*. If $T \ll T_g^*$ the latex particles will not sinter to form a coherent and continuous film. At the other extreme, $T \gg T_g^*$, a coherent and continuous film will readily form. However, in the latter case, the film may become tacky.

Implication to Spray Coating and Spray Drying. The film formation process for latex polymers discussed above and especially its sensitivity to the T_g, has a number of implications regarding their use as coating polymers for encapsulation. In discussing these implications we must consider the properties of the capsules as well as their processing.

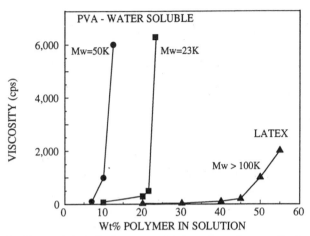

Figure 3. Comparison of the concentration dependance of viscosity between water soluble polymers and latex dispersion.

A. SOLUBLE POLYMER

B. LATEX DISPERSION

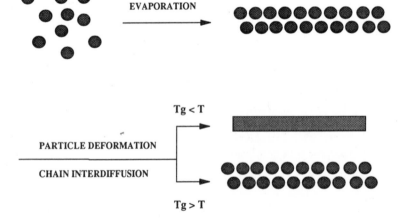

Figure 4. Simplified picture of evaporative film formation of water soluble polymer and latex dispersion.

The anticipated effect of latex T_g on spray coating is shown schematically in Figure 5A. If the process temperature, T_p, is far below T_g^*, the polymer will have difficulty coalescing with the solid active to be encapsulated and should not form a continuous coherent film. Here we expect poor coatings, and significant spray drying of the polymer solution which leads to dusting and excessive loss of material. At the other extreme, when the process temperature is much higher than T_g^*, the polymer becomes soft and readily deforms to coat the solid. However, in this case the coating is quite tacky and the particles readily clump which in the extreme case, can lead to caking of the bed. Thus, there must be a balance of T_p and T_g^* to form a coherent coating and a dry free flowing powder.

To be useful, the capsules so produced must have adequate storage and handling properties and to a first approximation, this should again be governed by the T_g of the dry polymer and the storage and handling conditions (Figure 5B). If the polymer has too low a T_g, the capsules can clump together and sinter under their own weight. At the other extreme, polymers that have too high a T_g, will form a brittle coating that does not adhere well to the core. In this case, the encapsulates will be friable and excessively sensitive to handling.

Similar considerations apply to the use of latex polymers in spray drying, however, there is an additional complication when the core material is an oil. Possible events that can occur during the formation of a microcapsule from an atomized droplet are shown in the diagram in Figure 6. As the droplet dries, it encounters other surfaces such as the wall of the drier or other particles at various stages of drying. The yield of free flowing powder that is produced by the process will depend on the strength of adhesion that results from these encounters. This will in turn depend on a variety of factors such as repulsive forces between aerosol particles, the area of the contact zone, the deformability of the particles and the relevant surface tensions. Four obvious possibilities can be distinguished.

If the droplets are sufficiently wet to still have a low viscosity when they encounter another surface, they will adhere to this surface. In this case, the particles produced are prone to coat the spray drying equipment and to form clumps. Thus, the yield of discrete free flowing powder will be low. As discussed above (Figure 3), non-swelling latex polymers require a much higher polymer concentration to build viscosity than do water soluble polymers. Such latexes should in general require either a lower water content or a higher drying temperature to achieve high capsule yield than do solution polymers.

Assuming that the solvent has completely evaporated, then the tendency of capsules to stick to other surfaces should broadly depend on the amount of exposed oil on the capsule surface and the surface properties of the dry polymer at the temperature of the process. The T_g should again have an important effect although other factors such as wetting, rheology, and emulsion stability no doubt make significant contributions. To a first approximation, if $T_p \ll T_g$, the latex will not coalesce to form a continuous matrix and significant migration of oil to the surface of the particle can take place. In this case we would expect a low yield of free flowing powder and significant coating of the equipment. This will be particularly severe when the oil phase has a positive spreading coefficient on the polymer surface, which will be favored by low surface tension oils such as silicones. In the extreme, the oil

A.

POOR COATING   CLUMPING

B.

ATTRITION   CAKING

Figure 5. Effect of T_g on the behavior of latex polymer in spray coating.
A. Coating and fluidization. B. Storage and handling characteristics.

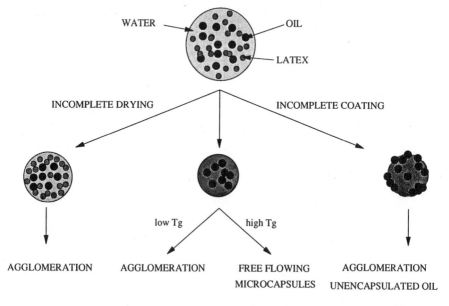

Figure 6. Possible events in the drying of an atomized droplet formed
during spray drying.

phase will in fact encapsulate the polymer particles to produce liquid bridges that will bind particles together.

When $T_p \ll T_g$ there is a much higher probability that the oil phase will be entirely sealed within the polymer matrix thus removing it as a source of surface tack. However, in this case, the surface of the polymer is deformable and mobile and can stick to other surfaces. Thus for $T_p \gg T_g$, we also anticipate a low yield of free flowing capsules but these should be well sealed with little free oil. These argument suggest that the spray drying temperature, water content, and glass transition temperature of the latex must be balanced to permit adequate sealing (film formation) without introducing excessive surface tack.

From the above discussion we conclude that the use of latex polymers for spray drying of oils (and probably soluble ingredients as well) is more tricky to control than when water soluble polymers are employed. The reasons stem from fundamental differences in rheology and film formation between these types of polymers. With latex polymers gelation occurs near to the random packing limit, and both film formation and surface tack are linked with T_g. The rheology and film formation of solution polymers are dominated by chain entanglements and these occur at much lower polymer concentrations than for latexes. Furthermore, film formation by chain entanglement is really independent on T_g and thus sealing efficiency (degree of encapsulation) and surface tack can be readily uncoupled.

Selection of Latex. Because film formation and surface tack both depend on T_g we can conclude that it can be tricky to balance the T_g of a uniform latex to achieve a high degree of encapsulation while maintaining a low tendency towards agglomeration. Although this can be achieved in practice (*18*), several routes can be identified to widen the process/composition window. For example, plasticizers are often required for a single T_g latex coating to tune the T_g to the process temperature of interest (*17, 22*). The problem is essentially to find ways to uncouple film formation from surface tack. We have explored two approaches to achieve this, namely the use of mixtures of latexes with different T_g, and the use of latexes with a core/shell morphology. These approaches are diagrammed in Figure 7.

It is known from the coating art (*17, 23*), that blends of a hard latex (high T_g) with a soft latex (low T_g) can yield tough coherent films that are nontacky. During the drying stage of evaporative coating, the soft latex particles deform and coalesce with themselves as well as with neighboring hard latex particles. This results in a coherent film. However, after drying, the hard segments maintain their spherical shape and prevent the soft segments from interacting with other surfaces. Thus, the films are nontacky when the ratio of latex particles is properly chosen.

The second approach is based on the use of latex polymers with a core-shell morphology. In this case, the core is selected to have a higher T_g than the process temperature while the shell has a relatively low T_g and is the soft segment. Such polymers are again known to form coherent and nontacky film (*24, 25*). During evaporative drying, the soft shells readily coalesces to form a coherent film. However, if the shell is sufficiently thin it will not be able to migrate to other surfaces and will not contribute to clumping of capsules.

Experimental

Materials. The latexes used for spray coating studies were a series of alkali soluble polymers and were provided by National Starch and Chemicals (*27*). Representative examples are listed in Table I. Four monomers were employed: methacrylic acid (MAA) for alkali solubility, butyl acrylate (BA) a soft hydrophobic monomer, and styrene (STY)/methyl methacrylate (MMA) a hard hydrophobic monomer used to control T_g.

Table I. Examples of Alkali-Soluble Latex Polymers Employed

Sample	T_g (°C)	BA	MMA	STY	MAA
A	35	65	5		30
B	52	60	10		30
C	52	60		10	30
D	60	50		20	30

The latexes used for spray drying were a series of styrene (Aldrich)/butyl acrylate (Rohm & Haas) copolymers. The compositions are shown in Table II. Latexes 1-5 were prepared by a semicontinuous emulsion polymerization (*24, 28*) employing potassium persulfate (Fisher) as initiator and a mixture of Triton X-100 (Union Carbide)/Siponate DS-10 (Alcolac) as the emulsifier. Triton X-100, DS-10 and deionized water were added to a 500 ml four-neck round bottom flask fitted with condenser, stirrer, nitrogen inlet, and temperature controller and heated to 80°C under stirring and flowing nitrogen. (5%) of the appropriate monomer mixture was charged to the reactor and emulsified for 10 minutes. Persulfate was added to start the polymerization and after 5 minutes, the remaining monomer was added over a period of 70 minutes. The reaction was held at 80°C for another 40 minutes and cooled to room temperature. Non-volatiles were determined gravimetrically.

Latex 6 was an inverted core-shell latex (*26, 29, 30*) with a hard polymer core (STY) and soft polymer shell (BA) morphology and was prepared using a similar procedure as described above except for the monomer addition. The BA/MAA monomer mixture was first fed to the reactor over 20 minutes and held another 10 minutes before neutralization with dilute ammonium hydroxide. After neutralizing the BA/MAA copolymer, STY monomer was metered into the reactor over a period of 45 minutes to make the core-shell latex.

The solid core material for spray drying studies was an organic chlorine bleach, sodium dichloroisocyanurate, Clearon CDB-56 (Olin) and had an average particle size of approximately 1000 µM. This and related oxidizing agents are of interest in a range of cleaning product applications (*27*).

For spray drying studies, silicone fluid L-45 (Union Carbide) was employed as the core material. Because of its spreading ability, silicone oil represents a sensitive test material to probe encapsulation efficiency.

Table II. Styrene/Butyl Acrylate Latexes Used for Spray Drying

Latex number	1	2	3	4	5	6
Styrene (wt%)	100	70	60	50	0	70
Butyl acrylate (wt%)	0	30	40	50	100	30
MAA (wt%)						2
T_g (°C)	100	58	44	30	-40	-

a) Latex 6 is a core-shell polymer (styrene core).

Procedures. The glass transition temperature of each of the alkali soluble polymers was measured by thermal analysis using a Perkin-Elmer DSC 7 Differential Scanning Calorimeter. The glass transition temperatures of styrene/butyl acrylate copolymers were calculated according to the Fox equation (*31*). As seen in Figure 8, the measured T_g of the alkali soluble polymers were consistently higher than those predicted from the Fox equation possibly because of formation of polymer micro-domains due to incompatibility.

Spray coating of CDB-56 was carried out using a Glatt Air Techniques fluidized-bed coater fitted with a Wurster bottom spray (*32*, see also Figure 1). The batch size was 800 to 1000 gm of bleach. The alkali soluble latex was applied from a 30-45% aqueous dispersion pumped at a rate of ~10 ml/min. A coating level of ~15 wt% was employed.

Spray dried silicone oil capsules were prepared as follows. 1:1 by weight silicone oil to latex dispersions for spray drying were prepared by adding preformed silicone emulsion to the latex under slow agitation. The silicone emulsion (50 wt% L-45) was prepared by a process previously described (*33*) using 3% Tergitol NP10 (Union Carbide) as the emulsifier and had an average droplet size of 0.64 microns as measured by dynamic light scattering using a Brookhaven BI-90 particle size analyzer (*33*). Before addition of the silicone emulsion, the latex was neutralized with dilute ammonium hydroxide to a pH ~8.3 because the silicone emulsion is not stable at lower pH due to complex formation of the nonionic surfactant with carboxylic acid groups on the surface of the polymer particle. Microcapsules were produced by spray drying the silicone-latex dispersion using a Yamato Pulvis Mini Spray Drier Model GA31 (See Figure 2). The conditions were: composition of feed dispersion - 20 wt% L-45, 20 wt% latex; feed rate - 2.5 ml/min; atomization pressure - 1.5kgf/cm²; air inlet temperature - 80°C; air outlet temperature - 50°C.

Capsules formed by spray coating were evaluated in terms of their release rate (bleach dissolution in water), uniformity of coating (electron microscopy), storage stability (% available chlorine after open cup storage at 27°C, 80% RH), resiliency (resistance to attrition by ball milling (*34*)), and caking tendency (storage at 95°C under 1 psi).

For spray dried silicone oil capsules the properties that were studied were the % yield of free flowing powder that was recovered from the cyclone and drying chamber of the spray drier, and the degree of encapsulation. The latter was determined by the

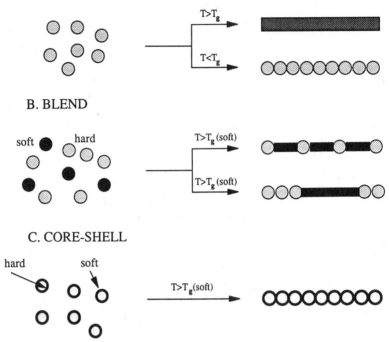

A. UNIFORM COPOLYMER

$T > T_g$

$T < T_g$

B. BLEND

soft   hard

$T > T_g$ (soft)

$T > T_g$ (soft)

C. CORE-SHELL

hard   soft

$T > T_g$ (soft)

Figure 7. Diagram depicting the effect of latex morphology on evaporative film formation.

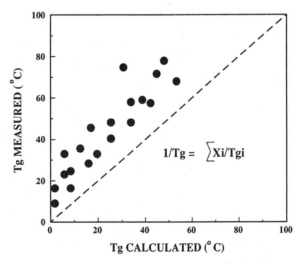

$$1/T_g = \sum X_i/T_{gi}$$

Tg MEASURED ($^\circ$C)

Tg CALCULATED ($^\circ$C)

Figure 8. Comparison of experimentally determined glass transition temperature (DSC) for alkali soluble latexes with theoretical values calculated from the Fox equation.

amount of silicone oil that could be extracted by hexane at room temperature and is commonly designated the percent "washable oil" or "surface oil".

Results and Discussion

Spray coating. Representative results of spray coating trials of chorine bleach, CDB-56, with alkali soluble latexes under different conditions of T_p and T_g are collected in Table III. The *target* latex coating level in all cases is 15%. The integrity of the coating was judged qualitatively by SEM evaluation (quality), and quantitatively by the time in minutes for 50% release of bleach under very mild agitation (in a Rotovac) in the presence of dilute phosphate buffer (pH 8.0). The results can be understood by comparing the glass transition temperature of the latex with the operating temperature of the fluidized bed.

When the T_g of the latex is more than about 10°C higher than T_p, as in Runs 1, 2, and especially 4 and 5, the resulting polymer coating contains many cracks which extend through the coating to the core surface. This leads to rapid release under mild alkaline conditions (R_{50} ~1 min) and poor barrier properties. We have also found that the actual amount of polymer deposited on the bleach particles is lower than the theoretical level anticipated based on the polymer charge. For example, the measured coating levels for Runs 4 and 5 is 7.4% and 8.3% respectively compared with a target of 15%. This suggests significant spray drying of the latex due to incomplete adhesion.

Table III. Highlights of Spray Coating Results With Alkali-Soluble Latexes

Run	T_g (°C)	Process Temp. (°C)	Coating[a] Quality (SEM)	Release Time (Min)	Process Problems
1	52	40	Fair	6.8	None
2	52	41	Fair	6.6	"
3	52	44	Good	6.4	"
4	60	35	Poor	1.0	Spray Drying
5	67	45	Poor	0.8	"
6	48	50	Excellent	8.4	None
7	35	38	Excellent	8.7	"
8	35	55	--	--	Clumping
9	25	45	--	--	"

a) Poor: Cracks extend to core and poor adhesion of coating
   Fair: Surface cracks and some holes in coating, good adhesion to core
   Good: Some holes in coating and good adhesion
   Excellent: Few imperfections detected

When the T_g of the latex matched the process temperature more closely, the coating becomes more uniform and its barrier properties increase as measured by an increase in R_{50} (Runs 3, 6, and 7). Photomicrographs illustrating this effect are shown in Figure 9. The series of photographs on the right show a capsule in which a latex having a T_g of 60°C was spray coated at 38°C. Note the poor coating characterized by cracks extending to the core and uplifting at the polymer/core junction. The capsule on the left was also processed at 38°C but the polymer employed in this case had a T_g of 35°C. Few cracks are evident and the coating has excellent adhesion to the core.

As seen by Runs 8 and 9 in Table III, when the T_g of the polymer falls to about 20°C below the operating temperature of the fluid bed, the latex act as a binder for the bleach particles. The particles clump together rapidly and fluidization ceases. The results in Table III suggest that for optimal spray coating with latex, the T_g of the polymer and the process temperature should be chosen to fall within the temperature range (T_g-10, T_g+20) and probably within the range (T_g-5, T_g+10).

The storage stability of the bleach capsules in open cup trials is shown in Figure 10. In this case Latex C (Table I) was employed and the capsules were stored at 27°C and 80% relative humidity. It is clear from these results that an alkali soluble latex with good barrier properties can be designed.

The attrition resistance of coated bleach particles in laboratory tests was used to predict handling characteristics, e.g., airveying, is shown in Figure 11. The test employed was a simple ball mill test (34) and the change in particle size was determined through sieve analysis. There is a good correlation between the % change in particle size and the glass transition temperature of the latex. For example, if a specific handling procedure requires a tolerance of no more than a 40% change in size based on this attrition test, the latex must have a T_g below about 45°C.

Caking of dry capsules can lead to serious problems in transport and storage. The tendency of bleach particles coated with alkali soluble latex to cake was simulated by storing a fixed quantity of capsules at different temperatures under an applied load of 1 psi. The results, shown in Table IV, illustrate the behavior observed. There is a critical T_g below which caking becomes increasingly severe. When the T_g is higher than the storage temperature there is some caking but these agglomerates readily disintegrate.

Table IV. Effect of Glass Transition Temperature of Latex Coating on Caking Properties of Spray Dried Capsules

Capsule No.	T_g (°C)	Caking[a] @ 52 °C	Caking[a] @ 45 °C
1	35	3	3
2	42	3	3
3	50	2	2
4	52	2	1
5	60	1	0

a) Rating: 0 - Free flowing; 1 - Loosens with tapping; 2 - Loosens with tapping but plug exists; 3 - Solid cake

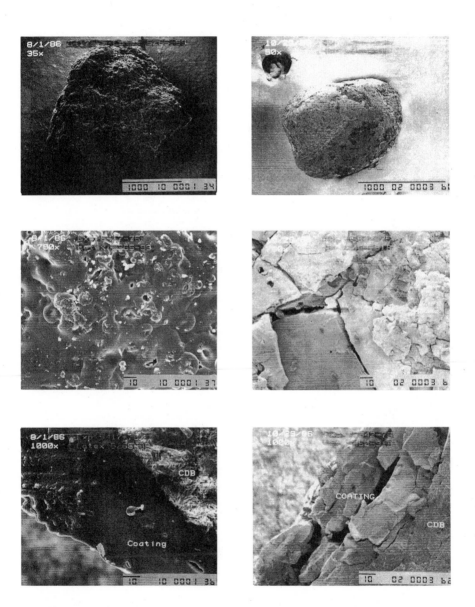

Figure 9. SEM micrographs of bleach capsules coated with alkali soluble latex polymers.

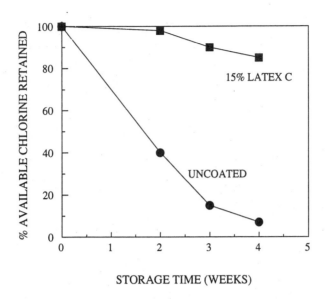

Figure 10. Storage stability of bleach capsules coated with an alkali soluble latex polymer (Latex C, Table 1) in open cup tests @ 27°C, 80% relative humidity.

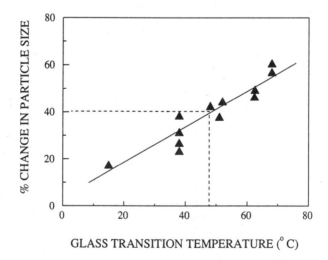

Figure 11. Influence of T_g on attrition resistance of spray coated bleach capsules.

The results presented above are consistent with the behavior anticipated for latex polymers as discussed in the previous section. Optimal manufacture of the capsule by spray coating, and adequate handling and storage characteristics requires a latex polymer having certain material properties. The glass transition temperature appears to be a good measure of these properties and the latex can be tuned so that its T_g falls as closely as possible within the required range. However, it may not always be possible to meet the process, storage and handling requirements with a single polymer having one T_g and other approaches must be utilized. Also, if the latex polymer has a complex composition, its components may be incompatible on a microscopic level making the thermal transition broad. This probably explains why the T_p limits in processing are not sharp, i.e., T_g-10 to T_g+20, since we have used the point of inflection as the measure of the T_g.

Spray Drying. The results described below relate to the spray drying of silicone oil (L-45) emulsions with the styrene/butyl acrylate latex polymers described in Table II. The polymers were not chosen with any particular application in mind but rather as a model system to test the concepts presented above. The capsule loading was 50% silicone oil in all cases and the Yamato spray drying conditions were held constant (see experimental). The study compared the use of a single T_g latex with a blend of latex polymers having different T_g's and with a latex designed to have a core/shell morphology (latex 6 - Table 2). We focused on the yield of free flowing powder produced and on the extent of encapsulation as determined by the ability to extract silicone from the capsules with a solvent in which the latex polymer is insoluble.

The spray drying results using single latex copolymers of varying T_g are shown in Figure 12. A moderate level of encapsulation of the silicone oil (> 40%) could only be achieved with a styrene/butyl acrylate latex having a T_g of less than about 40°C. However, these spray dried particles were very tacky and coated the spray drier. The yield of free flowing powder was very low in this case. The yield of capsules increased somewhat as the T_g was raised, but never reached more than about 40%.

The reason for the plateau in capsule yield observed in Figure 12 is due to the increase in unencapsulated oil that occurs as the T_g is raised. This unencapsulated oil migrates to the capsule surface and acts as a binder. This effect is particularly severe with silicone oil which, because of its low surface tension, is prone to migrate to the capsule surface. SEM photomicrographs of agglomerated microcapsules are shown in Figure 13. The photograph on the top shows untreated agglomerates of capsules while the photograph on the bottom is of capsules washed with hexane. Comparison of these photographs clearly show the presence of liquid bridges at interstices between particles that contribute to the agglomeration process.

The results in Figure 12 support our earlier conclusions that it can be difficult to simultaneously achieve high encapsulation efficiency and high yield using a single latex that has a uniform T_g. The use of a plasticizer and an increase in process temperature would probably be required.

The situation is greatly improved when blends of latex polymers are employed as seen by the results in Figure 14. Here we have used a mixture of polystyrene (T_g = 100 °C) and butyl acrylate (T_g = -50°C) homopolymer latexes. A 1:1 blend of these polymers simultaneously produces a reasonable capsule yield with about 60% of the oil phase encapsulated. It is likely that still better results could be obtained with this

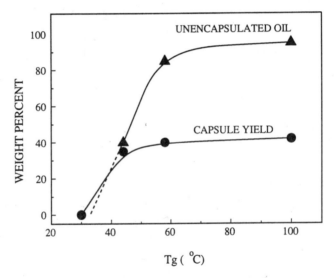

Figure 12. Influence of T_g on the yield of silicone oil microcapsules and the degree of encapsulation using a single styrene/butyl acrylate latex for each encapsulation.

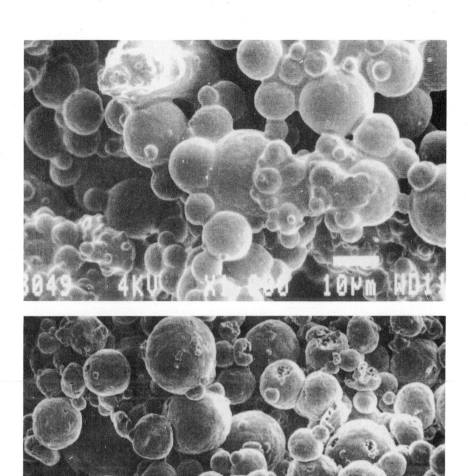

Figure 13. Scanning electron photomicrographs of silicone oil microcapsules formed by spray drying. The photograph on the top corresponds to unwashed microcapsules while the photograph on the bottom correspond to capsules washed with hexane. Note migration of oil to form liquid bridges.

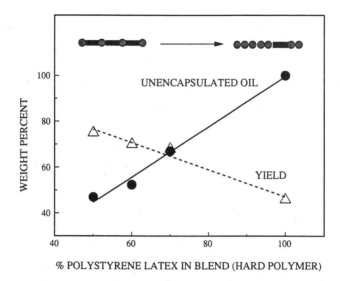

% POLYSTYRENE LATEX IN BLEND (HARD POLYMER)

Figure 14. Influence of blend composition on the yield and degree of microencapsulation of silicone oil for microcapsules formed by spray drying.

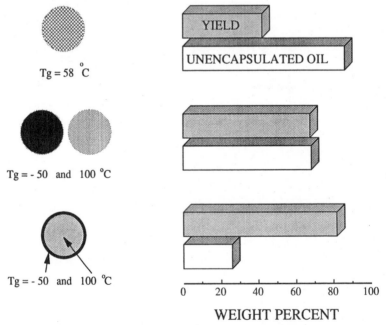

Figure 15. Comparison of core-shell latex with single T_g latex and blends for the microencapsulation of silicone oil by spray drying.

approach by optimizing the composition further. As the level of high T_g polymer increases, the yield decreases, reflecting an increase in the level of free oil because of poor film formation.

Figure 15 compares the spray drying performance of a core/shell latex (Latex 6, Table II) with both a single T_g latex and a blend of two latexes having different T_g's. The capsule was 50% silicone oil in the three cases and the composition of the polymer matrix was 30% butyl acrylate and 70% styrene. The result are averages of three spray drying trials for each polymer system. The core/shell morphology produced microcapsule in highest yield (80%) with consistently the lowest level of unencapsulated silicone oil (~25%). This finding is consistent with the concepts of film formation discussed earlier and suggest considerable scope for tailoring latex polymers for spray drying encapsulation.

Conclusions

The present study has considered some of the fundamental principles that govern the use of polymers in latex form for encapsulation by spray coating and spray drying. A simple, qualitative analysis of film formation suggested that the T_g of the polymer should be a key material variable that governs their behavior in processing, handling and storage of encapsulates made by these procedures. Experiments with bleach particles spray coated with alkali soluble latexes and microcapsules of silicone oil in styrene/butyl acrylate made by spray drying were carried out. The experimental results are consistent with conclusions drawn from the model and point to methods by which polymer latexes can be tailored to meet specific requirements.

Acknowledgements

The authors thank Mr. Joseph Moschetto for experimental assistance, Mr. R. Gursky, and Ms. K. Hoyberg for electron microscopy, Mr. Walter Samuel for photographic assistance, and Unilever Research for permission to publish this work.

Literature Cited

1. *Microcapsules and Other Capsule Advances Since 1975*, Gutcho, M. H., Noyes Data Corp., Park Ridge, New Jersey, 1981.
2. *Flavor Encapsulation*, Risch, S. J., and Reineccius, G. A., ACS Symposium Series 370, ACS, Washington, DC, 1988.
3. *Controlled Release Technologies: Methods, Theory and Applications*, Kydonieus, A. F., Vol. 1 and 2, CRC Press, Inc., Florida, 1980.
4. *Controlled Release Systems: Fabrication Technology*, Hsieh, D. S. T., CRC Press, Inc., Boca Raton, Florida, 1988; Vol. 1 and 2.
5. Jones, D. M., In *Flavor Encapsulation*; Editor, Risch, S.J., and Reineccius, G.A., ACS Symposium Series 370, ACS Washington, DC, 1988, P 158-176,.
6. Thacker, H., Forster, E., and Wallace, J., In *Controlled Release Systems, Fabrication Technology*, Editor, Dean S.T. Hsieh, CRC Press, Inc., Boca Raton, Florida, 1988 Vol. 1; P. 69-81 .

7.   Jones, D.M., *Pharm. Techn.* **1985**, Vol 9, 50-62.
8.   Mehta, M., Valazza, M. J., and Abele, S., E., *Pharm. Techn.* **1986**, Vol 10, 46-56.
9.   Mehta, M. In *Aqueous Polymeric Coatings for Pharmaceutical Dosage Forms*; Editor, McGinity, J. W., Marcel Dekker, Inc., New York, NY 1989, P267-302.
10. Master, K. *Spray Drying Handbook*, 4th Edition, Halsted Press, NY, NY, 1985.
11. Marrott, N. G., Boettger, R. M., Nappen, B. H., and Szymanski, C. D., U.S. Patent 3,455,838 (1969).
12. Reineccius, G. A. In *Flavor Encapsulation*, Editor, Risch, S.J. and Reineccius, G.A., ACS Symposium Series No. 370, Washington, DC, 1988; P55-66.
13. Trubiano, P. C., LaCourse, N. L., In *Flavor Encapsulation*; Editor, Risch, S.J. and Reineccius, G.A.; ACS Symposium Series No. 370, ACS, Washington, DC, 1988, P55-56.
14. Traae, J., and Kala, H., *Pharmazie*, **1984**, Vol 39, 233.
15. Blank, R.G., European Patent application EP 267,702, 1988.
16. Koppers Co. Inc., British Patent GB1,135,581, 1968.
17. Lehmann, K., In *Aqueous Polymeric Coatings for Pharmaceutical Dosage forms*, Editor, McGinity, J. W., Marcel Dekker, New York, NY, 1989 P153-245.
18. *Encyclopedia of Polymer Science and Engineer*, 2nd Edition, Mark, H., Bikales, N., Overberger, C. G., John Wiley, New York, NY, 1989, Vol 17, P. 734.
19. Vanderhoff, J., Bradford, E., Carrington, W., *J. Polym. Sci. Sym.*, **1973**, Vol 41, 155.
20. Bradford, E., and Vanderhoff, J. W., *J. Macromol. Chem.* **1966**, Vol 1, 335.
21. Vanderhoff, J. W., Tarkowski, H., Jenkis, M., and Bradford, E., *J. Macromol. Sci.*, **1966**, Vol 1, 361.
22. Banker, G. S., *J. Pharm. Sci.*, **1966**, Vol 55, 81.
23. Lehmann, K., *Acta Pharm. Technol.*, **1985**, Vol 31, 96.
24. Morgan, L. W., *J. Appl. Polym. Sci.*, **1982** Vol 27, 2033.
25. Min, T. I., Klein, A., El-Aasser, M. S., and Vanderhoff, J. W., *J. Polym. Sci., Polym. Chem. Ed.*, **1983**, Vol 21, 2845.
26. Amick, D., Melamed, S., and Novak R., US Patent 4,455,402, 1984.
27. Amer, G., Foster, J., and Iovine, C. US Patent 4,759,956, 1988.
28. *Encyclopedia of Polymer Science and Engineering*, 2nd Edition, Mark, H., Bikales, N.M., Overberger C.G., and Menges, G.M., John Wiley, New York, NY, 1989, Vol. 6, P1-51.
29. Okubo, M., Yamada, A., and Matsamoto, T., *J. Polym. Sci., Polym. Chem. Ed.*, **1980**, Vol 16, 3219.
30. Lee, D. I., and Ishikava, T., *J. Polym. Sci., Polym Chem. Ed.*, **1983**, Vol 21, 147.
31. Fox, T. G., *Bull Am. Phys. Soc.* **1956**, Vol 1, 123.
32. Wurster, H., US Patent 3,253,944, 1966.
33. Aronson, M. P., *Langmuir*, **1989**, Vol 5, 494.
34. Degussa Friability Test - Steel Ball Mill I422; Degussa Corporation.

RECEIVED August 14, 1992

Chapter 7

Modeling Swelling Behavior of Cellulose Ether Hydrogels

D. C. Harsh and S. H. Gehrke

Department of Chemical Engineering, University of Cincinnati, Cincinnati, OH 45221

Successful application of responsive gels will require knowledge of factors that control the swelling degree, transition temperature, and sharpness of the volume transition. As a result of the failure of the Flory network theory in predicting lower critical solution temperature (LCST) phenomena, more sophisticated theories accounting for the system free volume and specific interactions are required. In this work, the compressible lattice theory of Marchetti et al. is modified to account for ionic content and the elastic behavior of highly swollen networks. The resulting model is used to describe the observed swelling behavior of cellulose ether gels. The predicted and observed effects of controllable synthesis parameters are compared. The compressible lattice theory successfully describes LCST behavior, but is extremely sensitive to the parameter values, limiting its potential as a predictive tool. Accounting for non-Gaussian chain statistics results in the introduction of an additional parameter which does not greatly affect predicted swelling behavior.

Environmentally responsive hydrogels have been proposed for a variety of applications including artificial muscles and drug delivery devices (1,2). Successful applications will require both the magnitude and location of the volume change be well characterized and controllable. There has been much interest in relating the physical properties of the linear polymers to the swelling behavior of the crosslinked networks (3,4), but the predictive ability of such observations is limited beyond qualitative guidelines. Hence, it is desirable to make use of thermodynamic theories for gel swelling to model the swelling behavior. A successful theory would be able to predict gel swelling behavior based on the properties of the pure components (solvent and polymer). Development of thermodynamic theories is the subject of much current work (5-14), with particular interest in describing temperature-sensitive swelling behavior. In this chapter, the predictions of the classical Flory network theory will be presented, with discussion of how the controllable synthesis parameters affect these predictions. Then, the compressible lattice theory of Marchetti and Cussler (5-7) will be presented with modifications to account for ionic content and non-Gaussian network elasticity.

0097–6156/93/0520–0105$08.50/0
© 1993 American Chemical Society

Theoretical predictions of gel swelling with the modified Marchetti theory will then be presented in order to compare predictions of the two theories. Finally, the swelling behavior of cellulose ether gels will be modeled using the modified theory and the predictive utility of the modified theory will be discussed.

Equilibrium Swelling Thermodynamics

A gel will swell until the solvent chemical potential is equal in the gel and the free solution. For conceptual convenience, it is convenient to convert chemical potential to osmotic swelling pressure, which is zero at equilibrium. According to Flory, total osmotic swelling pressure is represented as the sum of the individual contributions of polymer-solvent mixing, network elasticity, ionic osmotic pressure, and electrostatic effects (15):

$$\Pi_{mix} + \Pi_{elas} + \Pi_{ion} + \Pi_{elec} = 0 \tag{1}$$

Π_{mix} represents the contribution from polymer-solvent interaction, which, in a good solvent, tends to increase gel swelling. Π_{elas} represents the contribution due to the network elasticity which arises from the restraints on swelling imposed by the crosslinks. In gel that contains charged or ionizable groups, it is necessary to include Π_{ion}, which represents the osmotic pressure of the counter-ions. In addition, it may be necessary to account for electrostatic effects, represented by Π_{elec}. The theories used to describe these contributions will now be reviewed.

Polymer Solution Theory. The Flory-Huggins theory uses a mean-field approach and introduces the interaction parameter, χ_1, to describe the solvent quality; χ_1 increases with decreasing solvency quality. However, the concentration and temperature dependence of χ_1 is highly non-linear so direct measurement is required. In addition, as originally defined, χ_1 decreases as temperature increases, failing to predict the inverse solubility behavior of temperature-sensitive gels.

Current theoretical work does not use classical polymer-solvent interaction parameters. Instead, by a corresponding states approach, equations of state have been developed based on pure component parameters which may be obtained from experimental data. These models attempt to account for the features that have been neglected in the Flory-Huggins theory, allowing description of volume changes of mixing and non-random mixing processes. Several different approaches have been taken, which will be now be described.

Flory et al., in an early attempt to correct the deficiencies of the Flory-Huggins theory, used a corresponding states approach to develop an equation of state accounting for non-zero volumes of mixing (16,17). The resulting theory has been useful in modelling of hydrocarbon solutions, although there has been limited success in application to aqueous systems due to strong orientation-dependent interactions such as hydrogen bonds.

Sanchez and Lacombe introduced the concept of a compressible lattice. The ternary version of the Flory-Huggins theory is adapted, adding holes to the lattice as a third component (18-21). The interaction parameter is defined as a function of the solvent and polymer cohesive energy densities. One adjustable parameter is added which characterizes the mixture: the deviation from the geometric mean mixing rule. The holes are treated as non-solvents for the polymer, with zero cohesive energy density. Solvent parameters are determined from P-V-T data; values for the polymer may be obtained by DSC measurements. This model will satisfactorily predict the phase behavior of some linear polymer solutions. Application to crosslinked systems has been accomplished theoretically (5); modelling experimental data has been accomplished by fitting the cohesive energy densities to observed phase behavior (6). In addition, this model predicts volume transitions in response to pressure changes, which has been experimentally confirmed (7).

Prausnitz and co-workers have taken a significantly different approach (8-12). Arguing that LCST behavior in aqueous systems is due to order-disorder transitions involving orientation dependent interactions such as hydrogen bonds, free-volume effects are neglected since the systems studied are at temperatures far from the solvent critical point. A partition function is derived based on a lattice model with three categories of interaction sites: hydrogen-bond donors, hydrogen-bond acceptors, and dispersion force contact sites. The resulting model qualitatively predicts LCST without the use of temperature dependent parameters in linear polymer solutions, although the parameter values must be fit to experimental data for accurate quantitative predictions. Although appealing from a conceptual standpoint, this theory has two major problems. First, free volume effects are totally neglected; volume changes in mixing and pressure induced volume changes are not described. Secondly, the model uses three "interaction energies" which are not easily accessible experimentally: the hydrogen bond strengths between like molecules, hydrogen bond strengths between unlike molecules, and the dispersion contacts. In addition, determining the solution of the equilibrium swelling equations is quite complicated numerically.

Although the theories discussed in the previous pages are the most commonly used, others have also been proposed. Otake et al. have introduced a separate contribution from hydrophobic interactions, which they use with a conventional virial expression to describe free energy of mixing (13). This approach, however, introduces three adjustable parameters to characterize the variation of hydrophobic interactions with temperature. A theory proposed by Painter et al. (14) accounts for the free energy change due to hydrogen bond formation by addition of a separate term based on self-association models used in alcohol-hydrocarbon systems. The extent of hydrogen bonding is quantified by an equilibrium constant, which must be determined experimentally.

Network Elasticity. Elastic contributions arise due to the configurational (entropic) changes in the swollen gels. The network entropy is decreased, since there are fewer possible configurations for the swollen network. At moderate deformations, the elasticity may be represented by an ideal network model. The ideal network assumes Hookean elasticity (retractive force proportional to deformation), with freely jointed polymer chains, tetrafunctional crosslinks of zero volume, crosslinked in the bulk state, and a Gaussian distribution of chain end-to-end distances. The elasticity may be expressed as follows for a network that swells isotropically (15):

$$\Pi_{elas} = -\rho_x RT [\phi_2^{1/3} - \phi_2/2]$$
(2)

ρ_x = Number of elastically effective network chains per unit volume dry polymer. Equation 2 neglects non-Gaussian contributions which arise due to physical restraints imposed by the crosslinks as swelling increases.

Equation 2 may be modified to account for crosslinking in solution; elastic force is relative to the unstrained state of the network. In this case, the "relaxed" state of the network is in solution, hence the introduction of the factor ϕ_{2f}, the polymer volume fraction at network formation (22):

$$\Pi_{elas} = -\rho_x RT \phi_{2f}[(\phi_2/\phi_{2f})^{1/3} - (\phi_2/\phi_{2f})/2]$$
(3)

Note that equation 3 reduces to equation 2 in the case of network formation in the absence of solvent, i.e. $\phi_{2f}=1$.

As the degree of swelling increases, the distribution of end-to-end distances of the network chains begins to deviate from a Gaussian distribution as the polymer chains begin to reach their maximum extensions. Hence, the validity of the Gaussian distribution becomes invalid, requiring use of non-Gaussian chain statistics. Alternatively, if the polymer chain is not freely jointed, then it may be necessary to account for the flexibility of the polymer backbone. Several models for non-Gaussian networks have been proposed to approximate the exact distribution of ideal chain extensions (23-25). These models use power series in ϕ_2 and require knowledge of the number of effective links per network chain, or equivalently, the number of structural units in an effective chain segment (N). Galli and Brumage have developed a function which closely matches the exact distribution of freely jointed random chains (25):

$$\Pi_{elas} = -RT\rho_x B \tag{4}$$

$$\text{where } B = \quad \phi_2/2 - \phi_2^{1/3}[1-1/N + (2/5)/N^2 + (8/25)/N^3]$$

$$+\phi_2^{-1/3}[-1/N +(13/5)/N^2 - (43/25)/N^3]$$

$$+\phi_2^{-1}[-(11/5)/N^2 + (221/25)/N^3] + \phi_2^{-5/3}[-(171/25)/N^3].$$

Use of non-Gaussian chain statistics results in a "levelling off" of predicted swelling at high swelling degrees (low ϕ) as opposed to the continued increase in swelling as solvent quality increases. This is due to stress increasing faster than strain at high swelling degrees. The non-Gaussian expression presented in equation 4 is written for networks prepared in the bulk state. Thus, the expression must be modified to account for the dilution at the time of crosslinking. Although there are other non-Gaussian models for solution crosslinked networks (23,24), the Galli and Brumage model (25) comes closest to describing the exact distribution of chain extensions of the models that have been proposed. All non-Gaussian models introduce an effective chain link, or an equivalent expression for the number of links per network chain. The expression developed by Galli and Brumage is a more complicated mathematical expression, but uses no more parameters than the other non-Gaussian models. The added complexity is not a concern for numerical calculations. The modification, as developed by Harsh (26), introduces the factor ϕ_{2f} in a manner analogous to equations 2 and 3 to account for the relaxed network state in equation 5.

$$\Pi_{elas} = RT\rho_x\phi_{2f}\{0.5(\frac{\phi_2}{\phi_{2f}})-(\frac{\phi_2}{\phi_{2f}})^{\frac{1}{3}}(1-\frac{1}{N}+\frac{2}{5N^2}+\frac{8}{25N^3})$$

$$+(\frac{\phi_2}{\phi_{2f}})^{\frac{1}{3}}(\frac{1}{N}+\frac{13}{5N^2}-\frac{43}{25N^3})+(\frac{\phi_2}{\phi_{2f}})^{-1}(\frac{11}{5N^2}+\frac{221}{25N^3})+(\frac{\phi_2}{\phi_{2f}})^{\frac{5}{3}}(\frac{171}{25N^3})\} \tag{5}$$

Ionic Osmotic Pressure and Electrostatic Effects. The presence of ionized groups will significantly affect equilibrium gel swelling, in some cases causing the gel to swell in otherwise poor solvents for the polymer. This is due to the osmotic pressure from the counter-ions of the charged groups, which must remain in the gel due to electroneutrality conditions. This ionic swelling pressure may be written as follows using an ideal solution model, with 'i' representing the fraction of charged repeat units in the network:

$$\Pi_{ion} = RT \ (i\phi_2/V_m) \tag{6}$$

where V_m = monomer specific volume. It is also necessary to account for the deswelling pressure which arises from solutes that are excluded from the gel network. At low ionic strengths, charged species will be excluded from the network by the Donnan effect, and may cause significant deswelling of a charged gel. This may be represented as follows for a binary salt:

$$\Pi_{ion} = RT \ [(i\phi_2/V_m) + 2(c_o-c_s)] \tag{7}$$

where c_o = concentration of mobile species in gel, and c_s = concentration of mobile species in free solution. In addition to charged species, any excluded solute may contribute to the deswelling pressure, such as high molecular weight linear polymers.

The mobile ion concentration in the gel may be calculated from the charge concentration in the gel and the external ion concentration. Donnan ion exclusion has been one the few quantitative successes in gel swelling theory. Changes in degree of gel swelling have been quantitatively predicted based on Donnan theory with no adjustable parameters (27-29).

Electrostatic effects are also present in a charged gel. However, in comparison with ionic osmotic pressure, electrostatics are a relatively minor component, especially in highly swollen gels in good solvents. Sample calculations using reasonable values of the significant parameters show little difference between calculated swelling equilibria with and without electrostatics (30). More importantly, the quantitative success of the ionic osmotic pressure term in interpretation of changes of gel swelling in salt solutions also indicates that this contribution may be neglected.

Flory Network Swelling Theory

Since all more sophisticated swelling theories make comparisons with the Flory theory, it is appropriate to examine the predicted swelling behavior in greater detail. As discussed previously, the Flory theory assumes the effects of mixing, elasticity, ionic osmotic pressure and electrostatics contribute to gel swelling independently. Therefore, equilibrium swelling is written as given by equation 8 for Gaussian network chains crosslinked in solution:

$$[\ln(1-\phi_2)+\phi_2+\chi_1\phi_2^2]/V_1-\phi_{2f}\rho_x B-\frac{i\phi_2}{V_m}=0 \tag{8}$$

While it is not possible to solve equation 8 explicitly for ϕ_2, it is possible to solve explicitly for χ_1 or i. This allows determination of predicted variations of swelling as a function of solvent quality (or indirectly, temperature) or network ionization (pH, for weakly ionizable networks). The solved equation for χ_1 is therefore:

$$\chi_1 = -\{\ln(1-\phi_2)+\phi_2+ V_1(\phi_{2f}\rho_x B+\frac{i\phi_2}{V_m})\}/\phi_2^2 \tag{9}$$

Equation (9) allows calculation of the interaction parameter as a function of swelling degree ($Q=1/\phi_2$) for fixed values of crosslinking and ionization. It should be noted that as originally defined by Flory, χ_1 decreases with increasing temperature; in this case it is treated simply as an adjustable parameter with a complex temperature dependence. Such calculations are easily performed on a spreadsheet, allowing the calculations to be rapidly performed for different values of the network parameters. By varying either parameter over a range of values, the predicted effects on swelling behavior may be determined.

Figures 1 and 2 show the effects of varied crosslink density and network ionization on predicted gel swelling behavior. Both of these figures show changes in swelling with changes in interaction parameter. Although there cannot be a direct comparison with temperature-sensitive swelling, qualitative comparisons can be made if χ_1 is considered an adjustable parameter that increases with temperature. The dotted lines represent unstable equilibrium states similar to those generated by cubic equations of state. As transition sharpness increases, a loop develops, and discontinuous transitions are predicted. In a manner analogous to the Maxwell's construction for real fluids, the point of transition is determined by the value of χ_1 that will divide the area between the lines $\chi_1=\chi_{1c}$ and $Q=f(\chi_1)$. This procedure has been discussed in greater detail elsewhere (*31*).

In Figure 1, swelling curves for gels with varying degrees of ionization are presented. As ionization increases, the degree of swelling increases, and the transition occurs at higher values of χ_1 (implying transition temperature increases). If crosslink density were reduced, discontinuous transitions would be predicted at lower levels of ionization.

Figure 2 presents the effect of crosslink density; decreased crosslinking increases the degree of swelling in the swollen state but does not significantly affect the point of transition. The sharpness of transition increases as crosslinking is decreased, until a discontinuous transition is predicted at a higher value of χ_1 than predicted for the continuous transitions. As ionic content increases, discontinuities are predicted at higher levels of crosslinking. These two figures demonstrate that the Flory-Huggins theory predicts a discontinuous transition for lightly crosslinked networks with partial ionic content.

The third parameter we can control at synthesis is the polymer type or relative hydrophobicity. This is not a prediction that is described from Figures 1 and 2. As the interaction parameter increases, the solvent quality decreases; at constant temperature this implies that the polymer is more hydrophobic. However, for a given polymer, the question that must be addressed is the variation of χ_1 with temperature. Qualitatively, we would expect χ_1 would increase at a lower temperature for a more hydrophobic polymer. Unfortunately, the actual values of χ_1 or the details of the variation with temperature or concentration are not known. Although χ_1 could be determined experimentally, measurements would be required across the range of concentration and temperature studied. Once obtained, the values would be restricted to the polymer sample (molecular weight and distribution) used. As a result, the ability to make quantitative predictions from these calculations is limited.

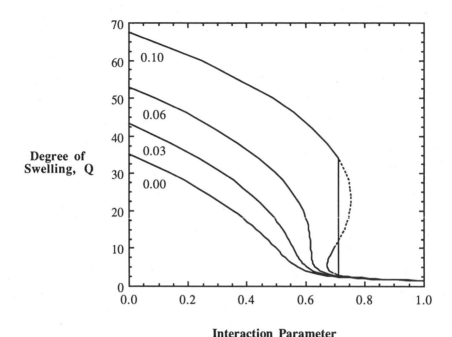

Interaction Parameter

Figure 1. Predicted swelling behavior of gels with varied ionic content with changing solvent quality according to Flory network theory using equation 9. As charge density of network increases, transition becomes more sharp until a discontinuous transition is predicted. V_m=90 cm³/mol, V_1=18 cm³/mol, ρ_x=1.0 × 10^{-6} mol/cm³, N=20, ϕ_{2f}=0.10. Values on the curves indicate the fraction of repeat units that are ionized.

Compressible Lattice Model for Gel Swelling

The model used for the description of polymer solvent mixing in this work is the compressible lattice model of Sanchez and Lacombe, as adapted to temperature-sensitive gels by Marchetti et al. (5-7). The Flory theory, as presented in the previous section, does not allow for compressibility, or free volume effects, in the lattice model. Such effects are known to be involved in LCST phenomena, as are orientation dependent interactions such as hydrogen bonding (32). Near the critical temperature of the solvent, there are large differences in the thermal expansion of the solvent and polymer, and the solvent expands more rapidly than the polymer. The free volume acts as a non-solvent for the polymer, thus solubility decreases until phase separation occurs. Although the use of quasi-chemical models that account for orientation dependent interactions such as hydrogen bonding is appealing, the free volume model is more tractable numerically and has been effectively applied to linear and crosslinked polymer systems displaying LCST behavior (6). More importantly, the quasi-chemical model requires three parameters that must be fitted

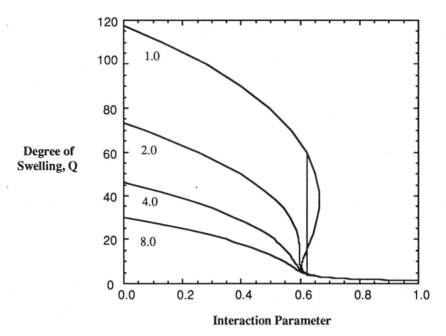

Figure 2. Predicted swelling behavior of gels with varied crosslink density with changing solvent quality according to Flory network theory using equation 9. As effective crosslink density decreases, transition becomes more sharp until discontinuous transition is predicted. Values on the curves indicate effective crosslink densities in 10^{-6} mol/cm^3. V_m=90 cm^3/mol, V_1=18 cm^3/mol, i=0.04, N=20, ϕ_{2f}=0.10.

to experimental data; the compressible lattice theory, by contrast, uses pure component parameters with only one adjustable parameter accounting for the mixture. By means of this parameter characterizing the mixture, hydrogen bonding and other non-dispersive interactions, are accounted for through a mean field approach.

The Sanchez and Lacombe theory for polymer-solvent mixing is a ternary version of the Flory-Huggins theory, with the vacant lattice sites, or "holes", representing the third component. In terms of solvent chemical potential, the free volume theory is written as follows, with the subscripts g, o, and s representing the gel, hole, and solvent, respectively (5):

$$\left(\frac{\mu_s}{V_s}\right)_{Gel} = RT\left\{\frac{1}{V_s}\left(\ln\phi_s + \left(1-\frac{V_s}{V_o}\right)\phi_o+\phi_g\right)+[\chi_{sg}\phi_g+\chi_{so}\phi_o](\phi_o+\phi_g)-\chi_{go}\phi_o\phi_g\right\} \qquad (10)$$

V_s, V_o = solvent and hole molar volumes, ϕ_s, ϕ_o, ϕ_g = solvent, hole and gel lattice volume fractions, and χ_{ab} = interaction parameter for species a with species b. The interaction parameters are determined from the pure component cohesive energy densities. The polymer-solvent interaction is calculated from the pure component cohesive energy densities, and a correction factor for deviation from the geometric mean mixing rule, Z_{sg}. The holes have zero cohesive energy density, so the three interaction parameters may be calculated as follows:

$$\chi_{sg}=\frac{1}{RT}[(\sqrt{P_s^*}-\sqrt{P_g^*})^2+2Z_{sg}\sqrt{P_s^*P_g^*}] \tag{11}$$

$$\chi_{so}=\frac{P_s^*}{RT} \tag{12}$$

$$\chi_{go}=\frac{P_g^*}{RT} \tag{13}$$

P_g^* = polymer cohesive energy density, P_s^* = solvent cohesive energy density, and Z_{sg}= correction for deviation from geometric mean mixing rule. Examination of equations 10 through 13 show that there are three parameters that affect the polymer phase behavior: gel cohesive energy density, P_g^*, solvent cohesive energy density, P_s^*, and the polymer-solvent mixing correction factor, Z_{sg}. Additional parameters include the solvent and hole molar volumes, V_s^* and V_o^*, respectively, which may be determined from experimental solvent vapor-liquid equilibrium data. The only additional parameter is Z_{sg}, the correction factor from the geometric-mean mixing rule. Examination of equation 11 shows that negative values of Z_{sg} indicate attractive interactions as compared to that obtained from the geometric mean mixing rule, thus resulting in lowered values of χ_{sg}. Although usually treated as an adjustable parameter characterizing the polymer-solvent pair of interest, Z_{sg} could be determined experimentally by fits of linear polymer cloud point data.

Swelling equilibrium is reached when the gel and solution phase chemical potentials are equal, hence we set $\phi_g=0$ in equation 10 to obtain the free solution chemical potential:

$$\left(\frac{\mu_s}{V_s}\right)_{Solution} = RT\left\{\frac{1}{V_s}\left(\ln\phi_s + (1-\frac{V_s}{V_o})\phi_o + \chi_{so}\phi_o^2\right)\right\} \tag{14}$$

Equation 14 for the solvent chemical potential in the free solution may be subtracted from equation 10 to obtain solvent swelling pressure. The resulting equation may be combined with equations 5 (non-Gaussian elasticity) and 6 (ionic pressure) to express the equilibrium swelling equation for the solvent. However, since the holes in the lattice are also capable of distributing between the gel phase and free solution, the swelling pressure for the holes must also be solved. The ionic and elastic terms are not changed, since ionic osmotic pressure is a function of the charge concentration, and elasticity is a entropic configurational phenomena. Therefore, equations analogous to equations 10 and 14 must be obtained for the hole chemical potential (5):

$$\left(\frac{\mu_o}{V_o}\right)_{Gel} = RT\left\{\frac{1}{V_o}\left(\ln\phi_o+(1-\frac{V_o}{V_s})\phi_s+\phi_g\right)+[\chi_{so}\phi_s+\chi_{go}\phi_g](\phi_s+\phi_g)-\chi_{sg}\phi_s\phi_g\right\} \tag{15}$$

$$\left(\frac{\mu_o}{V_o}\right)_{Solution} = RT\left\{\frac{1}{V_o}\left(\ln\phi_o + (1-\frac{V_o}{V_s})\phi_s\right)+\chi_{so}\phi_s^2\right\} \tag{16}$$

Similar subtraction of equation 16 from 15 gives the osmotic swelling pressure for the holes in the system.

We now have two equations for gel-solvent-hole equilibrium accounting for non-Gaussian elasticity and partial ionic content in the network. These equations, representing equality of chemical potentials for the holes and solvent, must be solved simultaneously subject to the restriction of all lattice sites being occupied. These equilibrium expressions may be written:

$$0=\frac{1}{V_s}(\ln\frac{\phi_s}{\phi_s^S}+\phi_g)+(\frac{1}{V_s}-\frac{1}{V_o})(\phi_o-\phi_o^S)+\chi_{sg}(\phi_o\phi_g+\phi_g^{\ 2})+\chi_{so}(\phi_o^{\ 2}+\phi_o\phi_g-(\phi_o^S)^2)$$

$$-\chi_{go}\phi_o\phi_g-\rho_x\phi_{2f}B'-(\frac{i}{V_m})(\frac{\phi_g}{1-\phi_o}) \qquad (17)$$

$$0=\frac{1}{V_o}(\ln\frac{\phi_o}{\phi_o^S}+\phi_g)+(\frac{1}{V_o}-\frac{1}{V_s})(\phi_s-\phi_s^S)+\chi_{so}(\phi_s\phi_g+\phi_s^{\ 2}-(\phi_s^S)^2)+\chi_{go}(\phi_g^{\ 2}+\phi_s\phi_g)$$

$$-\chi_{sg}\phi_s\phi_g-\rho_x\phi_{2f}B'-(\frac{i}{V_m})(\frac{\phi_g}{1-\phi_o}) \qquad (18)$$

where

$$B'=0.5(\frac{\phi_g}{\phi_{2f}})-(\frac{\phi_g}{\phi_{2f}})^{\frac{1}{3}}(1-\frac{1}{N}+\frac{2}{5N^2}+\frac{8}{25N^3})+(\frac{\phi_g}{\phi_{2f}})^{\frac{1}{3}}(\frac{1}{N}+\frac{13}{5N^2}-\frac{43}{25N^3})$$

$$+(\frac{\phi_g}{\phi_{2f}})^{-1}(-\frac{11}{5N^2}+\frac{221}{25N^3})+(\frac{\phi_g}{\phi_{2f}})^{\frac{5}{3}}(-\frac{171}{25N^3}) \qquad (19)$$

The superscript S in equations 17 and 18 denotes the lattice fraction in the free solution, all other composition variables refer to the gel phase. Solution for the lattice fractions was obtained using a Newton-Raphson algorithm. Validity of the calculations was verified using a homotopic continuation technique.

Solvent and Polymer Property Effects on Predicted Swelling

Once the validity of our calculations was established, the parameters were systematically varied in order to gain insight into what steps would be necessary to obtain valid fits on our experimental data. It was necessary to learn how to make coarse adjustments to the predicted transition temperature and sharpness. While it is known that degree of swelling, at least below the transition, is primarily controlled by the crosslink density, the effects of the gel and solvent parameters on transition temperature and sharpness were unclear. A series of simulations were performed around the values of the fit parameters for N-Isopropylacrylamide gels in water reported by Marchetti (6):

$P_g^* = 308.8$ cal/cm^3 $\qquad\qquad$ $P_s^* = 642.2$ cal/cm^3

$Z_{sg} = -.0633$ (dimensionless) $\qquad$ $\rho_x = 8.4 \times 10^{-6}$ mol/cm^3

It should be noted that the value of crosslink density reported by Marchetti et al. is based on the volume fraction as formed due to the nature of the elastic expression used in their work. The value above has been adjusted to the basis of dry polymer volume. Systematic variation of these parameters allowed determination of the type and extent of variation that could be expected by adjusting these values. As will be shown, the resulting phase diagrams are extremely sensitive. Small variations (<1%) in the interaction energy parameters can significantly change the degree of swelling, transition temperature, and the sharpness of the transition. The values listed above are used in all of the following graphs, with only one variable adjusted at a time. Although Marchetti

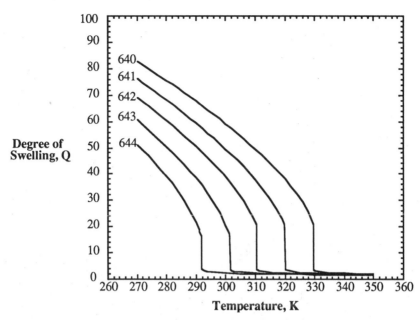

Figure 3. Predicted effect of solvent cohesive energy density according to compressible lattice theory. Decreased solvent quality (higher P_s^*) predicts decreased transition temperature and decreased volume change across transition. $P_g^*=308.8$ cal/cm^3, $Z_{sg}=-.0633$, $\rho_x=8.4 \times 10^{-6}$ mol/cm^3, N=20.

used an elastic expression that did not account for solution crosslinking or non-Gaussian chain statistics, the non-Gaussian elastic expression was used in these simulations for consistency with later modeling of cellulose ether gels with the value of N arbitrarily set at 20. The use of non-Gaussian chain statistics results in decreased degree of swelling below the transition, as the finite extensibility of the network chains is considered.

Solvent and Gel Cohesive Energy Density.   Changes in P_g^* and P_s^* primarily affect the quality of the solvent for the polymer. Closer values indicate better solvents, hence swelling degree and transition temperature increase, as predicted by the Flory theory presented previously. Figures 3 and 4 present the simulations for these variations, but it should be noted that Z_{sg} is held constant. This is an important point to consider, because as the solvent quality changes, it is likely that there will be changes in the specific interactions. As the polymer character changes, so will the nature of the interactions with the solvent. Thus it is difficult to draw conclusions for specific polymer-solvent systems from these figures, although they are helpful in determining appropriate steps to take in the process of fitting experimental data. For the remainder of this work, P_s^* has been left constant at 642.2 cal/cm^3, the value used by Marchetti et al. for water, originally reported by Sanchez and Lacombe (21). The molar volumes of water and the vacant lattice sites used, also taken from Sanchez and Lacombe, were 16.33 and 1.93 cm^3/mole, respectively.

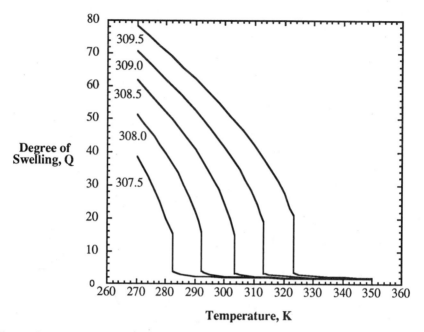

Figure 4.   Predicted effect of polymer cohesive energy density according to compressible lattice theory.   Increased polymer hydrophobicity (lower P_g^*) predicts decreased transition temperature and decreased volume change at transition. P_s^*=642.2 cal/cm^3, Z_{sg}=-.0633, ρ_x=8.4 × 10^{-6} mol/cm^3, N=20.

Tables I and II present the effect on the transition temperature and the volume change across the transition. In both cases, as the cohesive energy densities of the polymer and solvent become more similar the transition temperature increases, indicating a better solvent for the polymer. The transition sharpness, however, increases slightly as solvent quality improves for variations in both solvent and polymer cohesive energy density. This is in direct contradiction with the predictions of the Flory theory, although the trend is not as strong. Although not shown, the discontinuous transition shown in figures 3 and 4 gradually becomes smaller until a continuous transition is predicted. It should be noted that the variations in P_s^* presented are far less than would be encountered with different solvents. Sanchez and Lacombe have reported the molecular parameters for numerous organic solvents; values of P_s^* for solvents commonly used with gels are presented in Table III. Tabulated values of P_g^* are rare, especially for hydrophilic polymers used in hydrogel synthesis. As discussed previously, however, concomitant changes expected in Z_{sg} with changes in P_s^* and P_g^* limit the utility of these predictions.

Correction Factor.   The correction factor, Z_{sg}, represents the deviation from the geometric mean mixing rule as used in equation 11; negative values indicate attractive interactions between the species. Therefore, as Z_{sg} becomes more negative (smaller χ_{sg}) the degree of swelling increases, because of the favorable interactions between the

Table I. Predicted Effect of Solvent Cohesive Energy Density on Gel Swelling

P_s^* (cal/cm^3)	Transition Temperature (K)	Change in Swelling Across Transition (ΔQ)	$\chi_{SG} \times RT$ (cal/cm^3)
640	329.8	17.2	3.40
641	320.0	17.2	3.66
642	310.5	16.7	3.93
643	301.5	13.6	4.19
644	291.8	12.9	4.45

$P_g^* = 308.8$ cal/cm^3, $Z_{sg} = -.0633$, $\rho_x = 8.4 \times 10^{-6}$ mol/cm^3, N=20.

Table II. Predicted Effect of Polymer Cohesive Energy Density on Gel Swelling

P_g^* (cal/cm^3)	Transition Temperature (K)	Change in Swelling Across Transition (ΔQ)	$\chi_{SG} \times RT$ (cal/cm^3)
307.5	282.2	11.3	4.67
308.0	292.0	11.6	4.41
308.5	303.2	11.7	4.14
309.0	313.2	14.9	3.87
309.5	323.2	17.3	3.61

$P_s^* = 642.2$ cal/cm^3, $Z_{sg} = -.0633$, $\rho_x = 8.4 \times 10^{-6}$ mol/cm^3, N=20.

Table III. Equation of State Parameters for Selected Solvents

Solvent	P_s^* (cal/cm^3)	V_s^* (cm^3/mol)	V_o^* (cm^3/mol)
Water	642.2	16.33	1.93
Methanol	287.3	34.73	3.24
Ethanol	255.8	47.80	3.21
n-Propanol	212.2	61.78	3.93
2-Propanol	203.8	56.29	3.89

solvent and the gel. What it most significant is the degree to which minor variations in Z_{sg} can change swelling degree. Figure 5 shows the predicted swelling for values between -.0600 and -.0700. As can be seen, predicted swelling ranges from swollen at all temperatures across the range shown, to collapsed at all temperatures. Changes of ±.0025 result in transition temperature shifted by over 50 K. In addition, the transition sharpness increases significantly from a continuous transition at Z_{sg}=-.0625 to a discontinuity at Z_{sg}=-.0650.

Examining the effect over a smaller range of variation shows that, as expected, decreases in Z_{sg} (representing increased interaction) increases degree of swelling, transition temperature, and most significantly, sharpness of transition, shown in Figure 6. This is directly consistent with the work of Freitas, who inferred the importance of hydrogen bonding in swelling of NIPA gels by noting that swelling decreased in the presence of solutes that disrupt hydrogen bonds (*33*). As is the case with the cohesive energy densities, however, it is unlikely that changes in Z_{sg} could occur without simultaneous changes in P_s^* or P_g^*. Positive values of Z_{sg}, representing unfavorable interactions (larger χ_{sg}), result in prediction of a collapsed gel at all temperatures for the parameter values encountered in this work.

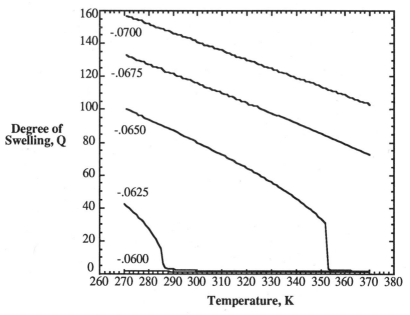

Figure 5. Predicted effects of geometric mean correction factor according to compressible lattice theory. Ten percent changes in Z_{sg} result in changes between a swollen and collapsed gel across the entire range of temperature studied. P_s^*=642.2 cal/cm^3, P_g^*=308.8 cal/cm^3, ρ_x=8.4 × 10^{-6} mol/cm^3, N=20.

Table IV. Predicted Effect of Geometric Mean Correction on Gel Swelling

Z_{sg}	Transition Temperature (K)	Change in Swelling Across Transition (ΔQ)	χ_{SG} × RT (cal/cm^3)
-.0628	295.8	9.6	4.42
-.0630	300.5	11.7	4.25
-.0632	306.0	14.3	4.07
-.0634	311.5	16.2	3.89
-.0636	316.8	18.1	3.71

P_s^*=642.2 cal/cm^3, P_g^*=308.8 cal/cm^3, ρ_x=1.8 x 10^{-6} mol/cm^3, N=20.

Table IV presents the effect of the correction factor on the changes of swelling across the transition. The degree of swelling above the transition increases significantly with more negative values of Z_{sg}, while swelling below the transition decreases. Thus, by comparison with Tables I and II, it can be seen that Z_{sg} exerts the greatest influence on the extent of the volume transition.

Figures 1 and 2 allowed direct comparison of the effects of hydrophilic or hydrophobic character by use of χ_1. Figures 3 through 6, however, involved variations of three parameters that interact to represent this same effect. Can the effects

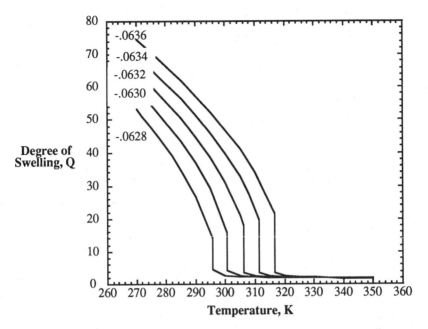

Figure 6. Predicted effects of geometric mean correction factor according to compressible lattice theory. More negative values of Z_{sg} (increased attractive interactions between polymer and solvent) result in increased transition temperature and sharpness of transition. $P_s^* = 642.2$ cal/cm^3, $P_g^* = 308.8$ cal/cm^3, $\rho_x = 8.4 \times 10^{-6}$ mol/cm^3, N=20.

of hydrophilic/hydrophobic character be related to changes in these three parameters? Since all experiments in this work used water as a solvent, the value of P_s^* will be assumed constant for this discussion. As P_g^* decreases, the difference between the polymer and solvent cohesive energy density increases, thus the hydrophilic character of the polymer is decreasing. This does not address the variation in Z_{sg}; there is no reason to expect a direct variation (increase or decrease) with the hydrophilic or hydrophobic character. If the changes in the polymer increase the ability of the polymer to hydrogen bond with the solvent (a favorable interaction), we would expect a decrease in Z_{sg}, although only qualitative estimates may be made.

Crosslink Density and Statistical Segment Length. In discussing the effects of crosslinking, it is necessary to account for the effects of both the crosslink density, ρ_x, and the number of effective links per network chain, N. The effective crosslink density represents the number density of elastically active chains in the network, while the number of units per chain segment gives an indication of the polymer chain flexibility. As N increases, there are more repeat units per segment, hence chain flexibility decreases.

The effect of crosslinking on volume transition behavior has been the subject of much discussion, both experimental and theoretical. The Flory theory does not predict discontinuous transitions in non-ionic gels, except at extremely high crosslinking values two to three orders of magnitude higher than typically seen in hydrogels which are not experimentally feasible (*34*). Prange et al., however, predict discontinuous transitions at low levels of crosslinking using the quasi-chemical theory (*8*).

Experimentally, both discontinuous and continuous transitions have been reported in N-isopropylacrylamide gels. These conflicting reports have involved gels synthesized from the same nominal content and under similar conditions. Since NIPA gels are produced by simultaneous copolymerization and crosslinking, minor variations in synthesis conditions may be responsible for these differences. Temperature variations during synthesis can influence the rate of both initiation and propagation of the polymerization reaction. In addition, varied initiator levels will affect both the number and lengths of the network chains. These effects have recently been addressed by Gehrke et al. (*35*).

The compressible-lattice theory used in this work predicts discontinuous volume transitions at low levels of crosslinking. The predicted effect of crosslinking on the swelling behavior of NIPA gels is shown in Figure 7. As crosslink density is decreased, the predicted transition shifts from continuous to discontinuous. These effects are reported in Table V. Note that even in the range of crosslink densities where continuous transitions are predicted, the sharpness increases as crosslinking is decreased. In addition, the degree of discontinuity increases; this is expected, since a low level of crosslinking produces increased swelling at low temperatures. Although increasing crosslinking changes the overall polymer character (and thus P_g^*), the small amounts of crosslinker in these gels is not likely to have a significant effect.

Figure 8 and Table VI present the effects of the number of units per statistical segment on predicted gel swelling. Note that while decreases in N (corresponding to decreased chain flexibility) have no effect on transition temperature, the degree of swelling at high temperature decreases. This may be interpreted in terms of the number of effective repeat units per network chain. As N decreases, there are fewer effective "chain links" per network chain, resulting in an increased crosslink density. Therefore, with increased crosslinking, the swelling decreases as would be expected. Interestingly, however, the transition sharpness increases as the network chain becomes less flexible; the change is similar in magnitude to that predicted with changes in crosslink density. This is more evident in gels that display gradual transitions than

Table V. Effect of Effective Crosslink Density on Predicted Gel Swelling

ρ_x (10^{-6} mol/cm^3)	Transition Temperature (K)	Continuous or Discontinuous	Sharpness (%/°C) or ΔQ at Transition
1.0	308.8	Disc.	86.6
2.5	308.8	Disc.	53.6
5.0	308.0	Cont.	23.2 %
7.5	308.0	Cont.	15.5 %
10.	307.0	Cont.	12.2 %

P_s^*=642.2 cal/cm^3, P_g^*=308.8 cal/cm^3, Z_{sg}=-.0633, N=20.

Table VI. Effect of Statistical Segment Parameter on Predicted Gel Swelling

N	Transition Temperature (K)	Sharpness of Transition (%/K)
2	308.0	28.5
4	308.0	19.0
6	308.0	15.8
10	308.0	13.5
50	308.0	11.3

P_s^*=642.2 cal/cm^3, P_g^*=308.8 cal/cm^3, Z_{sg}=-.0633, ρ_x=1.0x10^{-5} mol/cm^3.

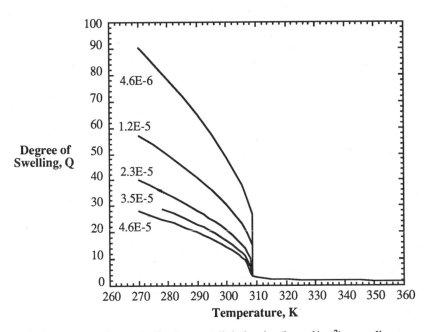

Figure 7. Predicted effects of effective crosslink density (in mol/cm³) according to compressible lattice theory. Decreased crosslink density produces discontinuous transitions and increases swelling below the transition. There is little effect on transition temperature. P_s^*=642.2 cal/cm³, P_g^*=308.8 cal/cm³, Z_{sg}=-.0633, N=20.

in near-critical gels, as will be discussed later. The effect of polymer chain flexibility has not been studied previously in any great detail, although Post and Zimm had observed discontinuous changes in crosslinked DNA with changes in solvent composition, which they attributed to the rigidity of the helical structure (*36,37*).

Ionic Content. As discussed previously, the Flory theory predicts discontinuous volume transitions in networks with partial ionic content. Such predictions were first made by Ilavsky et al.; these predictions suggest that partial ionic content is necessary for discontinuous volume transitions to be observed (*24*). Several researchers, however, have reported discontinuous volume transitions in non-ionic NIPA gels (*38,39*). Gehrke at al. (*35*) have identified significant quantities of acrylic acid monomer in samples of N-isopropylacrylamide monomer as received, hence the reported discontinuities might have been due to an ionic impurity.

The predicted effects of ionic content with the compressible lattice theory are much greater than those predicted by other theories. All theories predict increased transition temperature and sharpness of transition with increased ionic content. However, the compressible lattice theory predicts a greater increase and at much lower levels than predicted by the other theories. Figure 9 presents these trends for NIPA gels. Table VII presents the transition temperature and sharpness for the results in Figure 9 and the corresponding values from the results of Beltran et al. obtained from the quasichemical theory. The compressible lattice theory predicts greater increase in

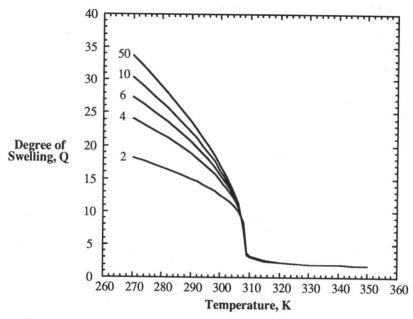

Figure 8.   Predicted effects of statistical segment parameter according to compressible lattice theory. As number of effective links per chain decreases, the swelling below the transition is decreased and transition sharpness increases. The transition temperature is unaffected. P_s^*=642.2 cal/cm^3, P_g^*=308.8 cal/cm^3, ρ_x=8.4 × 10^{-6} mol/cm^3, Z_{sg}=-.0633.

Table VII. Predicted Effect of Network Ionic Content on Gel Swelling for Compressible Lattice and Quasichemical Theories (9)

i	Transition Temperature[†] (K)	Sharpness (%/°C) or ΔQ[†]	Transition Temperature[‡] (K)	Sharpness (%/°C) or ΔQ[‡]
0.00	307.	12%/°C	306.	7.5 %/°C
0.02	326.	90. Δ	316.	25. Δ
0.04	342.	163. Δ	332.	58. Δ

[†] Compressible Lattice Theory
[‡] Quasi-chemical Theory

Δ - change in swelling across a discontinuous transition.

transition temperature and sharpness than that predicted by the quasichemical theory (9); also, the predicted effects are significantly greater than observed experimentally. It should also be noted that neither Beltran et al. (9) or Akhtar (40) have observed the discontinuous transitions predicted by the theories discussed here.

In an effort to explain the discrepancies in the predictions, the individual contributions to gel swelling from mixing, elasticity and ionic osmotic pressure were examined for ionic and non-ionic gels, using both the Flory network theory and the compressible lattice theory described in this work. The results are as expected; the

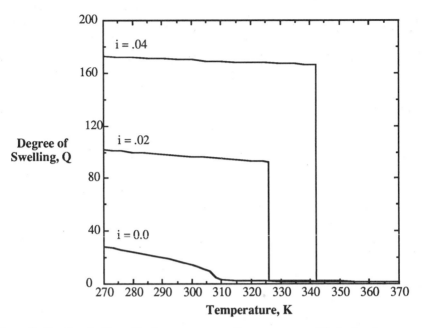

Figure 9. Predicted effect of ionic content according to compressible lattice theory. Ionic content increases degree of swelling below the transition; swelling above the transition is unaffected. Transition temperature and sharpness increase with ionic content and discontinuous transitions are predicted. P_s^*=642.2 cal/cm^3, P_g^*=308.8 cal/cm^3, ρ_x=1.0 $\times$ 10^{-5} mol/cm^3, Z_{sg}=-.0633., N=20.

non-ionic gel shows equal and opposite mixing and elastic swelling pressures across the entire range of swelling studied. Introduction of ionic content results in the ionic and elastic contribution dominating in the swollen gel; in the swollen gel, the swelling pressure due to polymer solvent mixing is minor. In the collapsed gel, however, the elastic contribution is greatly reduced and equilibrium is dominated by the ionic and mixing terms.

Swelling pressures as calculated for the Flory and compressible lattice theories for gels with ionic content are presented for comparison in Figure 10; predicted swelling behavior (as a function of interaction parameter and temperature, respectively) is shown in Figure 11. The curves in Figure 10 are restricted to high swelling degrees because it is here that the reasons for the large swelling can be found. Note that for the Flory theory, the swelling pressure due to polymer-solvent mixing is still negative immediately above the transition. In contrast, the bold lines show the corresponding values for the compressible lattice theory, which is much less above the transition, thus the degree of swelling is increased. The values for all contributions are similar for a given value of Q because the ionic and elastic contributions are only dependent on degree of swelling and are unaffected by the polymer-solvent interactions. One point of note is the continued increase in polymer-solvent mixing pressure according to the Flory theory. This is an artifact, due to the value of χ_1 becoming increasingly

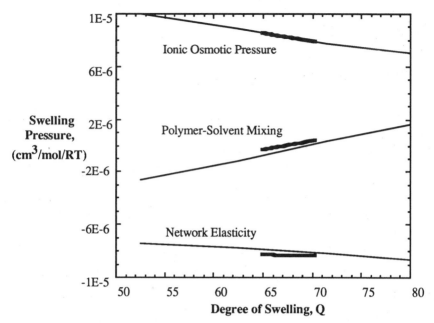

Figure 10. Swelling pressure contributions from mixing, elasticity, and ionic osmotic pressure according to Flory and compressible lattice (in bold) theories for a gel with 5% of repeat units ionized. Region shown is at degrees of swelling above the transition; the compressible lattice theory predictions are represented by the bold lines. Note that the mixing pressure from the Flory theory is negative just above the transition; thus the degree of swelling is less than that predicted by the compressible lattice theory. Swelling pressure is not converted into pressure units because variation of χ_1 with temperature is unknown. V_m=90 cm³/mol, V_1=18 cm³/mol, ρ_x=1.0 × 10⁻⁵ mol/cm³, N=8.

negative. As this value continues to decrease, the mixing pressure increases continuously. This is obviously unreasonable, as in a good solvent, this contribution should eventually taper off to zero as swelling increases. This is one of the problems that can be encountered if theoretical predictions are made using adjustable parameters beyond their range of physical meaning.

It is difficult to criticize the validity of the ionic expression used and the separation of the contributions of mixing, elasticity and ionic pressures, as the ionic osmotic pressure term has had the greatest quantitative success in describing gel swelling. As will be discussed in the next section, there may be problems with the elastic expression, but not significant enough to cause this large a deviation. The ionic expression assumes an ideal solution, which assumes the activity of the charged groups is equal to their concentration. Thus, the actual activity of the charged groups will be overpredicted by this expression, but by no more than a factor of 2. In addition, other factors that would reduce the effect of the charges such as counter-ion condensation and solvent dielectric effects have been neglected. If the presence of the ionized groups affects the polymer character, it would improve the solvent quality, thus predicting increased swelling (through modification of P_g^* and Z_{sg}). Since this is

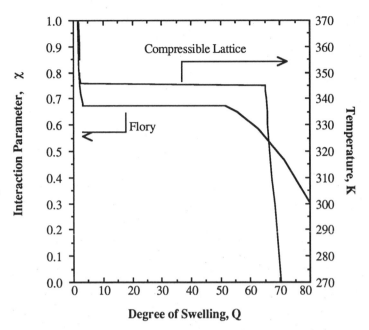

Figure 11. Predicted swelling behavior as a function of interaction parameter and temperature according to the Flory and compressible lattice theories, respectively for a gel with 5% of repeat units ionized. Continued increase in swelling predicted by the Flory theory is due to the decrease in values of c_1 used for the swelling calculations. P_s^*=642.2 cal/cm^3, P_g^*=308.8 cal/cm^3, ρ_x=1.0 x 10^{-5} mol/cm^3, Z_{sg}=-.0633., N=8.

neglected, this should in fact be underpredicted if this were the cause of the discrepancy. Because of the problems encountered with the ionic predictions in this work, the experimental data of ionic gels was not modeled in the subsequent section. Also, the variations in polymer character produced by including carboxymethyl cellulose in the gels would make meaningful comparison of the gel parameters obtained difficult.

Fit Swelling Behavior of Cellulose Ether Gels

The synthesis and equilibrium swelling behavior of the gels modelled in this work has been described elsewhere (4). Cellulose ethers are alkyl and hydroxyalkyl substituted polymers whose properties have been extensively reviewed (41). Cellulose ethers are designated "Generally Recognized as Safe" by the Food and Drug Administration, making them ideal for usage in controlled release applications. In addition, the available variations in substitution levels make them ideal for use as model systems in examination of thermodynamic models as in this work. The substitution levels of the polymers used in this work are presented in Table VIII. DS and MS indicate the number of substitute groups per glucose unit on the polymer chains; MS may be greater than three due to the hydroxyl group contained in the hydroxypropyl group, which may also be substituted.

Table VIII. Substitution Levels of Cellulose Ethers Used in This Work

Polymer	Methyl DS	Hydroxypropyl MS	M_n
MC	1.8	0.0	70,000
HPMC-E	1.9	0.23	70,000
HPMC-K	1.8	0.21	70,000
HPC	0.0	4.0	370,000

Method of Fit. The first steps in fitting the experimental data were to determine what portions of the swelling curves were to be fit and how the quality of the fit would be evaluated. Initially, qualitative comparisons of fit values may be made to judge the quality of fit to the experimental data, but this is not satisfactory from a standpoint of determining the "best" fit.

There were several factors involved in choosing what portion of the swelling curve should be fit. First, as temperature increases, the stiffness of the cellulose ether polymer backbone decreases, resulting in a higher value of N, the number of effective links per network chain (*42*). Also, in fitting acrylamide derivative gels, the actual value of N is not known and becomes another parameter to fit; the appropriate range may be estimated for cellulose ethers by measurement of intrinsic viscosity. More importantly, we are most interested in controlling the transition region; theoretical work indicates that this is not affected as much with changes in elasticity. Since the interaction parameters (obtained from P_g^*, P_s^*, and Z_{sg}) have the strongest effect on the characteristics of the transition, it was decided to focus on the transition region, where the sharpest collapse in gel swelling occurs. The transition temperature and sharpness information was determined from normalized swelling curves:

$$\frac{Q - Q_{min}}{Q_{max} - Q_{min}} \quad \text{vs. Temperature}$$

Q = degree of swelling at a given temperature, Q_{min} = minimum degree of swelling observed, and Q_{max} = maximum degree of swelling observed. From the normalized curves, only points below 0.5 normalized swelling were used; that is, from this temperature and higher. This maintained our focus in the transition region, as discussed above. In some cases it was necessary to use one point above 0.5, or else no points below the transition would have been included in the fit. Therefore, the points used stressed the volume transition region, minimizing any deficiencies in the elastic term as swelling increases.

Parameter fits were performed in pairs: polymer cohesive energy density with the geometric mean correction factor, and crosslink density with the statistical segment parameter. The pairs were not chosen arbitrarily: both crosslink density and the statistical segment parameter have the same general effect on the predicted swelling, and both appear only in the elastic expression. Similarly, both the polymer cohesive energy density and the correction factor are in the polymer-solvent mixing term of the gel equation of state and determine the location and extent of the volume transition. Similarly, the crosslink density and segment parameter have the most influence on the degree of swelling at low temperatures, although the sharpness of the volume transition is also affected. Therefore, it makes sense to modify these parameters together, especially since P_g^* and Z_{sg} are functions of the polymer character. After obtaining the best fit for the cohesive energy density and correction factor, the elastic parameters were fit using the interaction parameters determined. Best fits of all parameters were obtained by alternating fits of the two pairs of parameters, until best fits were obtained for all four values. Fits of gels with ionic content were not performed because of the

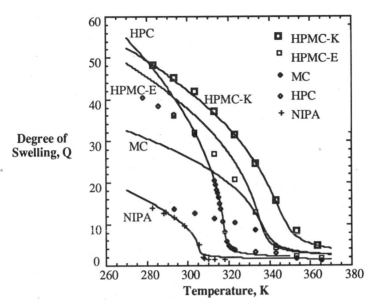

Figure 12. Fit of swelling data for gels of different polymer type. Values of fit parameters are given in Table IX.

Table IX. Values of Fit Network Parameters for Gels of Varied Polymer Type

Polymer	P_g^* (cal/cm^3)	Z_{sg}	N	ρ_x_fit (10^{-5} mol/cm^3)	ρ_x-meas. (10^{-5} mol/cm^3)
NIPA	306.7	-.0648	30	12.	13.
HPC	314.0	-.0598	8	0.97	1.1
HPMC-E	322.0	-.0547	6	2.9	2.5
HPMC-K	324.8	-.0531	4	5.1	2.0
MC	319.9	-.0562	3	1.5	2.1

problems encountered with the ionic term in the theoretical calculations in this work. Finally, the quality of the fits and the significance of the fit values of the network parameters obtained is discussed in terms of predictive ability of the compressible lattice theory.

Polymer Type. The fit swelling curves for gels of varying polymer type are presented in Figure 12; the fit values of P_g^* and Z_{sg} obtained are summarized in Table IX. The value of P_g^* obtained for the fit increases with the observed gel transition temperature, and thus with the increasing hydrophilic character of the polymer. This is consistent with the theoretical predictions of the model presented previously, although the effect on the sharpness of transition predicted with changes in polymer cohesive energy density is not observed. It is also seen that the sharper volume transitions correspond to more negative values of Z_{sg}. This is directly consistent with the predictions of the theory, although the effects on transition temperature are not observed due to the simultaneous large changes in P_g^*. If all parameters are held constant, more negative values of Z_{sg} indicate more hydrophilic polymers. This contradicts the trends obtained from the fits of the swelling data; the fit values of Z_{sg} for the most hydrophobic polymers were the most negative.

It is necessary, therefore, to discuss the fit values obtained as pairs, not as individual parameters. Past work has postulated that specific interactions, in particular hydrogen bonds, must be present for a discontinuous volume transition to be observed in a non-ionic gel (33). Figure 4 showed that more hydrophobic polymers swelled less and showed more gradual transitions at constant values of Z_{sg}. A gel of a hydrophobic polymer that did not interact favorably with water would not swell; as shown in Figure 5. Thus, if a hydrophobic gel can swell, there must be attractive interactions with the solvent. As the hydrophobic character of the polymer increases, there would have to be more interaction with the solvent for solubility to be maintained, hence the trend of more negative Z_{sg} with increased polymer hydrophobicity is reasonable.

Of particular interest is the widely varying quality of the fits obtained for the different polymer types. Note for example, that the HPMC, HPC and NIPA gels are fit very well over the entire swelling range, while the MC gel is fit quite poorly below the transition. It is expected that there would be some overprediction for the cellulose ether gels at low temperatures since the fits were performed on the upper portion of the curves and the value of N obtained for these fits would be larger than the true value at low temperatures.

As mentioned, certain gels were fit very well over the entire range of swelling, although the fits were performed on the values near the transition temperature. The only failure of this theory is the MC gel. Because the degree of swelling is so far off at high temperatures, a failure in the elastic term is likely, since this has the most effect on swelling below the transition temperature. The Galli and Brumage expression used in this work comes closest to describing the exact distribution of chain extensions for ideal networks, so the problem is not likely to be from the inaccuracy of the model. Instead, it is more likely that the applicability of the model is suspect. Network models assume uniform crosslinking without imperfections; that is, there are no regions of increased crosslink density. For a polymer such as MC, which has methyl substitution only, regions of concentrated crosslinking could have occurred near associated hydrophobic (methyl) groups. In contrast, the relatively hydrophilic HPMC-K shows excellent agreement across the entire curve; indicating that hydrophilic polymers that would be thoroughly solvated may form networks that are more appropriately described by the ideal model.

It could be argued that the MC gel should have a higher crosslink density or smaller N value since the low swelling values are overpredicted significantly. However, at low values of N, the sharpness is affected significantly with changes of N. The theoretical predictions presented previously used parameter values that were very near critical; away from values that predict discontinuous transitions (i.e. fit values for MC gels), changes in N affect the sharpness as shown in Figure 13. If the network is in fact non-uniform as discussed above, quantitative predictions of swelling at low temperatures may not be possible.

Crosslink Density. Gels of varying crosslink density were synthesized from HPC and HPMC-E. In both cases, the best fit values of P_g^* and Z_{sg} were the same for all of the crosslinking levels of a particular polymer type. In addition, the same value of N was obtained for a particular polymer type regardless of the crosslink density. The fits for the HPC gels of varied crosslink density is shown in Figure 14, with the fit values of the network parameters tabulated in Table X. The fits of the HPMC-E gels are presented in Figure 15, with comparison of experimental and fit crosslink densities given in Table XI. For the HPC gels, the fit crosslink densities overpredict swelling at low temperatures, although the relative degrees of swelling show the same qualitative trend. The gel with 5% DVS is both observed and predicted to swell significantly less

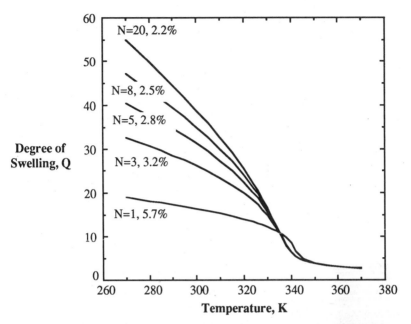

Figure 13. Effect of statistical segment parameter on transition region for Methyl cellulose parameters. For a gradual transition such as MC, increased chain stiffness results in sharper transitions and slight increases in transition temperature. Parameters listed are the value of N in the simulation and the sharpness of the volume transition.

than the other gels. Crosslink density was not determined experimentally, hence a comparison with independent values is not possible, although the theoretical range is 2.0 to 5.0 x 10⁻⁴ mol/cm³. The efficiencies calculated from the fit values are on the order of 5% (for all but the 2% DVS gel), which is a reasonable range.

The crosslink densities fit to the swelling data of the HPMC-E gels compare very well at moderate crosslink densities, as shown in Table X. However, at extremely low levels of crosslinking (0.5% DVS), there is a great difference between the experimental and the fit crosslink densities. It is unlikely that the equilibrium swelling theory is at fault here. Since the crosslink density was determined experimentally using a compression test that relied on ideal rubber elasticity for crosslink density determination, it is likely that the extreme degree of swelling at room temperature (~250) is beyond the validity of the ideal network model used for the calculations. The model used for the crosslink density is based on a Gaussian distribution of chain extensions. At such extremely high degrees of swelling, this is clearly invalid. In fact, the valid region of linear deformation for ideal rubber elasticity is generally limited to 10%-20% (*43*). Thus, for a gel crosslinked in solution ($\phi_{2f}=0.1$), this corresponds to a maximum degree of swelling of near 17. An improved model for the distribution of network chain extensions may be necessary for accurate determination of effective crosslink density.

Another significant point is better illustrated in Figure 16, which shows the HPMC-E fits and data for degrees of swelling below 60. Note that above the transition

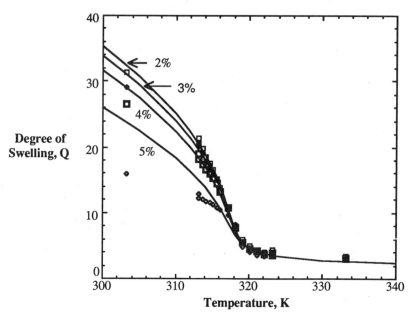

Figure 14. Swelling behavior of Hydroxypropyl cellulose gels with varied nominal crosslinker dosage. Values of fit crosslink densities and segment parameters are given in Table X.

Table X. Values of Fit Crosslink Density and Statistical Segment Parameter for Hydroxypropyl Cellulose Gels

Nominal DVS Dosage	ρ_x-fit (10^{-5} mol/cm^3)	N-fit
2%	9.7	8
3%	1.0	8
4%	1.2	8
5%	1.9	8

$P_g^*=314.0$, $Z_{sg}=-.0648$.

Table XI. Values of Fit Crosslink Density for Hydroxypropylmethyl Cellulose Gels

Nominal DVS Dosage	ρ_x-meas. (10^{-6} mol/cm^3)	ρ_x-fit (10^{-6} mol/cm^3)
0.5%	2.4	0.31
1.0%	6.5	6.2
2.0%	23.	21.
4.0%	44.	27.

$P_g^*=322.0$, $Z_{sg}=-.0547$, N=3.

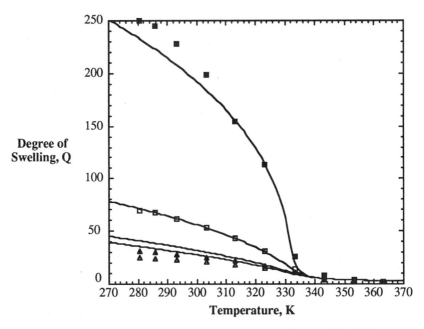

Figure 15. Fit swelling behavior Hydroxypropylmethyl cellulose (HPMC-E) gels with varied nominal crosslinker dosage. Values of fit crosslink densities and segment parameters are given in Table XI.

temperature, the degree of swelling is overpredicted consistently. This is also seen for the NIPA gel in Figure 12. This overprediction arises from the elastic expression which accounts for solution crosslinking. When the degree of swelling is greater than the degree of dilution at network formation, there is a retractive force. However, at low degrees of swelling, when the gel is collapsed and water is not a good solvent for the polymer, the solution crosslinked expression exerts a positive swelling pressure, thus the swelling is overpredicted. In a good solvent, this would be reasonable, but in a poor solvent, such as warm water, this contribution is suspect. Some expansive force would be expected below the degree of swelling at formation, but not to the extent as predicted here.

Elastic response of networks in collapsed states has not been examined. This is a potential problem for the application of responsive gels in controlled delivery devices. Many devices have been proposed that would completely block release of the drug used, based on a size exclusion mechanism. Thus the swelling in the collapsed state must be known. Although the error in prediction is relatively small on an absolute basis, it could make a significant difference in the observed versus predicted release rate for a small solute.

Conclusions

The compressible lattice theory presented in this work is able to accurately model the temperature-sensitive swelling behavior exhibited by cellulose ether gels. The model uses only one adjustable parameter, the deviation from the geometric mean

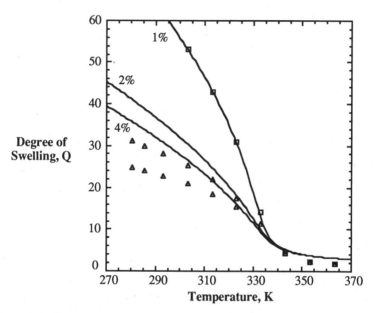

Figure 16. Fit swelling curves of HPMC-E gels at low degrees of swelling. Note that the degree of swelling is consistently overpredicted above the transition; this is due to the solution crosslinked elastic expression which increases predicted swelling in the collapsed gel.

mixing rule; the remaining parameters are pure component properties. The fit values of the polymer and solvent parameters are reasonable, and can be explained in light of variations in the hydrophilic or hydrophobic character of the polymer. The Flory network theory can also describe this swelling behavior, but the temperature dependence of the interaction parameter is not consistent with its definition. The compressible lattice model is extremely sensitive to minor variations in the parameter values, requiring that the pure component parameters be determined to within 1% of their true value for the model to be effective. As a result, the predictive capability of this model for use with responsive hydrogels is limited at the present time.

The modified non-Gaussian elasticity expression introduces an additional parameter to describe the number of effective links per network chain. The predicted swelling behavior was relatively insensitive to variations of this parameter, with the exception of degree of swelling at low temperatures. Controlling the degree of swelling at these temperatures could be accomplished empirically, based on the nominal crosslinker dosage. Furthermore, the values of the segment parameter should have varied with crosslinking, but were found to be independent of the crosslink density.

The modification for solution crosslinking produces a model that overpredicts gel swelling in the collapsed state, due to the positive swelling pressure at degrees of swelling below that at network formation. This implies that a network that is collapsed due to poor solvent quality would exert a positive swelling pressure. This situation has not been examined experimentally and indicates that the elastic theory used for gels synthesized in solution may be inaccurate at low degrees of swelling.

Acknowledgments. The authors wish to thank Dr. Marcello Marchetti and Dr. Ed Cussler for providing a copy of their computer code and advice which quickened the pace of this work.

Literature Cited

1. Stinson, S.; Chem. Eng. News **1990**, *69*, 30-31.
2. Hoffman,A.S.; J.Cont. Rel. **1987**, *6*, 297-305.
3. Bae, Y.H.; Okano, T.; and Kim, S.W.; J. Polym. Sci. **1990**, *28*, Part B, 923-936.
4. Harsh, D.C.; and Gehrke S.H.; J. Cont. Rel., **1991**, *17*, 175-186.
5. Marchetti, M.; Prager, S.; Cussler, E. L.; Macr. **1990**, *23*, 1760-1765.
6. Marchetti, M.; Prager, S.; Cussler, E. L.; Macr. **1990**, *23*, 3445-3450.
7. Lee, K. K.; Cussler, E. L.; Marchetti, M.; McHugh, M. A.; Chem. Eng. Sçi. **1990**, *45*, 766-767.
8. Prange, M.M.; Hooper, H.H.; Prausnitz, J.M.; AIChE J. **1989**, *35*, 803-813.
9. Beltran, S.; Hooper, H.H.; Blanch, H.W.; and Prausnitz, J.M.; J. Chem. Phys. **1990**, *92*, 2061-2066.
10. Hooper, H.H.; Baker, J.P.; Blanch, H.W.; and Prausnitz, J.M.; Macr. **1990**, *23*, 1096-1104.
11. Hooper, H.H.; Blanch, H.W.; and Prausnitz, J.M.; Macr. **1990**, *23* 4820-4829.
12. Beltran, S.; Baker, J.P.; Hooper, H.H.; Blanch, H.W.; and Prausnitz, J.M.; Macr. **1991**, *24* 549-551.
13. Otake, K.; Inomata, H.; Konno, M.; and Saito, S.; J. Chem. Phys. **1989**, *91* 1345-1350.
14. Painter, P.C.; Graf, J.; and Coleman, M.M.; J. Chem. Phys. **1990**, *92* 6166-6174.
15. Flory, P.J., *Principles of Polymer Chemistry*, Cornell University Press, Ithaca, NY, 1953.
16. Flory, P.J.; Orwoll, R.A.; and Vrij, A.; J. Amer. Chem. Soc. **1964**, *86* 3515-3520.
17. Eichinger, B.E.; and Flory, P.J.; Trans. Farad. Soc. **1968**, *64* , 2035-2052.
18. Sanchez, I.C.; and Balazs, A.C.; Macr. **1989**, *22*, 2325-2331.
19. Sanchez, I.C.; and Lacombe, R.H.; Macr. **1978**, *11*, 1145-1156.
20. Lacombe, R.H.; and Sanchez, I.C.; J. Phys. Chem. **1976**, *80*, 2568-2580.
21. Sanchez, I.C.; and Lacombe, R.H.; J. Phys. Chem. **1976**, *80*, #21, 2352-2362.
22. Peppas, N.A.; and Merrill, E.W.; J. Polym. Sci. Polym. Chem. **1976**, *14*, 441-457.
23. Peppas, N.A.; and Lucht, L.M.; Chem. Eng. Comm. **1984**, *30*, 291-308.
24. Hasa, J.; Ilavsky, M.; and Dusek, K.; J. Polym. Sci. Polym. Phys. Ed. **1975**, *13*, 253-262.
25. Galli, A.; and Brumage, W.H.; J.Chem. Phys. **1983**, *79*, 2411-2418.
26. Harsh, D.C.; *Controlling Swelling Behavior of Novel Cellulose Ether Hydrogels*, Ph.D. Dissertation, University of Cincinnati, Cincinnati, OH, 1992.
27. Ricka, J.; and Tanaka, T.; Macr. **1984**, *17*, 2916-2921.
28. Siegel, R. A.; Firestone; Macr. **1988**, *21*, 3254-3259.
29. Vasheghani-Farahani, E.; Vera, J.H.; Cooper, D.G.; and Weber, M.E.; Ind. Eng. Chem. Res. **1990**, *29*, 554-560.
30. Gehrke, S.; and Lee, P.; In *Specialized Drug Delivery Systems: Manufacturing and Production Technology*; Tyle, P., Ed.; Marcell Dekker, New York, NY, 1989, pp. 333-392.
31. Ilavsky, M.; Polymer **1981**, *22*, 1687-1691.
32. Saito, S.; J. Chem. Eng. Jap. **1989**, *22*, 215-228.

33. Freitas, R.F.S.; *Extraction with and Phase Behavior of Temperature-Sensitive Gels*, Ph.D. Thesis, University of Minnesota, Minneapolis, MN, 1986.
34. Hsu, T-P.; Cohen, C.; J. Polym. Sci.: Polym. Letters Ed. **1985**, *23*, 445-451.
35. Gehrke, S.H.; Palasis, M.; and Akhtar, M.K.; Int. Polym. J. **1992**, In Press.
36. Post, C.B.; and Zimm, B.H.; Biopolymers **1979**, *18*, 1487-1501.
37. Post, C.B.; and Zimm, B.H.; Biopolymers **1982**, *21*, 2123-2137.
38. T. Tanaka; Phys. Rev. Lett., **1978**, *40*, 820-823.
39. Cussler, E.L.; Stokar, M.R.; Varberg, J.E.; AIChE J. **1984**, *30*, 578-582.
40. Akhtar, M.K.; *Volume change kinetics in near-critical gels*, M.S. Thesis, University of Cincinnati, Cincinnati, OH, 1990.
41. Doelker, E.; In *Hydrogels in Medicine and Pharmacy, Vol. II*, Peppas, N., Ed., CRC Press, Boca Raton, 1987, pp. 115-154.
42. Conio, G.; Bianchi, E.; Ciferri, A.; Tealdi, A.; and Aden, M.A.; Macr. **1983**, *16*, 1264-1270.
43. Erman, B.; and Mark, J.E.; Macr. **1987**, *20*, 2892-2896.

RECEIVED October 1, 1992

Chapter 8

Diffusional Delivery of Oligonucleotides and Proteins from Gel-in-Matrix Devices

T. Chad Willis, Richard B. Provonchee, and Francis H. Kirkpatrick

FMC BioProducts, 191 Thomaston Street, Rockland, ME 04841

A novel immobilization/delivery system is demonstrated to be potentially useful for small DNA probes
and other molecules of potential pharmacological
interest. Diffusion of small molecules from fractured
agarose immobilized in foam matrices is essentially
first order, with effective diffusion constants similar to
unfractured materials. The gel-in-matrix device may
be useful for drug delivery.

Improved control of delivery of drugs has been of high interest
for several decades (*1*), and has been reviewed extensively (*2-
5*). Both the drugs and the coatings or matrices initially used for
this technology were hydrophobic or compatible with organic
solvents, and this approach is still of interest in some situations
(*6,7*), such as the current nicotine patch.

However, many newer drugs in development are proteins (*8*)
or oligonucleotides (DNA or RNA). These, like more classical
protein drugs such as insulin, are not easily delivered via the
traditional organic polymer approaches. Hydrogels are thus of
increasing interest for delivering these molecules because of their
known compatibility with newer biological reagents (*9-13*).
There is a long history of using hydrogels as inert anticonvective
agents, and it is known that diffusion proceeds in a normal and
reproducible manner in hydrogels, for a large variety of proteins
(*14, 15*). In particular, Ackers and Steere (14) were able to
observe retardation of diffusion by a gel layer, and to extrapolate

0097–6156/93/0520–0135$06.00/0
© 1993 American Chemical Society

the effective diffusion constant to zero agar concentration with reasonable agreement with literature values.

GEL-IN-MATRIX DEVICES

We have recently discovered a new method of making composite structures containing hydrogels which are very inert towards biological molecules, yet can be manufactured as monolithic articles on a large scale. *(16,17)*. These objects can be made from a variety of materials by the simple steps of impregnating a porous material with a hydrogel-forming mixture; gelling the mixture; and then fracturing the gel by mechanically stressing the porous matrix – for example, by squeezing or centrifuging. The resulting particles of fractured gel are about the size and shape of the pores in the matrix, and are trapped in the pores. The fracturing operation creates a large surface area, which is exposed to the channels between gel beads created by the fracturing process. In addition, the gels are porous, and there is an additional internal volume available to sufficiently small molecules by diffusion. We have called such a composite a "gel-in-matrix", or GIM.

Structure of a Gel-in-Matrix Device. A gel-in-matrix device is a composite with several levels of structure. These are shown diagrammatically in Figure 1. Panel 1 shows individual polymer strands of the gel. Drugs for delivery could be tied to the polymer strands by reversible links, as described in the literature of controlled release *(5)*. Panel 2 shows the gelled polymer level, with gel domains connected by polymer strands. The resulting network has a characteristic porosity, which can be controlled by varying the concentration of the gel *(14,18,19)*. If the characteristic pore size is smaller than the size of the molecule, then diffusion will be restricted. However, as shown below, pores only slightly larger than the molecule give effective diffusion rates similar to very large pores.

Panel 3 illustrates a GIM device after fracturing. Small chunks or beads of gel are separated by gel-free fracture channels, caused by the mechanical stresses of the fracturing process. The dimensions of the beads are determined by the cell size of the porous matrix; the channels are believed to follow the matrix, in preference to the interior of the gel phase.

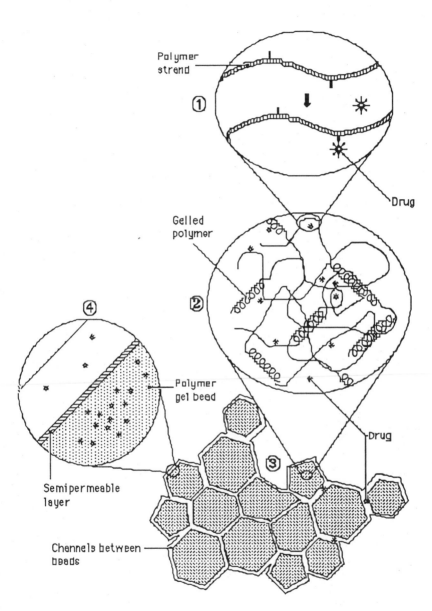

Figure 1 : Gel-in-Matrix Structure

Panel 4 illustrates a further diffusion-restricting option: after fracturing, the beads can be coated to produce a low-permeability layer on their surface, substantially reducing permeability for smaller molecules (20).

Potential Uses. Most materials of interest can be included in the gel during fracturing. Large materials, such as cells and particles, will be retained quantitatively in the gel phase during fracturing, despite the loss of about 80% of the aqueous phase and of small ions and molecules (16). Molecules and particles of intermediate size, which are small enough to pass through the pores of the gels, but large enough to experience sieving (reduction of effective diffusion constant), will be partially but not completely retained.

The fractured, partially dehydrated gel can then be treated in a variety of ways. It behaves like a sponge, and will readily recover the volume lost on fracturing if exposed to water. In the experiments described below, fracturing was done by squeezing in small containers (20 mm wells in tissue-culture plates), and the fractured gel was allowed to re-imbibe the liquid expressed during fracturing.

Fractured, partially-dehydrated gels are easily dried, due to the high surface area. They can also be coated with a surface layer at this state. In experiments to be described elsewhere, the gels were made of carrageenan, fractured, and then coated with chitosan. Properties of carrageenan-chitosan composites are described by Rha (20).

One possible use for a GIM is as a separation device, such as a chromatography column. We have described such applications elsewhere (17); in summary, the devices do not have sharp resolution at analytical scale, but may be useful for larger scale separations. In all cases, the molecules tested were capable of diffusing through the gel beads, apparently with no substantial alteration from the presence of the matrix or the effects of fracturing.

Diffusion is also crucial to performance in controlled drug delivery. Since the materials used in chromatography are necessarily compatible with the molecules being separated, we have begun to evaluate the suitability of the gel-in-matrix technology for drug delivery. As a first step, the effects of the fracturing and of the matrix on the diffusion of model molecules has been determined.

EXPERIMENTAL TECHNIQUES

The gel phase can be any gel with sufficient rigidity to break before flowing. This excludes many of the "gel" formulations found in shampoos and the like. We have used rigid natural thermosetting gels in most of our experiments, such as agar, agarose, and kappa carrageenan. These are easy to handle, inert (or having known absorptive properties) and non-toxic. The matrix can be any porous material which will give beads of a desired size. Our preferred materials for experimentation have been flexible reticulated polyurethane foams and rigid polypropylene or polyethylene foams. Since most of these are hydrophobic, we have wet them by immersion in alcohol (87% isopropanol) followed by rinsing in water or buffer, finally followed by immersion in the hot ungelled polymer. With kappa carrageenan, it is important to choose a KCl concentration at which thermogelation can still be obtained; this is typically about 0.1 M (*21*).

In the experiments reported here, discs (16 mm diam. by 3 mm thick) were cut from 1/8 inch thick sheets of 100 pore per inch flexible polyurethane foam (Scott). Since the foam was hydrophobic, the discs were pre-wetted with alcohol and rinsed with water. Excess water was shaken out, and discs were placed individually in wells of a cell culture tray. The gelling solution was 1.0% agarose (SeaKem LE) or 3% low-melting agarose (NuSieve agarose) in TBE buffer (*22*). Agarose was added to saturate the discs (about 0.6 ml) and allowed to gel. Molecules to be tested for diffusion were included in the gelling mixture when required.

Fracturing has been done primarily by squeezing of gels cast in flexible polyurethane foams, having a pore size of about 100 pores per inch. We have also used various rigid foams, including both materials with uniform pore sizes such as Porex, and materials with variable pore sizes such as generic polypropylene water filters. Both rigid and flexible materials can be fractured by centrifugation at a few hundred G's. Drying can also be used to fracture rigid foams (but not flexible foams), but is slow.

In these experiments, gels were fractured to form GIMs in the wells by pressing with a syringe plunger. Controls were handled similarly but not fractured. After allowing at least 30 min. (or overnight) for reabsorption of fluid and macromolecules by the fractured GIM, diffusion experiments were begun by addition of

buffer to the well. Typically, the buffer was the same volume as
the GIM; in some cases, larger volumes were used. The samples,
in their wells in the tray, were agitated gently on a rotary shaker
table (at about 10 rpm/ 0.16 Hz) to help stir the solution above
the gels. Samples were taken from the fluid layer in one of two
patterns: either a large, single aliquot at a fixed time, or in small
repeated aliquots from the same sample, optionally with
replacement of the volume removed with buffer. An
approximately exponential equilibration between the two
compartments was observed with both techniques.

RESULTS

Results of initial experiments are presented in Figures 2, 3 and 4,
in which the concentration in the liquid compartment is plotted
vs. a logarithmic time axis. Data points are averages of several
samples from the same experiment; variations in measured CPM
are smaller than the symbols. The curves on the figures are least-
square fits to the data, and give no evidence of significant
differences in flux among different prototype devices loaded with
a given probe. The linear fit cannot be a completely accurate
picture of the efflux, since the concentration is zero (by definition)
at zero time; however, the efflux at longer times is approximately
logarithmic as it approaches the limiting value of about 1/2 of the
initial concentration of probe.

Figure 2 shows the relative diffusion (during the first 8 hours
of an experiment) of a small molecule (adenosine triphosphate,
ATP, labelled with 32P) from a fractured GIM and from an
unfractured control, at two agarose concentrations. Equilibrium
was obtained by 40 hours (data not shown.) As expected, there is
little difference in the efflux rate, with or without fracturing or
change in gel concentration. The average pore radius in 1%
agarose gels is about 0.2 micron (23; compare 14), which will
not significantly retard the movement of small molecules.

Figure 3 shows the relative diffusion of a larger molecule, a
25-base oligonucleotide probe which was end-labelled with 32P,
in two different agarose concentrations and with and without
fracturing. Significant retardation of the probe molecule
(compared to ATP) is observed, as expected. (Compare the overall
duration of the experiments). No statistically significant
differences were seen between the agarose concentrations,
although the higher concentration (3%) is nearly enough to
produce sieving at this molecular weight during electrophoresis.

Figure 4 shows the efflux of the protein beta-lactoglobulin from a similar system. The values at 100 hours are at equilibrium. Again, efflux acts as if from a single compartment, and is not markedly affected (at these concentrations) by agarose concentration, or by fracturing.

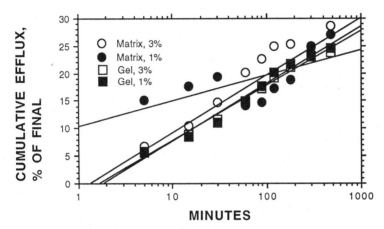

Figure 2. Efflux of ATP from GIM devices.
Devices prepared as described in text. "Matrix", GIM after fracturing; "gel", unfractured device; %, gel concentration.

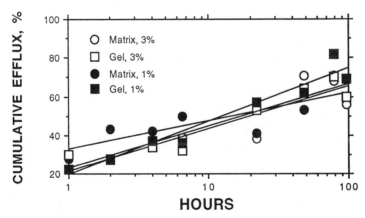

Figure 3. Efflux of DNA Oligomers from Gel in Matrix.
Legend: as in Fig. 2

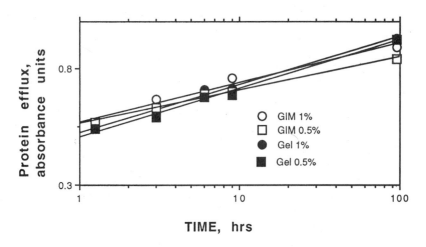

Figure 4. Efflux of beta-lactoglobulin from GIM devices.
Legend as in Fig. 2.

DISCUSSION

These studies demonstrate that a fractured Gel-in-Matrix device
can deliver small to medium sized molecules, with kinetics that
are not substantially different from those observed in intact gel
blocks.   The geometry of these studies is experimentally simple
and practical, but makes calculation of meaningful effective
diffusion constants very difficult.

     The results show that when used in a purely diffusive mode,
there is little difference in diffusional delivery rate between a
fractured gel-in-matrix and an intact gel.   In either case,
diffusional exit from the device can be markedly slowed for
medium-sized molecules by the presence of a gel.   However, in
these studies we see less effect of the gel concentration on efflux
rate than expected, especially for the largest molecule, beta-
lactoglobulin (Fig. 4).   From the literature (14), there should have
been an observable slowing (ca. 10%) of the efflux rate in 1% gels
vs. 0.5%, which was not detected.   The most likely explanation is
that our simple efflux measurement system is not sensitive
enough to detect variations of less than 20% or so in effective
diffusion constant.

     We have found, as expected from the geometry of the

provide new methods to achieve a zero-order drug delivery system.   The two most common approaches to this objective, erodable leaching and the osmotic pump approach, can be used in GIM devices, but the devices themselves do not contribute to pseudo-zero order diffusion.

One utility of a fractured gel-in-matrix device is the ability to load the device after it is formed.   Although not shown, we achieved identical delivery curves from devices loaded by fracturing to remove water, followed by soaking in the molecule to be delivered.   This can be considerably faster than loading an intact gel block by diffusion, since only the 200-micron gel bits need to be at diffusional equilibrium, compared with the 3000-micron thickness of the entire device.   While further studies would be required to completely characterize the speed of loading, our minimum reequilibration time after fracturing (30 minutes) is about 0.5% of the equilibrium time during efflux (100 hours) for the DNA probe and lactoglobulin, as expected from the relative thickness ratio (1:15, implying 1:225 in diffusion time).

A final utility of the system is its compatibility, as an aqueous system, with materials known to be compatible with the molecules to be delivered.   Agarose, the particular gel used here, is widely used in gel electrophoresis and in immunodiffusion, in large part because it does not specifically interact with the vast majority of biological molecules.   It is thus an obvious choice for a gel system for drug delivery.   In addition, it is non-toxic, and derived by purification from a food-grade material; the system is thus also likely to be non-irritating to the recipient of the delivered molecule.

Simple demonstration of predictability and probable non-toxicity is only a first step in the development of a delivery system.   We do not, as a company, make complete drug delivery devices; but we are happy to work with organizations who are actually constructing such devices.

LITERATURE CITED:

1. Williams, A. *Sustained Release Pharmaceuticals*, Noyes, Park Ridge, NJ **1969**
2. Langer, R. *Science* **1990**, *249*, 1527
3. Langer, R *Pharm. Tech*, **1989**, Aug., 18
4. Graham, N B; Wood, D A *Polymer News* **1982** *8* 230

5. Langer, R; Peppas, N *Rev. Macromol. Chem. Phys.* **1983** *C23(1)*, 61
6. Cohen, J; Siegel, R A; Langer, R *J Pharm Sci.* **1984** *73* 1034
7. Chien, Y W *Med. Res. Rev.* **1990** *10*, 477
8. Chien, Y W; Siddiqui, O; Sun,Y; Shi, W M; Liu, J C *Ann. N Y Acad Sci*, **1987**, *507*, 32
9. Peppas,N A; Lustig, S R *Ann N Y Acad Sci* **1985**, *446*, 26
10. Salzman, W M; Langer, R *Biophys J* **1989** *55* 163
11. Nair, M; Tan, J S US Patent 5,078,994
12. Lee, P I US Patent 4,749,576
13. Wheatley, M A; Langer, R S; Eisen, H N U S Patent 4,933,185
14. Ackers, G K; Steere, R L *Biochim Biophys Acta* **1962** *59* 137
15. Young, M E; Carroad, P A; Bell, R L *Biotech, Bioeng.* **1980**, *22* 947
16. Provonchee, R B and Kirkpatrick, F H European Patent Application 0 316 642
17. Kirkpatrick, F H , Provonchee, R B and Willis, T C *Proceedings of the 1991 Membrane Conference*; Business Comm.: Norwalk, CT, **1991**; pp 116-121
18. Serwer, P. *Electrophoresis* **1983** *4* 375
19. Kirkpatrick, F H in *Electrophoresis of Large DNA Molecules;* Cold Spring Harbor, NY; **1991**, pp 9-22
20. Rha, C. U S Patents 4,744,933; 4,749,620
21. Guiseley, K B; *Enzyme Microb. Technol.* **1989**, *11*, 706- 716
22. Sambrook, J; Fritsch, E F; Maniatis, T; *Molecular Cloning* (2nd ed); Cold Spring Harbor, NY; **1989**; Vol. 1, ch. 6.6
23. Greiss, G A; Moreno, E M; Easom, R A; Serwer, P; *Biopolymers* **1989** *28*, 1475-1484
24. Cussler, E L; *Diffusion* Cambridge: Cambridge, **1984**

RECEIVED October 1, 1992

Chapter 9

Centrifugal Suspension–Separation
Coating of Particles and Droplets

R. E. Sparks, I. C. Jacobs, and N. S. Mason

Microencapsulation and Granulation Laboratory, Department of Chemical
Engineering, Washington University, One Brookings Drive,
St. Louis, MO 63130

A new process has been developed for coating particles and
droplets, based on forming a suspension of the core particles
in the coating, then using a rotating disk to remove the excess
coating liquid in the form of small droplets, while a residual
coating remains around the core particles. All particles are
then solidified by cooling or drying. The fine pure-coating
particles are then separated from the larger coated particles
and recycled. Core size can be from about 30 micrometers to
several millimeters. Many wall materials are suitable and the
coating can be applied from a solution or from a melted material.
The process is fast, has high production capacity and is
inexpensive.

There are perhaps thirty processes for coating small particles
which might be termed "microencapsulation" processes, depending
upon the criteria employed (1,2,3). Most of these processes can
be placed in three basic categories.
 The first category could be called "spray processes," in which
the coating liquid is sprayed directly onto the particles as they
are being tumbled, mixed or fluidized. A second category would be
"wall deposition from solution", in which the core particles are
first suspended in a solution containing all or part of the
components needed to form the wall. The wall material is then
caused to come out of solution by reaction, phase separation, etc.,
after which the separated wall phase deposits onto the core particles
as a coating. A third category might be called "chemical reaction",
in which the wall is formed directly on the core particles by
chemical reaction of the wall-forming components.
 In the practice of these methods, a number of problems often
occur which make them difficult or impractical to use for a
particular application. Among these difficulties are

 Poor wetting of core particles by the wall material
 Aggregation of core particles by the wall formation or
  hardening

0097–6156/93/0520–0145$06.00/0
© 1993 American Chemical Society

Limited choice of wall materials
Tedious control of process steps
Solvent handling and removal from product
Limited production rate
Cost of processing (typically $1.50 - $20 per kilogram)

A New Approach to the Problem

The operational concept behind the available methods is that one
starts with a particle and then devises ways to place a coating
around it. Some of the difficulties of these processes might be
avoided, and some new directions in thinking about microencapsulation might be stimulated, if the problem of coating a particle
were conceived differently.

For example, imagine that a process begins by immersing the
core particles in the coating liquid, forming a coarse suspension.
The processing equipment is then designed to remove excess liquid
from between the suspended core particles, leaving the residual
liquid on them as a coating, as illustrated in Figure 1. Coated
particles are made virtually by default in the process of
separating the unwanted excess liquid from the suspension.

With this new concept for producing a coated particle, the
thought processes take a new direction, focusing on methods of
removing the excess liquid rather than on how to place a coating
on particles.

Centrifugal Suspension-Separation

A rotating disk is a particularly useful piece of equipment for the
process. Over a wide range of operating conditions, passing the
suspension over such a disk causes simultaneous separation of coated
core particles from each other while the excess liquid is being
removed (4). The rotating disk must be operated such that the
separated excess liquid is converted into droplets which are appreciably smaller than the coated particles, permitting them to be
easily separated from the product with cyclones or sieves. The
operation of the process is illustrated in Figure 2, where the
apparatus is a simple rotating disk. (Such a simple disk is
effective within a narrow range of operating variables, but is
not the apparatus of choice for most applications.)

Since all the liquid which is not needed in the final coating
must be removed and recycled as fine particles, the suspension
should contain the highest concentration of core particles for
which the suspension still flows. This is often 20-35% of the
core particles by volume.

Suspension Behavior on the Disk. As indicated in Figure 2, the
core/coating suspension is poured onto the central region of the
rotating disk where it begins to spread outward under the influence
of the centrifugal force. The importance of the centrifugal force
has caused the process to be called "Centrifugal Suspension-Separation" (CSS).

The liquid film becomes thinner as the suspension spreads
outward over the increasing area of the disk. It is critical that

SUSPENSION OF CORE
PARTICLES IN COATING

COATED CORE PARTICLES

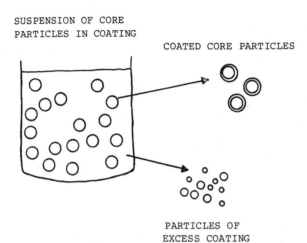

PARTICLES OF
EXCESS COATING

Figure 1. General method of separating a suspension to obtain coated particles.

CONCENTRATED
SUSPENSION

THIN FLUID FILM

COATED CORE
PARTICLES

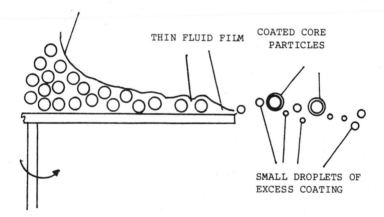

SMALL DROPLETS OF
EXCESS COATING

Figure 2. The process of centrifugal suspension-separation.

the disk be operated such that the thickness of the film of liquid
coating at the edge of the disk is appreciably less than the
diameter of the core particles.

As the core particles move into the region where the coating
film thickness is less than the diameter, the liquid layer covering
them is pulled into the surrounding thin film, leaving a gradually
thinning coating layer on top of the core particles.

Under these conditions the core and coating components of the
original suspension leave the periphery of the disk under two
different conditions. The excess coating liquid, which is present as
a thin film between the core particles, leaves the disk as droplets,
fine filaments or a thin sheet of liquid, breaking up into droplets
as if only the coating liquid has been fed to the disk for an atomization process. At the same time, the core particles are thrown
outward on the disk, appearing near the edge of the disk as relatively large "rocks" protruding upward from the thin liquid film. However, these large core particles still have a residual coating of
liquid covering them.

As the cores leave the disk, the coating film surrounding them
pulls away from the thin liquid layer on the disk, forming a tether-
ing thread. The thread then breaks, leaving the coated core particle with a "tail" which collapses back around the core particle
through the action of surface tension. This gives the final product
from the process, the coated or microencapsulated particles, which
are larger than the atomized droplets of excess coating liquid.

Influence of Wettability. An initial concern was the possibility
that a coating liquid which wetted the core particles poorly would
"de-wet" from the particles, leaving uncoated core particles to come
off the disk. This possibility was first tested by coating particles
of potassium chloride with a melted waxy mixture. The polar surface
of the KCl, which contains some adsorbed water molecules, often does
not wet well with hydrophobic coatings. However, KCl particles
approximately 700 microns in diameter have been easily coated with
several hydrophobic coatings by centrifugal suspension separation.

In a further test, the same core particles were coated with
molten Wood's metal, an alloy of bismuth, lead, tin and cadmium
which melts near 70°C. This material has high surface energy, low
viscosity and a poor tendency to wet anything except other metals
(and even then, only if the surface is sufficiently clean). However,
a coating of Wood's metal was easily placed on the potassium chloride
particles by the new coating method.

Apparently, the spreading of the liquid coating film on the
disk occurs so rapidly that there is not sufficient time for the
weak surface forces to cause the liquid film to pull away from the
core particles, even when the core is poorly wetted by the coating
liquid. This is occasionally an advantage for centrifugal suspension
separation over spray-coating processes. Since the core particles
are first totally immersed in the coating liquid, they are surrounded
by the liquid even if they are poorly wetted. On the disk, there is
not enough time for the liquid to pull away. However, in processes
where the coating is sprayed onto the core particles, the droplets of
sprayed coating must collide with the core particle surface and
spread over it instantaneously. In this case, wettability is a
strong determinant of whether a good coating can be formed.

Subsequent Particle Processing. After leaving the disk, both the coated core particles and the droplets of pure coating must be solidified to permit subsequent handling. This is accomplished by cooling of the coating liquid, if it is a melted material such as a wax or fat, or by drying the coating, if it is a solution or suspension. Alternatively, the coated particles can be caught in a liquid bath in which the coating is hardened by the various treatments or reactions.

If air is used as the cooling medium, then chambers, air-handling equipment and particle-handling equipment such as that used in spray-drying can be employed. The designs must be matched to the unique requirements of the centrifugal suspension separation coating process to account for the differences in air distribution, heat exchange regions and particle behavior. The use of such modified commercial equipment makes it possible to move rapidly from the preparation of laboratory samples to large-scale production with no need to design and test radically different equipment.

This is also an indication of the potential production capacity of centrifugal suspension-separation. Standard spray driers with rotating disk atomizers are routinely used in the ceramics industry to produce 20 tons/hour of product. This is equivalent to 120,000 tons per year on a three-shift basis. The production capacity of CSS should be comparable for well-behaved systems.

Differences From Spray-Drying

The suspension-separation process should be placed in perspective relative to spray-drying and spray-chilling, since rotating disks and drying chambers are used, and these are traditionally associated with spray-drying.

In spray-drying or spray-cooling of a suspension, the solid is in the form of suspended fine particles. This permits the suspension to flow over the entire disk as if it were a simple liquid, atomizing at the edge of the disk much the same as in simple liquid atomization. All the droplets which are formed contain the fine solid particles of active ingredient, and these atomized droplets are the product of the process. There are not two distinct size distributions, one for coated core particles and another for pure coating material, as in centrifugal suspension separation. Any subsequent separation, e.g. by cyclones, is for the purpose of meeting the size specification on the product, not to separate unused pure coating for recycle as in CSS.

In spray-drying or spray-cooling, a change in atomization conditions leads to a change in the size of the product which is formed. By contrast, in CSS a change in atomization conditions causes a change only in the size of the small pure coating particles to be recycled, but has a negligible effect on the size of the coated product (only changing the wall thickness slightly.) In CSS, size and size distribution of the coated product are determined largely by the size and size distribution of the core particles, since they are coated singly in the normal operation.

As a practical result of this difference in the physics of the

two processes, the size of the particles coated conveniently by CSS
is usually much larger than the size of product droplets typical of
spray-drying and spray-chilling. The particle size from spray-
chilling compares more closely to the size of the small recycled pure
coating droplets which are made in CSS, since both sets of particles
are the result of atomization.

Product Characteristics

Particle Size. Core particles from 30 microns up to 2 mm can usually
be coated. Particles up to several millimeters in diameter have been
coated, but coating such large particles commercially often requires
large and expensive equipment.

Wall Thickness and Capsule Payload. Wall thickness can be varied
from a fraction of a micron (for solutions) up to a few hundred
microns, depending on the size of the core particles, the viscosity
of the coating liquid and whether the coating is a melted material or
a solution.
      The capsule payload (fraction of core material in the coated
particle) can be well over 90%, but can also be very low if the
active material has been formed into a core particle containing a
high fraction of inert material.

Acceptable Core Materials. Since the wettability of the core
particle by the coating has little effect on the coating process, CSS
is effective for core particles which are hydrophilic, hydrophobic,
granular, porous, multi-component, etc.
      The coating of liquid core droplets requires considerable
modification of the process. However, droplets of aqueous liquids
can usually be coated and it is often possible to coat droplets of
hydrophobic liquids for many applications.
      The speed with which materials are handled in the process makes
it well suited for coating thermally labile materials. For example,
even if a melted wall material is employed at 115°C, highly unstable
molecules such as enzymes can be safely coated because the core
material is exposed to the high coating temperature for short periods
of time, perhaps 2-20 seconds, and little degradation can occur. A
number of enzymes have been successfully coated with negligible
degradation.

Acceptable Wall Materials. Melted coating materials handle well in
the process, partly because they often have low viscosity. Since no
solvent must be removed, the use of melts also gives the lowest
operating costs. A variety of melts have been used, including waxes,
fats, glycerides, stearates and polyethylene glycol. Figure 3 is a
scanning electron micrograph of porous granules (mean diameter 700
microns) coated with a multi-component wax based coating. Figure 4
is a scanning electron micrograph of a portion of the cross-section
through a coated particle. The coating has a solid appearance,
exhibiting no layering or gross porosity.
      Particularly useful wall materials are those containing high
polymers. Since the CSS process only has to form excess coating
particles which are separably smaller than the coated particles,

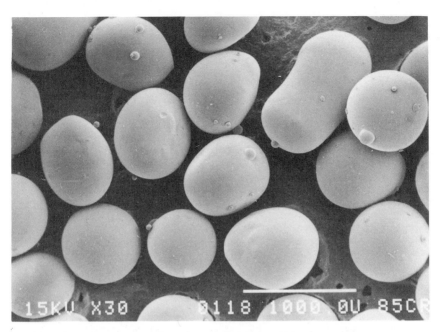

Figure 3.   Porous granules (mean diameter 700 micrometers) with a
wax-based coating produced by centrifugal suspension-separation.

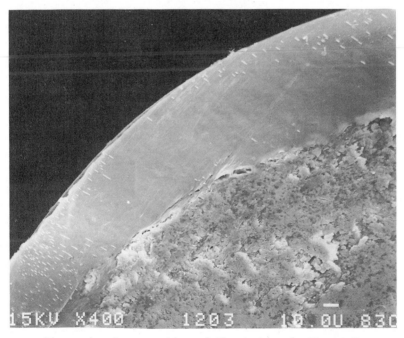

Figure 4.   Cross-section of the coating in Figure 3.

there is no requirement to form such small droplets as are needed in spray-coating methods. Hence, coatings of higher viscosity can be easily handled. Many melted coating formulations having viscosities up to 5000 centipoise have been applied by CSS.

A number of solutions have also been employed as coating liquids, including solutions of polymers in organic solvents and aqueous solutions of gelatin, food gums and other soluble materials. Water-based latexes behave well as coatings in the process.

Effect of Particle Shape. From the description of the mechanisms involved in CSS, it is apparent that the final shape of the coated particle will be determined by surface tension. When the core particle has corners or irregular protrusions, the coating liquid will fill the valleys between the protrusions and leave relatively thin coatings over the ends of the protrusions. This gives poor protection where the coating is thin. The best core particles for the process are rather round and smooth, but they need not be spherical. Granules made by pan-granulation, spray-drying or fluid-bed granulation are often acceptable. Crushed materials are usually not acceptable.

Advantages of CSS

Experience in applying the new process to a wide variety of applications has shown it to have a number of practical advantages:

> It is a simple, one-step process.
> A wide variety of core and coating materials can be used.
> Meltable coatings can be handled well.
> Aggregation is avoided.
> The process is very fast (seconds).
> Thermally sensitive materials can be easily coated.
> There are no tedious control problems.
> The process is continuous.
> High production rates are possible.
> The process is inexpensive (operating costs typically in the
> range of $0.25-2.50/kg, including the coating material).

Difficulties For CSS

Experience with the new process has also highlighted the difficulties for some applications.

> The lower limit in core diameter is 30-50 microns for most
> coatings. The coating viscosity is the determining factor.
> The less viscous the coating, the smaller the particle
> which can be coated easily.
> Solvent is required to obtain high loading in particles below
> 100 microns in diameter. This is required to decrease the
> viscosity sufficiently to spread the coating liquid into
> a sufficiently thin film.
> Thin uniform coatings cannot be applied to tablets and highly
> irregular particles. There are always thin, leaky spots
> in the coating.

Droplets are sometimes more difficult to coat, because their
viscosity must be increased enough for them to behave as
"solids" on the disk, allowing the major effect of the
forces to be upon the coating liquid.
Coatings having a long solidification time are impractical
to use. They stick together at the bottom of the tower.
Insoluble, non-melting coatings cannot be applied.

Conclusions

A method of coating particles and liquids has been developed which
offers the possibility of extending the fields of practical uses of
coated particles. These extensions could come from the ease of
applying melt coatings, the possibility of coating with very viscous
coatings, the safe handling of labile materials or the ability to
coat inexpensively particles below 150 microns in diameter.
      The process offers the engineer a process with much higher
production capacity than that of other coating processes. These
economies of scale, when coupled with the ability to coat with melts
lead to low operating costs.

Literature Cited

1.   Sparks, R. E. in Kirk-Othmer Encyclopedia of Chemical Technology,
Third Edition; John Wiley and Sons, Inc.: New York, NY, 1981; Vol.
15; pp. 470-493.
2.   Deasy, P. Microencapsulation and Other Drug-Related Processes;
Marcel Dekker: New York, NY, 1984.
3.   Sparks, R. E. In Encyclopedia of Chemical Process Technology;
McKetta, J., Ed.; Marcel Dekker, Inc.: New York, NY, 1989.
4.   Sparks, R. E.; Mason, N. S. Method and Apparatus for Coating
Particles and Liquid Droplets; U.S. Patent No. 4,675,140 (1987).

RECEIVED October 9, 1992

Chapter 10

Design of Biodegradable Polymer Systems for Controlled Release of Bioactive Agents

Barbara J. Floy, Gary C. Visor[1], and Lynda M. Sanders

Syntex Research, Palo Alto, CA 94304

Major advances in therapeutic regimens are now possible with biodegradable polymer systems for controlled delivery of bioactive agents. Therapies involving peptide and protein drugs will be most applicable in the future; however, these types of therapies are limited in use due to their low oral absorption potential and relatively short half-lives. Parenteral (subcutaneous or intramuscular) administration of a "depot" preparation of a peptide/protein drug in a controlled release polymer matrix can provide very effective continuous therapy. Many of these delivery systems use microspheres composed of biodegradable homo- and copolymers of lactic and glycolic acid. Extensive experience with these materials has established their safety, biocompatibility, and subsequent biodegradation to non-toxic products. Short-term therapeutic regimens can be designed based on the selection and optimization of a number of drug/polymer variables. This paper examines the technology involved in the design of a controlled release injectable product for peptides and proteins, with particular emphasis on the pharmaceutical aspects of quality control and regulatory compliance.

Parenteral controlled release technology is rapidly developing and offers convenient and successful approaches to the therapeutic treatment of a number of conditions. It is particularly attractive for delivery of peptide and protein compounds due to their poor oral absorption and requirement for frequent injectable dosing regimens. The technology is complex and development of a commercially viable system is challenging. Development requires significant time, capital, and manpower allocations. Although formulation design aspects of these products are relatively well understood, process development and scale-up factors have received less attention, yet are important to product success. The additional requirement of sterility for these products places process, equipment and facility design in a new realm.

[1]Current address: Gilead Sciences, 346 Lakeside Drive, Foster City, CA 94404

0097–6156/93/0520–0154$06.00/0
© 1993 American Chemical Society

Therapeutic Rationale for Controlled Release

Parenteral controlled release delivery systems offer a number of advantages in therapeutic treatment. The relatively short half life of peptides and proteins requires that these compounds be administered as often as several times each day to achieve therapeutic effect. However, parenteral controlled release systems of these compounds offer a prolonged duration of delivery, enabling dosing to be reduced to a monthly, or less frequent, interval. Therefore, the agent's pharmacological activity is actually prolonged.

Controlled delivery eliminates the "peak and valley" effects observed with frequent, pulsatile administration of traditional parenteral systems. The widely varying plasma levels can result in the development of undesirable side effects at peak plasma levels while periods of insufficient treatment may occur at trough levels. Therefore, controlled release delivery systems, in theory, can result in better drug utilization by delivering the drug at a desired rate resulting in a narrower range of plasma levels.

Parenteral controlled release delivery systems offer the advantage of improved patient compliance. The patient is not required to remember to take doses one or more times per day. Instead, the patient receives a dose during a routine physician visit, every four to six weeks, depending on the formulation delivery design. Alternatively, the patient (or a family member) could be trained to self-administer these formulations on a routine basis if desired.

Even though parenteral controlled release delivery systems offer several advantages, there are some limitations that must be considered in determining an agent's potential for these systems. An agent with minimal potency is not an optimal candidate for controlled release systems. The lower the potency, the greater the drug quantity required for a specific delivery period. Large quantities of drug are not compatible with a reasonably sized injectable or implantable system. Low potency will require high drug loading levels, which may result in unacceptable or unpredictable release profiles.

Agents with a narrow therapeutic index must receive careful scrutiny if a parenteral controlled release system is desired. For these agents, the dose at which side effects or even toxicity occurs, is very close to the dose required for therapeutic effect. The controlled release system, therefore, must perform very reliably and reproducibly to prevent overdosage or therapeutic failure. Little or no initial burst of drug from the delivery system is a strict requirement. Additionally, in systems designed with a biodegradable polymer, release of drug during the polymer degradation phase must be carefully controlled.

Controlled release systems designed with a biodegradable polymer are desirable because they do not need to be retrieved when drug delivery is complete. However, they do pose problems if there is a need to terminate therapy (due to some adverse event). In the time shortly after injection or implantation, little or no polymer degradation has taken place and the systems will be reasonably intact and may be retrieved. Microsphere systems, of course, will be more difficult to retrieve than solid matrix implants. However, once polymer degradation begins, all systems will become fragmented and smaller, making retrieval difficult, if not impossible.

Formulation Design Considerations

Formulation design decisions must be made in parallel to process design strategies for any controlled release system. A formulation prepared under one process design may have a very different release performance when the process is partially or completely modified. This is a critical aspect when developing a product for the commercial market. It is, therefore, extremely important to keep the overall formulation and process design strategies in mind throughout the product development cycle, from the earliest stage of feasibility evaluation to final product scale-up.

Certain formulation considerations will be required for any parenteral controlled release system. These are polymer selection, drug loading level, and additional excipients required for processing or incorporated to alter product performance.

Polymer Selection. Polymer selection is one of the most important factors in designing a parenteral controlled release product, because polymer is the major component in the final product and generates fundamental system behavior. The polymer can either be a non-degradable system requiring surgical removal or a biodegradable system requiring no removal. Biodegradable polymers, in the presence of an aqueous environment, degrade to soluble fragments which are eliminated from the body.

The number of biodegradable polymers available for drug delivery is small. The most popular and well understood class of polymers is the polyesters, although other systems such as polyanhydrides and poly(ortho)esters have potential application as well. Several general reviews of the biodegradable polymer systems are available (*1-3*).

Within the class of polyesters, homo- and copolymers of lactic and glycolic acid are most widely studied and used. Their commercial availability, in a variety of copolymer ratios and molecular weights, make them desirable candidates for product development. These polymers were initially studied and used for biodegradable sutures (*4*), resulting in a long and well accepted safety profile. They have now been incorporated into at least two commercially available (on the U.S. market) delivery systems containing peptides. The primary polymer degradation route is hydrolysis (*5*), although macrophage penetration of the delivery system may alter biodegradation (*6*). The degradation products, lactic acid and glycolic acid, are found endogenously and/or are renally excreted.

Homopolymers of lactic acid can be prepared from either the D or L isomers, resulting in poly(D-lactide), poly (L-lactide), or the racemic poly (D,L-lactide). Copolymerization with glycolide will result in a similar series. The resultant polymers have differing physical characteristics (e.g. glass transition temperature, crystallinity, and hydrophobicity), resulting in differing biodegradation profiles, from as little as two months to two years or more, depending on molecular weight and device geometry. Polyester degradation has been described as a biphasic process (*7*). In the initial or lag phase, polymer mass remains constant. The time course of the lag phase is decreased by reduction in polymer molecular weight and decreased lactide content. An erosional phase follows the lag phase during which an exponential reduction in polymer mass occurs. The erosional phase is also reduced in duration by molecular weight reduction and by decreased lactide content (from 100% to 40% lactide). In general, copolymers nearer a 50:50 lactide:glycolide ratio will degrade more rapidly than polymers composed of a single monomer or a high ratio of one monomer. A higher molecular weight will result in a longer total degradation time.

Drug release profiles from microspheres prepared from poly (D,L-lactide-co-glycolide) (PLGA) typically exhibit a triphasic release profile (*8-10*). An initial, rapid release of compound occurs during the first few days. This release is due to diffusion of drug from the superficial regions of the microsphere. Thereafter, a latent period with little drug release occurs (secondary phase). During this time, polymer molecular weight is decreasing, however, copolymer mass remains relatively constant (*11*). After this latent phase, bulk erosion of the polymer occurs (tertiary phase), resulting in loss of copolymer mass and the major phase of drug release (*12*).

The in-vivo duration of the tertiary phase, total duration of efficacy, and median onset of the tertiary release phase have been shown to be a function of the polymer lactide ratio (*13*). All three responses were reduced when polymer lactide content was reduced from 100% to 80% while keeping molecular weight constant. This behavior is due to reduced hydrophobicity of the polymer, thereby resulting in an increased hydrolysis rate. The corresponding in-vitro evaluation of these systems followed the same trends, although the responses were accelerated compared to in-vivo data.

Polymer molecular weight also affects biodegradation and the resultant release profile. For a series of PLA polymers, increasing molecular weight (from 0.36dL/g to 1.02 dL/g intrinsic viscosity) resulted in an increased time between the initial release phase of nafarelin (an LHRH agonist decapeptide) and the tertiary or bulk erosional phase (*13*). Release rate of leuprolide (another LHRH agonist) from microcapsules increased with decreasing PLA molecular weight (*14*).

Polymer molecular weight also affects the amount of drug released in the initial release phase (*15*) for protein-containing microspheres. An increase in polymer molecular weight from 5000 daltons to 10,000 daltons resulted in a substantial reduction in burst. However, an additional increase to 14,000 did not result in further burst reduction. For thyrotropin-releasing factor, a similar reduction in initial release was achieved through increased polymer molecular weight (*16*). Increased polymer molecular weight also resulted in prolonged drug release.

Desired release rates from controlled release systems may be accomplished through rational selection of polymer molecular weight and copolymer ratio. A single polymer may be selected and provides for a simplified system. Release profile possibilities can be expanded by employing mixtures of two or more polymers (*17, 18*), however, this approach adds additional complexity to an already complex system.

Drug Loading. Drug loading in the controlled release system must be considered along with the polymer selection, since there may be an interaction between the two variables. The loading for a system will depend on the agent's potency, desired duration of therapy, limitations on the system's physical dimensions or injection volume, and acceptable release profile. For water soluble peptides in microcapsule systems, drug loading is generally not more than 10% (*19*).

The effects of drug loading on in-vivo release was studied in Rhesus monkeys receiving nafarelin-loaded PLGA microspheres (*20*). Increasing drug loading from 2.3% to 8.4% resulted in a decrease in total duration of release, increase in the magnitude of the initial diffusional phase and an unexpected increase in bioavailability.

The alteration in release profile due to different drug loading levels of nafarelin-containing microspheres (2-7%) in humans has been described (*21*). For microspheres loaded with 2, 4, and 7% nafarelin, plasma profiles exhibited overall similar behavior. However, release during the first 10 days after injection increased with increasing drug loading while release beyond day 20 decreased with increasing loading. Theoretical calculations of plasma levels for various mixtures of the differently loaded microspheres were also presented. This approach may be useful for achieving desired profiles, however, the assumption that release profiles from mixtures is an additive function of the individual components is, as yet, unproven.

The effect of drug loading level on release has also been described for TRH (*16, 22*). An optimal TRH loading for minimizing initial burst (24 hour release) was approximately 5-10%, depending on polymer molecular weight.

Additional Excipients. A number of additives may be used to alter release profile, enhance entrapment efficiency or for process improvements. The effects of the additives on release behavior, entrapment efficiency and process parameters must be studied systematically to make confident selections of appropriate levels.

A solvent evaporation encapsulation method for leuprolide acetate, prepared through a multiple emulsion (w/o/w) process, incorporated gelatin into the internal aqueous drug solution (*23*). The gelatin resulted in increased internal aqueous phase viscosity, improving the entrapment efficiency from as low as 1.9% (without gelatin) to as much as 70.7%, enabling up to 20% drug loading. The external aqueous phase incorporated polyvinyl alcohol, presumably to stabilize the emulsion during the drying process.

In an effort to increase release rate of leuprolide from PLA microcapsules, several compounds were added to the oil phase of the multiple emulsion to induce pore, or aqueous channel formation (*14*). However, the compounds only increased the initial release, failing to increase the overall release rate.

Encapsulation techniques utilize one or more organic solvents to solubilize the polymer or aid in polymer deposition and microsphere hardening. The solvents are then removed during the encapsulation process. Solvent volume and polymer concentration in the solvent have been found to affect entrapment efficiency (*23*) and microsphere sphericity and size (*24*).

The solvent volume required for processing should be carefully evaluated to ensure optimal processing and product performance. Additionally, environmental and safety issues for handling solvents must be considered, particularly in scaling up the process and in designing manufacturing facilities. Safety issues regarding residual solvent levels in the product must also be addressed in the product development process.

Process Design and Strategy

Process design and strategy is as important as formulation in developing a parenteral controlled release product. The process must be scaleable to meet commercial manufacturing needs, capable of reproducibly producing product within specifications, and must be capable of producing a sterile product. These considerations are enhanced when developing potent or expensive peptide or protein agents. At each stage of product and process development, the overall goals of the commercial development program must be well understood and kept in mind to ensure satisfactory progress.

Microsphere Processes. Three processes are most frequently considered for preparation of microspheres, although newer processes are being developed. These processes are phase separation, solvent evaporation (also called the in-water drying process) and spray drying.

The phase separation process (Figure 1) has been described for the encapsulation of LHRH analogues (*25*). This process requires emulsification of drug solution, or suspension of drug particles, in a solution of biodegradable polymer (PLGA). Addition of a second polymer (silicone oil), causes precipitation of PLGA around the aqueous droplets or solid particles. Hardening of the microspheres is accomplished through addition of a non-solvent (heptane), from which the microspheres are isolated.

There are a number of process variables in this method which must be adequately characterized and controlled for satisfactory final product. These include addition rates of solutions and solvents, polymer concentration, drug concentration, solvent volumes, mixing rates, mixing times, process temperature, vessel and stirrer geometries, and isolation techniques. Some of these variables may have greater or lesser effects on the final product, but the only approach to determining the extent of these effects is through experimentation One way of approaching the complex optimization process is through statistical experimental design strategies. Several computer software packages are available for this purpose, and are best used along with the advice of an experienced statistician.

Solvent evaporation techniques (Figure 2) for microencapsulation have been described for leuprolide (*26*). In this process, a triple emulsion is formed (w/o/w) with drug contained in the inner aqueous phase, polymer in an organic solvent as the oil phase, and an external aqueous phase. The solvent is removed under reduced pressure, usually at a reduced temperature to increase internal phase viscosity, followed by microsphere isolation. Process variables similar to those of the phase

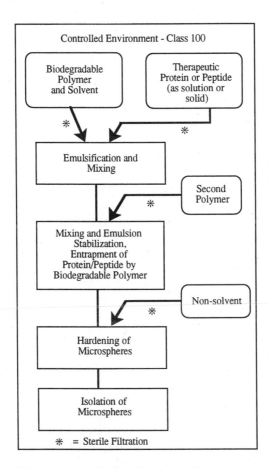

Figure 1. Process schematic for phase separation microencapsulation.

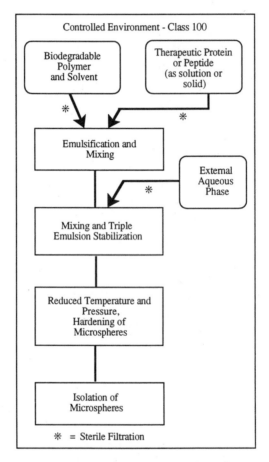

Figure 2. Process schematic for solvent evaporation microencapsulation.

separation technique must be considered in this process in addition to pressure control during evaporation.

The spray drying process (Figure 3) also offers a potential process for microencapsulation. In this approach, drug, in aqueous solution or as solid particles, is dispersed in a solution of polymer in solvent. This mixture is pumped through the atomizer of the spray drier into the drying chamber where a heated carrier gas dries the particles as they are carried to the jet separator for collection. Typical process variables are polymer concentration, peptide concentration, solution volumes, atomizer characteristics, fluid pumping rate, inlet and outlet gas temperature, and gas flow rate.

A number of publications have evaluated some process variables for the various encapsulation techniques (15, 27-30). However, these process variables must be assessed, preferably in combination with formulation variables, for each system and process. Statistical experimental design may be the most direct method for evaluating independent factors as well as determining interactions between factors in order to develop an optimized and satisfactory process.

Process and Product Evaluation

During process and product development, controls and specifications must be in place for raw material, process steps, bulk microspheres and finished product to ensure product consistency and to evaluate effects of process changes. The specifications and controls may evolve during development to accurately reflect product requirements. In addition to product performance (e.g. release rate), careful attention must be paid to the sterility and endotoxin content of the final product since it is intended for parenteral administration.

Typical raw material control tests and methodologies for PLGA polymer are listed in Table I. The specifications for each of the tests must be set to ensure product performance reproducibility; some critical specifications may be quite narrow while others may be relatively wide. Bioburden and endotoxin content of raw materials must be established in order to adequately assess the sterilization process and sterility of the final product.

The microsphere product will require a number of control tests at different processing stages as well as appropriate stability testing. A candidate list of tests and appropriate stages for testing is shown in Table II. The bulk stage refers to microspheres after isolation and drying processes have been completed, while the finished stage refers to product that has been filled into unit dose containers. Microspheres may be filled as a dry powder into vials, although this process is prone to significant particulate contamination. Additionally, because of the small microsphere size, static charge may cause difficulty in powder transfer, resulting in content uniformity problems. An alternative approach involves suspension of the microspheres in an appropriate suspending vehicle to enable a liquid filling process. After filling, the product is lyophilized to remove the water, forming a solid cake that is resuspended prior to use. If this approach is selected, careful attention must be paid to uniformity of the suspension and to evaluation of the lyophilization process effects on release behavior.

In-vitro release evaluation of the product is an important aspect of product assessment, both in early development and as a regular quality control test. Ideally, the in-vitro release test procedure will correlate to the in-vivo release of the product, and thus become a predictive test useful in evaluating formulation and process design. Practically, in-vitro release media will not completely model the complex biological milieu, so reliance on in-vitro release profiles has some risk. In-vitro release of leuprolide (7) was shown to be comparable to in-vivo release, while for nafarelin (13) the tertiary phase of release occurred earlier in-vitro than in-vivo. Variation in release

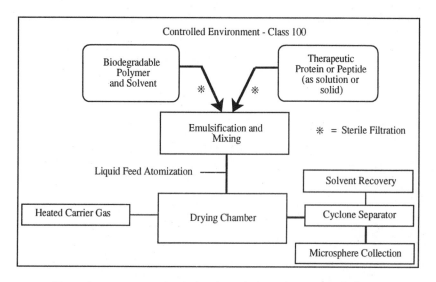

Figure 3. Process schematic for spray drying microencapsulation.

Table I. PLGA Raw Material Controls

Test	Method
Identity	Infrared Spectroscopy
Molecular Weight (Dispersity)	Size Exclusion Chromatography, Viscometry, Gel Permeation Chromatography
Monomer Ratio	NMR Spectroscopy
Residual Monomer	Gas Chromatography
Water Content	Karl Fischer Titration
Residual Solvent	Gas Chromatography
Heavy Metals	Atomic Absorption
Bioburden	Membrane Filtration
Endotoxin Content	LAL (Limulocyte Amebocyte Lysate)

profiles between animal species has also been noted (20), and, therefore, predictions of the clinical performance of these systems from animal data may not always be accurate.

Particle size may or may not be an important test procedure for the final product, depending on how significant an effect it has on the product performance. In the case of nafarelin (8), large particles (80-150μm) produced an in-vivo duration of action

Table II. Microsphere Specifications

Test	Bulk Stage	Finished Stage	Stability
Identity	--	✔	--
Product/Container Appearance	--	✔	✔
Molecular Weight Distribution	✔	✔	✔
Peptide Loading and % Label Strength	✔	✔	✔
Degradation Products	✔	✔	✔
Content Uniformity	--	✔	--
In-Vitro Release	✔	✔	✔
Particle Size and Syringeability	--	✔	✔
Dose Delivery and Suspendability	--	✔	✔
Apparent pH	--	✔	✔
Water Content	--	✔	✔
Residual Solvent	✔	✔	--
Bioburden	✔	✔	--
Sterility	--	✔	✔
Endotoxin Content	✔	✔	--

similar to small particles (30-50μm). However, for leuprolide *(23)* entrapment efficiency was greater for larger particles within a batch. This result suggests that shifts in particle size distribution may result in a change in entrapment efficiency. Particle size may also be a factor in product syringeability, resulting in needle clogging or difficult injection *(31)*.

Sterility of the product must be determined for each lot of finished product. In addition, internal sterility of the microspheres must be evaluated, if not for every lot, at least in the validation of the process for production of a sterile product. Evaluation of internal sterility can be difficult due to the fact that most solvents for PLGA are organic and have bacteriostatic or bacteriocidal effects.

Aseptic Processing or Terminal Sterilization

In developing an encapsulation process for a parenteral product, a decision must be made relatively early in the development cycle on whether to rely on aseptic processing techniques or terminal sterilization to achieve a desired level of sterility assurance. In aseptic processing, individual components used in the process or product are sterilized and then processing and packaging activities occur under controlled, aseptic conditions. Guidance in the aseptic processing of drug products has been provided by the FDA *(32)* and has received further discussion *(33, 34)*.

Terminal sterilization processes are performed on the final, filled product to destroy remaining microbiological contamination. The main difference in the two approaches is the sterility assurance level (SAL) of the final product. Typically, aseptic processes result in an SAL of 10^{-3}, or that there is a chance that as many as 1 unit per 1,000 units of final product is contaminated. Terminal sterilization processes generally result in an SAL of 10^{-6}, or the chance that up to 1 unit per one million units of product is non-sterile. To show its desire for improved sterility assurance, the U.S. FDA has proposed that manufacturers must supply supportive data to justify preparing a sterile drug product without using terminal sterilization *(35)*.

The tradeoff for improved sterility assurance for controlled release systems is the deleterious effects of the terminal sterilization process. Although heat and steam procedures are appropriate for other sterilization processes, irradiation (by gamma, electron beam or cobalt) is the most appropriate approach for controlled release systems. The degradative effect of gamma irradiation on PLGA molecular weight in microspheres containing nafarelin has been described *(8)*. The molecular weight decrease resulted in a shortening of the secondary, or lag, phase of release and a reduction in the time of onset of the tertiary release phase. Degradation of the polymer during product storage was shown to be enhanced due to gamma irradiation and to depend on polymer glycolide content *(36)*. In addition to polymer degradation, the drug may also be degraded by the irradiation process or may form adducts with the polymer.

The usual radiation sterilization dose is 2.5Mrad, and is considered as on "overkill" dose, and will likely cause some level of product degradation. In order to reduce the radiation dose (to reduce product degradation) while still providing an improved SAL (10^{-4} - 10^{-6}) over aseptic processing, an "ultra clean" manufacturing approach may be used. This approach is outlined for medical devices in an AAMI publication *(37)*.

The "ultra clean" approach yields a product with a low level of contamination, preferably less than 10 cfu per finished unit. The contamination must be defined in terms of its radiation resistance in order to set an appropriate radiation dose (<2.5Mrad) to assure the desired SAL, preferably $>10^{-3}$.

Validation Considerations

Validation is the process of establishing documented evidence that provides assurance that a process consistently results in a product of desired specifications. A protocol, or written plan, is developed explaining how the studies will be conducted, what equipment will be used, what test parameters will be monitored and, importantly, a description of acceptable test results. Validation procedures typically make use of "worst case" conditions; process parameters at their upper and lower limits (within standard operating procedures). These limits should pose the greatest chance of process or product failure when compared to "ideal" conditions, yet do not necessarily induce process or product failure.

Validation exercises may be conducted prospectively or retrospectively. Prospective studies are conducted before product made with the defined process is released for use. Retrospective validation is based on the accumulation of testing information for an ongoing process (ie. the product is already in distribution). Retrospective validation is not appropriate for aseptic processes, and particularly not for a complex micro-encapsulation process. Therefore, validation must be considered as an important component of the product development cycle to ensure a streamlined development process.

Validation of the processes for manufacturing a parenteral controlled release product requires close collaboration of a number of groups: manufacturing/process development, quality control, microbiology, quality assurance, formulation development, validation, and engineering. The key elements for validating the entire process include equipment qualification (installation qualification and operation qualification for utilities and process), equipment calibration, process development and documentation, qualification of the process, instituting a change control process and, finally, a revalidation plan. Additionally, the facility and environmental validation and control must be considered.

There will be a large number of validation studies for parenteral controlled release products, and the best approach is to break validation studies into smaller units to make the overall validation process achievable. Table III lists some possible validation units within the overall validation process.

Table III.   Validation Starting Points

Equipment preparation, cleaning, calibration, sterilization

Component preparation, sterilization

Aseptic filtration operations

Encapsulation process

Microsphere isolation and drying process

Suspension or powder filling operation

Lyophilization process

Sterilization process

Analytical methods, including sterility and endotoxin testing

Documentation of bactericidal nature of solvent systems

Conclusion

Microencapsulation processes are typically multi-step, complex designs with requirements for specialized equipment, well trained personnel, and dedicated facilities. Development of parenteral controlled release products will, therefore, require significant time, capital, and manpower expenditures considerably above those associated with typical development programs.

The utilization of these systems for systemic delivery of peptides and proteins is certainly an attractive and viable goal. However, the application of the technology for commercial purposes is not yet routine, and will benefit with advances in clean room design, robotics, and encapsulation techniques.

Literature Cited

1. Holland, S. J.; Tighe, B. J.; Gould, P. L. *Journal of Controlled Release* **1986**, *4*, pp. 155-180.
2. Chasin, M.; Langer, R. *Biodegradable Polymers as Drug Delivery Systems*; Marcel Dekker, Inc.: New York, 1990; Vol. 45, pp 1-347.
3. Smith, K. L.; Schimpf, M. E.; Thompson, K. E. *Advanced Drug Delivery Reviews* **1990**, *4*, pp. 343-357.
4. Schneider, A. K.; U.S. Patent 3,636,956.
5. Lewis, D. H. In *Biodegradable Polymers as Drug Delivery Systems*; M. Chasin and R. Langer, Ed.; Marcel Dekker, Inc.: New York, 1990; Vol. 45; pp 1-41.
6. Csernus, V. J.; Szende, B.; Schally, A. V. *Int. J. Peptide Protein Res.* **1990**, *35*, pp. 557-565.
7. Ogawa, Y.; Okada, H.; Yamamoto, M.; Shimamoto, T. *Chem. Pharm. Bull.* **1988**, *36*, pp. 2576-2581.
8. Sanders, L. M.; Kent, J. S.; McRae, G. I.; Vickery, B. H.; Tice, T. R.; Lewis, D. H. *J. Pharm. Sci.* **1984**, *73*, pp. 1294-1297.
9. Singh Hora, M.; Rana, R. K.; Nunberg, J. H.; Tice, T. R.; Gilley, R. M.; Hudson, M. E. *Bio/Technology* **1990**, *8*, pp. 755-758.
10. Hutchinson, F. G.; Furr, B. J. A. *Biochem. Soc. Trans.* **1985**, *13*, pp. 520-523.
11. Kenley, R. A.; Lee, M. O.; Mahoney, T. R., II; Sanders, L. M. *Macromolecules* **1987**, *20*, pp. 2398-2403.
12. Sanders, L. M.; McRae, G. I.; Vitale, K. M.; Kell, B. A. *J. Controlled Release* **1985**, *2*, pp. 187-195.
13. Sanders, L. M.; Kell, B. A.; McRae, G. I.; Whitehead, G. W. *J. Pharm. Sci.* **1986**, *75*, pp. 356-360.
14. Ogawa, Y.; Yamamoto, M.; Takada, S.; Okada, H.; Shimamoto, T. *Chem. Pharm. Bull.* **1988**, *36*, pp. 1502-1507.
15. Cohen, S.; Yoshioka, T.; Lucarelli, M.; Hwang, L. H.; Langer, R. *Pharm. Research* **1991**, *8*, pp. 713-720.
16. Heya, T.; Okada, H.; Ogawa, Y.; Toguchi, H. *Int. J. Pharm.* **1991**, *72*, pp. 199-205.
17. Tice, T. R.; Labrie, F.; McRae-Degueurce, A.; Dillon, D. L.; Mason, D. W.; Gilley, R. M. *Polym. Prep. (Am. Chem. Soc. Div. Polym. Chem.)* **1990**, *31*, pp. 185-186.
18. Asano, M.; Fukuzaki, H.; Yoshida, M.; Kumakura, M.; Mashimo, T.; Yuasa, H.; Imai, K.; Yamanaka, H.; Kawaharada, U.; Suzuki, K. *Int. J. Pharm.* **1991**, *67*, pp. 67-77.
19. Maulding, H. V. *J. Controlled Release* **1987**, *6*, pp. 167-176.
20. Sanders, L. M.; Vitale, K. M.; McRae, G. I.; Mishky, P. B. In *Delivery Systems for Peptide Drugs*; S. S. Davis, L. Illum and E. Tomlinson, Ed.; Plenum Press: New York, NY, 1986; pp 125-137.

21. Burns, R. A., Jr.; Vitale, K.; Sanders, L. M. *J. Microencapsulation* **1990**, *7*, pp. 397-413.
22. Heya, T.; Okada, H.; Tanigawara, Y.; Ogawa, Y.; Toguchi, H. *Int. J. Pharm.* **1991**, *69*, pp. 69-75.
23. Ogawa, Y.; Yamamoto, M.; Okada, H.; Yashiki, T.; Shimamoto, T. *Chem. Pharm. Bull.* **1988**, *36*, pp. 1095-1103.
24. Sato, T.; Kanke, M.; Schroeder, H. G.; DeLuca, P. P. *Pharm. Research* **1988**, *5*, pp. 21-30.
25. Kent, J. S.; Lewis, D. H.; Sanders, L. M.; Tice, T. R.; U.S. 4,675,189.
26. Okada, H.; Ogawa, Y.; Yashiki, T.; U.S. 4,652,441.
27. Arshady, R. *J. Controlled Release* **1991**, *17*, pp. 1-22.
28. Jalil, R.; Nixon, J. R. *J. Microencapsulation* **1990**, *7*, pp. 297-325.
29. Watts, P. J.; Davies, M. C.; Melia, C. D. *Crit. Rev. Ther. Drug Carrier Syst.* **1990**, *7*, pp. 235-259.
30. Wang, H. T.; Schmitt, E.; Flanagan, D. R.; Linhardt, R. J. *J. Controlled Release* **1991**, *17*, pp. 23-32.
31. Sanders, L. M. In *Peptide and Protein Drug Delivery*; V. H. L. Lee, Ed.; Marcel Dekker, Inc.: New York, NY, 1991; pp 785-806.
32. "Guideline on Sterile Drug Products Produced by Aseptic Processing," Center for Drugs and Biologics and Office of Regulatory Affairs, Food and Drug Administration, Rockville, Maryland, 1987.
33. Akers, J. E.; Agalloco, J. P.; Carleton, F. J.; Korczynski, M. *J. Parenter. Sci. Technol.* **1988**, *42*, pp. 53-56.
34. Akers, J. E.; Agalloco, J. P.; Carleton, F. J.; Korczynski, M. S. *J. Parenter. Sci. Technol.* **1988**, *42*, pp. 114-117.
35. *Federal Register* **October 11, 1991**, *58*, pp. 51354-51358.
36. Spenlehauer, G.; Vert, M.; Benoit, J. P.; Boddaert, A. *Biomaterials* **1989**, *10*, pp. 557-563.
37. *AAMI Recommended Practice Process Control Guidelines for Radiation Sterilization of Medical Devices*; Association for the Advancement of Medical Instrumentation: Arlington, VA, March 1984.

RECEIVED September 24, 1992

Chapter 11

Microcapsules Containing Water-Soluble Cyclodextrin Inclusion Complexes of Water-Insoluble Drugs

Thorsteinn Loftsson and Thórdís Kristmundsdóttir

Department of Pharmacy, University of Iceland, Reykjavik IS–101, Iceland

One of the most common practiced microencapsulation in the pharmaceutical industry is to form slow-release oral dosage forms of lipophilic water-insoluble drugs. In general, water-insoluble drugs possess inherent slow-release properties, but in many cases their release rate can be increased by formation of water-soluble drug-cyclodextrin inclusion complexes. Micro-encapsulation of such water-soluble complexes will result in drug-delivery forms with release rates independent of their water solubility.

In recent years, microencapsulation has been used successfully in many technological fields. In the pharmaceutical industry microcapsules have for example been used to improve the bioavailability of drugs, to protect drugs from decomposition, to convert liquids to free-flowing powder, and to mask unfavourable odour and taste (1). However, one of the most common usage of microencapsulation in the pharmaceutical industry is to form slow-release oral dosage forms and in the literature there are numerous reports of successful developments of such dosage forms. Most of the drugs tested so far have a lipophilic character. Lipophilic water-insoluble drugs are rather easily microencapsulated and they possess inherent slow-release properties in the pure form as well as when they are incorporated into polymeric matrices. The preparation of microcapsules containing hydrophilic water-soluble drugs is a greater challenge and there are fewer reports on such forms of microcapsules. We believe in many cases the release rate of water-insoluble drugs from microcapsules can be better controlled if, water-soluble drug-cyclodextrin inclusion

0097–6156/93/0520–0168$06.50/0
© 1993 American Chemical Society

complexes of the drugs, rather than the pure drugs themselves are incorporated into the microcapsules. In this chapter we describe a method for producing microcapsules of hydrophilic water-soluble cyclodextrin complexes containing lipophilic water-insoluble drugs.

Cyclodextrins

Cyclodextrins form a group of structurally related cyclic oligosaccharides with a hydrophilic outer surface and a rather lipophilic cavity in the center. They are natural products formed by the action of enzymes on starch. Various cyclodextrin derivatives have also been synthesised. Following is a short review of cyclodextrins and their properties. For further information excellent reviews have been published in recent years (2-7).

History. Cyclodextrins were first isolated from starch degradation products by Villiers in 1891 (8). The foundations of the cyclodextrin chemistry were laid down by Schardinger in the period 1903 to 1911 (9, 10), and many of the old literature refer to cyclodextrins as Schardinger's dextrins. Cyclodextrins have also been called cycloamyloses and cycloglucans. Cyclodextrins are formed by enzymatic cyclization of starch by a group of amylases called glycosyltransferases. The enzymes convert partially prehydrolysed starch to a mixture of cyclic and acyclic dextrins from which the pure homogeneous, crystalline cyclodextrins are isolated. Until 1970, only small amounts of cyclodextrins could be produced in the laboratory and the high production cost prevented the usage of cyclodextrins in industry. The recent biotechnological advancements have resulted in dramatic improvements in the cyclodextrin production which has lowered their production cost (11). This has made industrial applications of cyclodextrins possible. Since 1960, close to 1000 patents have been filed on both the production of cyclodextrins and their usage. The first patent on the preparation of drug-cyclodextrin complexes was registered in 1953 (12).

Chemistry of cyclodextrins. The important structural characteristics of the cyclodextrin molecules are their cylindrical shape, somewhat hydrophobic central cavity and the hydrophilic hydroxyl groups on the outer surface. The polarity of the cyclodextrin cavity has been estimated to be similar to that of aqueous alcoholic solution (13). Due to lack of free rotation about the bonds connecting the glucopyranose units, the cyclodextrins are not perfectly cylindrical molecules but to some extent cone shaped. All the primary hydroxyl groups are located on the narrow side while all the secondary hydroxyl groups are located on the wider side. The most common cyclodextrins are α-cyclodextrin (or cyclohexaamylose), β-cyclodextrin (or cycloheptaamylose) and γ-cyclodextrin (or cyclooctaamylose), consisting of 6, 7 or 8 α-1,4 linked glucopyranose units, respectively (Table I).

Cyclodextrins are capable of forming inclusion complexes with many molecules by taking up a whole molecule, or some part of it, into the cavity (5, 6). The cyclodextrins become the "host" molecules and the molecules which go into the cavity the "guest" molecules. No covalent bonds are formed or broken during the complex formation and in aqueous solutions the complexes are readily dissociated and free guest molecules are in a rapid equilibrium with the molecules bound within the cyclodextrin cavity. The driving force for the complex formation is thought to be the release of enthalpy-rich water from the cyclodextrin cavity. The water molecules located inside the cavity cannot satisfy their hydrogen bonding potentials and therefore exist in a higher energy state compared to the water molecules out in the aqueous cyclodextrin solution (3, 13). These enthalpy-rich water molecules are readily replaced by suitable guest molecules less polar than water. The size and chemical structure of the guest molecule are also important. Only relatively apolar molecules of appropriate size can go into the cyclodextrin cavity. Too large or too polar molecules will not form inclusion complexes with cyclodextrins. However, it is often sufficient for complex formation that some apolar part of a molecule fits into the cavity. For example, many polypeptides and proteins contain some amino acids carrying hydrophobic aromatic moieties capable of forming inclusion complexes with cyclodextrins (6, 14). The size of the cyclodextrin cavity is also very important (Table I). For instance, the α-cyclodextrin cavity is too small for naphthalene and only the γ-cyclodextrin cavity can accommodate anthracene (5). α-Cyclodextrin can be used for small molecules or slim side chains of larger molecules. β-Cyclodextrin is the most useful for complexation of average size molecules such as most drugs. γ-Cyclodextrin can be used for complexation of large molecules such as macrolide antibiotics. Unfortunately, β-cyclodextrin has low aqueous solubility (only 1.85 g/100 ml) and causes severe toxic effects upon systemic administration (15, 16).

To improve their physicochemical and biological properties, the molecular structure of the parent cyclodextrins, i.e. α-, β- and γ-cyclodextrin, have been modified. Branched cyclodextrins can be obtained by reacting a cyclodextrin with glucose or maltose in the presence of pullulance enzyme [13]. Other common cyclodextrin derivatives are formed by alkylation (e.g. methyl- and ethyl-β-cyclodextrin) or hydroxyalkylation (e.g. hydroxypropyl- and

Table I: The structure of β-cyclodextrin and physical properties of α-, β-, and γ-cyclodextrin[1]

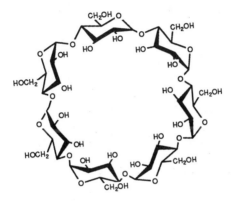

β-Cyclodextrin

	α-Cyclodextrin	β-Cyclodextrin	γ-Cyclodextrin
Molecular weight	972	1135	1297
Number of glucopyranose units	6	7	8
Internal cavity diameter (Å)	4.7-5.3	6.0-6.5	7.5-8.3
Approx. cavity volume (Å^3)	174	262	472
Molecules of water in cavity	6	11	17
Water solubility at 25°C (g/100 ml)	14.5	1.85	23.3
Water of crystallization (%)	10.2	13-15	8-18
Melting range (°C)	255-260	255-265	240-245

[1]Modified from references 7 and 13

hydroxyethyl-derivatives of α-, β- and γ-cyclodextrin) of the hydroxyl groups (5). These manipulations frequently transform the crystalline cyclodextrins into amorphous mixtures of isomeric cyclodextrin derivatives and, thus, the aqueous solubility of the derivatives is usually much larger than that of the parent cyclodextrin. For example, 2-hydroxypropyl-β-cyclodextrin is obtained by treating a base-solubilized solution of β-cyclodextrin with propylene oxide. The aqueous solubility of 2-hydroxypropyl-β-cyclodextrin is over 60 g/100 ml and it shows virtually no toxic side effects upon parenteral administration (7, 17). Both the molar substitution, i.e. the average number of propylene oxide molecules that have reacted with one glucopyranose unit, and the location of the hydroxypropyl groups on the β-cyclodextrin molecule will affect the complexation properties of the 2-hydroxypropyl-β-cyclodextrin molecule (18).

Various methods have been used to detect formation of cyclodextrin inclusion complexes. In the solid state thermal analysis, X-ray analysis, infrared spectroscopy, rate of dissolution, vacuum sublimation and vacuum drying have been used. In aqueous solutions spectral changes (e.g. NMR, ultraviolet, fluorescence, phosphorescence and circular dichroism), and the effect of the cyclodextrin molecules on the solubility and stability of the guest molecules have been used (19). Solubility and kinetic methods are mainly used for determining stability constants of the cyclodextrin inclusion complexes (19, 20).

Cyclodextrin encapsulation of a hydrophobic guest molecule will affect many of its physicochemical properties. In the solid state, the cyclodextrin complexes frequently increase the rate of dissolution of the guest molecule, increase its chemical stability, and reduce its volatility and sublimation. In aqueous solutions, cyclodextrin complexes can both increase the solubility and stability of the guest molecule, as well as reduce its volatility and absorption into or adsorption on surfaces and membranes. For these reasons various industrial applications of cyclodextrins and their derivatives are now being tested.

Usage of cyclodextrins in pharmaceutical formulations. In the pharmaceutical industry cyclodextrins are mainly used to increase the aqueous solubility, stability and bioavailability of drugs but they have also been used to convert liquid drugs into microcrystalline powders, reduce gastro-intestinal irritation, and reduce or eliminate unpleasant taste and smell. Following are few examples of usage of cyclodextrins in pharmaceutical formulations.

Cyclodextrins have been used to improve both the solubility and chemical stability of anticancer drugs. Many anticancer drugs owe their pharmacological activity to their instability in aqueous solution, therefore it is often difficult to introduce these drugs into aqueous formulations. For example, the solubility of the anticancer drug chlorambucil in water is only 0.3 mg/ml and its shelf-life in aqueous buffer solution at pH 7.4 and 25°C is only about 11 minutes. The desired intravenous dose would be about 10 mg and to give this dosage one would have to inject about 35 to 40 ml of pure aqueous solution of the drug within 11 minutes from its dissolution, which is quite impossible. Chlorambucil forms an inclusion complex with 2-hydroxypropyl-β-cyclodextrin. In aqueous isotonic 2-hydroxypropyl-β-cyclodextrin solution (approximately 23% w/v) the solubility of chlorambucil is about 17 mg/ml and, thus, 10 mg dose of the drug could easily be dissolved in one ml of isotonic 2-hydroxypropyl-β-cyclodextrin solution (21). The complexation will also result in 2- to 3-fold increase in the shelf-life. Cyclodextrins have also been shown to increase the aqueous solubility and/or the stability of following anticancer drugs: Daunorubicin (22), doxorubicin (22, 23), estramustine (24), lomustine (25), melphalan (21) and tauromustine (Loftsson, T.; Baldvinsdóttir, J. *Acta Pharm. Nord.*, in press).

The non-steroidal anti-inflammatory drugs (NSAIDs) are one of the most commonly used group of drugs in the world. However, most NSAIDs cause some form of gastro-intestinal irritation especially at the high sustained dosages necessary for treatment of arthritis. Several approaches have been tested for reducing or preventing this side effect of the NSAIDs including prodrug formation, microencapsulation, addition of neutralising excipients and co-prescription of anti-ulcer agents. A recently tested method for reduction of gastrointestinal irritation is complexation of NSAIDs with cyclodextrins (26). The cyclodextrin complexation of NSAIDs frequently results in more rapid absorption of the NSAIDs after oral administration which again can reduce the potential for gastric lesions due to shorter time of contact between the NSAIDs and the mucosa. It has for example been shown that gastric tolerance of piroxicam can be increased by β-cyclodextrin complexation (27). β-Cyclodextrin also increases the chemical stability of acetylsalicylic acid in aqueous solutions (Loftsson, T., unpublished data). NMR studies of the complexation of the acetylsalicylic acid with β-cyclodextrin have shown that during the complex formation the benzene ring of the drug molecule goes into the cavity from the wider side of the β-cyclodextrin molecule. The NMR studies also show that in the inclusion complex the benzene ring is located well inside the cavity but the acetyl group is standing out of the cavity (Fig. 1). The complexation increases the stability of the drug by protection of the chemically weak spot, i.e. the ester bond, of the molecule. Cyclodextrins have also been shown to increase the aqueous solubility and/or the stability of following NSAIDs: Diclofenac (28), ibuprofen, indomethacin (28, 29), naproxen (30) and tenoxicam.

Steroids as a group consist of relatively large water-insoluble molecules. The molecules are too large to fit completely into the cyclodextrin cavity but since only some lipophilic part of the molecule, e.g. the A-ring, has to fit into the cavity many steroids form cyclodextrin inclusion complexes. Thus, steroids with a flat unsaturated and unhindered A-ring, like dexamethasone and 17β-estradiol, generally have a good solubility in aqueous 2-hydroxypropyl-β-cyclodextrin solutions. Steroids with a hindered A-ring, like ethynylestradiol 3-methyl ether, have less solubility (31).

Microencapsulation

Microencapsulation can be defined as the coating of small particles or droplets of liquids, where the size of microencapsulated particles can range from a fraction of a μm to several hundred μm. The term microcapsule has not only been applied to coated particles or droplets but also to dispersions in a solid matrix (32). Microencapsulation makes it possible to alter the properties of the encapsulated product; controlling release characteristics, improving stability by protecting from the environment and converting liquids into solids (33).

Preparation of microcapsules. Microcapsules can be prepared by several processes, each has its advantages and disadvantages, and no one process is suitable for all substances. The choice of a process is dependent upon the intended size of the microencapsulated product, and the physicochemical properties of both the substance to be encapsulated as well as the coating material. The structure of the microcapsule, whether the encapsulated substance is surrounded by a thin coating or is dispersed throughout a polymeric matrix, is also dependent upon the process used. The various microencapsulation techniques are (34): coacervation phase separation; interfacial polymerization; solvent evaporation; spray drying and spray congealing; pan coating; multiorifice centrifugal process and air suspension. For a description of the microencapsulation techniques the reader is referred to the many excellent reviews available (32-24).

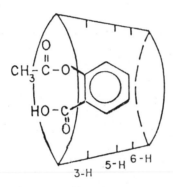

Figure 1.   Schematic drawing of a acetylsalicylic acid - β-cyclodextrin inclusion complex in aqueous solution (Loftsson, T. *et al.*, unpublished data).

The shape and the morphology of microcapsules depend on the core material, the type of coating material used and the method of microencapsulation. When applying a thin surface coating on drug crystals (e.g. by coacervate phase separation or by air suspension) the shape of the microcapsule will be governed by the crystal shape, whereas when the coating is thicker or when the method of microencapsulation involves dissolving the core material during the process (e.g. by solvent evaporation), spherical or nearly spherical microcapsules will be obtained. The size and the size distribution of microcapsules is influenced by the method of encapsulation and by the process parameters used. Since the size of the microcapsules affects the release rate of the encapsulated substance, an accurate characterization of the size distribution is important. Sieving and Coulter Counter methods have been used for the size determination of microcapsules (32).

Usage of microencapsulation in pharmacy. The first practical application of microencapsulation was in the printing industry in the 1950s when it was used in production of carbonless copying paper (1). Microencapsulation has since been used for a variety of applications in many other fields, e.g. in the food, chemical, agriculture, cosmetic and pharmaceutical industries. Within the pharmaceutical industry microencapsulation has mainly been used for: changing liquids into solids, masking unpleasant taste or odor, protecting drugs from the environment (moisture, oxidation), modification of the drug dissolution profile, decreasing the evaporation of volatile substances, preventing incompatibility, improving the powder flow characteristics and obtaining controlled/sustained release of drugs.

In a sustained release formulation, the microcapsules can be presented as powder, capsules or tablets, or they are suspended in liquid formulations for parenteral use. Microcapsules usually exhibit high mechanical resistance and are able to withstand abrasion and other damage during the manufacturing procedure and subsequent storage. However, there has been some concern that microcapsules might loose their integrity during tableting causing increased release rate of the active ingredient (35, 36). On the contrary, several researchers have demonstrated that in many instances compression of microcapsules results in substantial prolongation of the release of the active ingredient (37-39). One of the determining factors in drug release from tableted microcapsules is the structure of the microcapsules, with matrix type microcapsules being more resilient than microcapsules consisting of thinly coated drug crystals. Other determining factors are compression pressure (40, 41), core to wall ratio and size distribution (40), the excipients used (39-41), flow characteristics (43) and other manufacturing conditions (37) .

One of the first drugs to be microencapsulated was aspirin and other NSAIDs, including indomethacin (44), ibuprofen (45, 46) and naproxen (47). Other examples of microencapsulated drugs are propranolol (48), theophylline (49, 50), isoniazid (51) and potassium chloride (52). Steroids have been encapsulated for prolonged release, especially for the release of contraceptive steroids (53) and parenteral usage of microencapsulated drugs have shown clear advantages targeted delivery systems and in parenteral sustained drug (54-56). Anticancer drugs which have rapid clearance and toxic side effects are obviously good candidates for microencapsulation (57, 58).

Release mechanism for microencapsulated drugs can be diffusion through the coating material, release through the pores of a matrix or by disruption or erosion of the coating. In many cases an initial "burst effect" is seen, which is probably caused by dissolution of free drug on or near the surface of the microcapsule (56). The release of drugs from microcapsules usually follow first order kinetics but zero order release kinetics are sometimes obtained (32).

Coating materials used for microencapsulation. To a large extent the polymeric coating material controls the microcapsule properties. The choice of coating material is governed by the purpose of encapsulation. If the aim is to obtain sustained/controlled release, then the coating material should facilitate a sufficient reduction in the dissolution rate of the drug. A considerable variety of polymers have been used for coating in microencapsulation. A prerequisite for the coating material is its good stability, and formation of a continuous film which adheres to the core. Ethylcellulose is the polymer which has been most commonly used in the microencapsulation of drugs for oral use. It forms a strong, flexible film stable against both heat and light, and is therefore a most suitable candidate for microencapsulation (59, 60).

For the microencapsulation of drugs intended for parenteral use (injections or implanting) biodegradable polymers have been used (61). Release from biodegradable polymers is by diffusion through the polymer and by polymer degradation. If the polymer degradation is rapid, then the release of drug could become erratic and it is therefore desirable the polymer is degraded at a considerably slower rate than the drug release (1).

A wide range of natural and synthetic biodegradable polymers have been investigated but most attention has been focused on poly(lactic acid), poly(glycolic) acid and poly(lactide-co-glycolide)

(54, 61-63). The properties of poly(lactic acid) depend to a large extent upon whether the polymer is synthesized from the optically active L-lactic acid or the optically inactive racemic mixture DL-lactic acid. Poly(L-lactic acid) is a semicrystalline polymer and poly(DL-lactic acid) is amorphous but the crystallinity of poly(lactide-co-glycolide) is dependent upon the ratio of the two components, with less than 70 % glycolide resulting in an amorphous polymer (62). The release rate of drugs from poly(lactic acid) is affected by the crystallinity of the polymer and its molecular weight. Jalil and Nixon (64) reported that the molecular weight of poly(L-lactic acid) affected the microcapsule structure, with the high molecular weight poly(L-lactic acid) forming a more porous structure than low molecular weight poly(L-lactic acid). Bodmeier and Chen (65) found, when working with mixtures of high and low molecular weight poly(DL-lactic acid), an increase in the release rate with an increase in the low molecular weight fraction in the mixture.

Microencapsulation of water-soluble 2-hydroxypropyl-β-cyclodextrin complexes. We have tested several methods for preparing microcapsules containing drug-cyclodextrin complexes, including microencapsulation of the lyophilised drug-cyclodextrin complex. However the most convenient method for preparing microcapsules containing water-soluble drug-cyclodextrin complexes is the solvent evaporation process. The original (O/W) version of the solvent evaporation process involves the making of an emulsion where the disperse phase consists of a solution of the coating material and the drug to be encapsulated in an organic solvent (methylene chloride or chloroform). The solution is emulsified in an aqueous phase containing an emulsifier. The organic solvent is then removed by heating or under vacuum whereafter the microcapsules can be separated by decantation and filtration. The solvent evaporation method of microencapsulation gives microcapsules containing the encapsulated drug dispersed throughout a polymeric matrix. The particles obtained may present various internal structures depending upon the nature of the components and on the ratio of drug to coating material. Mostly three physical states are distinguishable i.e. metastable molecular dispersion, stable molecular dispersion and crystalline dispersion (45).

The amount of drug in the microcapsule depends both on the ratio of drug to coating material and the solubility of the drug in the processing medium, i.e. unless the drug is relatively insoluble in the processing medium a proportion of the drug will be lost from the microcapsules during preparation. Several process parameters (e.g. type and concentration of emulsifier, rate of stirring, ratio of organic phase to aqueous phase) can be adjusted to control the characteristics of the microencapsulated product. The solvent evaporation method of microencapsulation has been used extensively for the preparation of poly(lactic acid) microspheres (66) but it has not been commonly used for other coating materials. The solvent evaporation method is particularly suitable for the encapsulation of slightly-soluble materials in a hydrophobic polymer.

The microencapsulation of water-soluble drugs is not practical using the O/W version of the solvent evaporation technique described above as the drug would to a great extent be lost to the aqueous phase. Tsai et al. reported using a W/O version of the solvent evaporation technique where the drug and the coating polymer were dissolved in acetonitrile and then dispersed in liquid paraffin (67). Huang and Ghebre-Sellassie reported the use of ethanol as the dispersed phase (68). This method, using ethanol for the disperse phase, was applicable for the microencapsulation of the water-soluble 2-hydroxypropyl-β-cyclodextrin complexes with ethylcellulose. However when it came to the microencapsulation using poly(DL-lactic acid) the method had to be modified due to the low solubility of the coating polymer in ethanol. By using methylene chloride as the dispersed phase it was possible to dissolve the poly(DL-lactic acid) but the 2-hydroxypropyl-β-cyclodextrin complex was suspended in that solution.

Preparation of microcapsules

The microcapsules containing water-soluble drug-cyclodextrin complexes were prepared by the emulsion-solvent evaporation method (69) and following is a short description of this method. The coating vessel described in Fig.2 was used. When using ethylcellulose as coating material it was dissolved in ethanol, then diethyl phthalate, cyclodextrin and finally the drug were added and dissolved in the polymeric solution, which formed the disperse phase of the emulsion (Fig. 2). The exact drug-cyclodextrin ratio depends on how much drug is incorporated into the cyclodextrin derivative used in the experiment, and the incorporation was determined from drug-cyclodextrin phase-solubility diagram obtained in aqueous solutions (31). In our experiments we mainly used hydrocortisone as a sample drug and the cyclodextrin derivative was 2-hydroxypropyl-β-cyclodextrin. When using poly(DL-lactic acid) as coating material the polymer was dissolved in methylene chloride and after suspending the lyophilized drug-2-hydroxylpropyl-β-cyclodextrin complex in the solution, it was dispersed in the continuous phase.

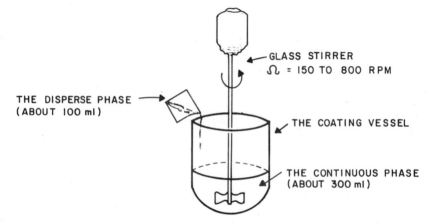

Figure 2.   Apparatus for preparation of microcapsules by the emulsion-solvent evaporation method.

The continuous phase of the emulsion was composed of n-heptane (90 ml), paraffin preliquidum Ph. Eur. (180 ml) and sorbitan monooleate (2.7 to 8.1 ml). The two phases were mixed in the coating vessel and the mixture stirred continuously at a rate of 150-800 rpm under ambient conditions until all solvent had evaporated (19-48 h). The microcapsules formed were then filtered, washed with n-heptane, and dried for one hour in a vacuum oven at 55°C.

Release studies

The release studies were carried out in the USP XXII described paddle apparatus for dissolution rate determination. The release rate was determined at $37\pm1°C$ and 100 rpm by adding microcapsules equivalent to 5 mg of the drug to 500 ml of simulated gastric fluid TS USP XXII pH 7.50 (without enzyme), 1% 2-hydroxypropyl-β-cyclodextrin or 0.02% polysorbate 80 in simulated gastric fluid. Samples (1.5 ml) were withdrawn at various time intervals, filtered through 0.45 mm membrane filters (Millex-HV; Millipore, U.S.A.) and analysed by high-performance liquid chromatographic (HPLC) methods.

Results and discussion

We have successfully microencapsulated several water-soluble drug-cyclodextrin complexes by the described method, including a 2-hydroxypropyl-α-cyclodextrin complex of carboplatin and 2-hydroxypropyl-β-cyclodextrin complexes of dexamethasone, 17β-estradiol, hydrocortisone and methylprednisolone. The preparation was very simple and the drug-cyclodextrin complex was formed during the evaporation of the disperse phase. The products consisted of regularly shaped spherical microparticles with a smooth surface (Fig. 3). The average drug loading of the microcapsules was determined to be between 2 and 4%.

Fig. 4 shows the effect of a plasticizer, i.e. diethyl phthalate, on the release of hydrocortisone from ethylcellulose microcapsules. The release rate of hydrocortisone from microcapsules consisting only of a hydrocortisone - 2-hydroxypropyl-β-cyclodextrin complex and the ethylcellulose polymer is very slow. About 30% of the drug was released within the first 150 min. Similar results were obtained with ethylcellulose microcapsules containing a 17β-estradiol - 2-hydroxypropyl-β-cyclodextrin complex (Fig. 5). Addition of a plasticizer significantly increases the release rate and it could in fact be controlled by the amount of plasticizer added during the microcapsule production. Addition of plasticizers, which improves the flexibility of the ethylcellulose polymer chains and also reduces the

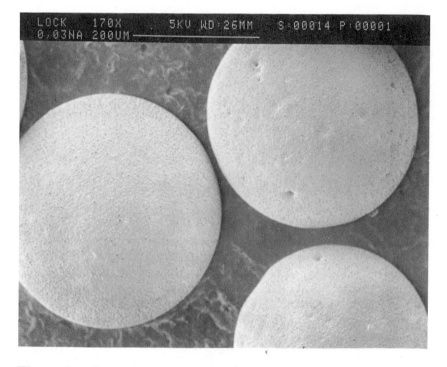

Figure 3. Scanning electron micrographs of microcapsules containing a hydrocortisone - 2-hydroxypropyl-β-cyclodextrin complex, 170x enlargement. The coating material was ethylcellulose.

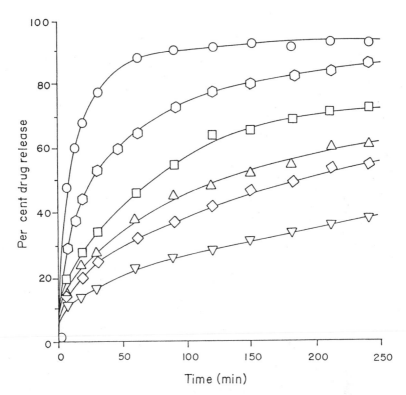

Figure 4. The effect of the polymer-plasticizer (i.e. ethylcellulose - diethyl phthalate) ratio on the release of hydrocortisone from microcapsules containing a hydrocortisone - 2-hydroxypropyl-β-cyclodextrin complex: 1:3, ○ ; 1:1, ⬡ ; 2:1, □ ; 3:1, △ ; 9:1, ◇ ; only ethylcellulose, ▽ . The total amount of the polymer and plasticizer used was kept constant at 2.00 g per 1.00 g hydrocortisone - 2-hydroxypropyl-β-cyclodextrin complex. (Reproduced with permission from ref. 69. Copyright 1992 Taylor & Francis.)

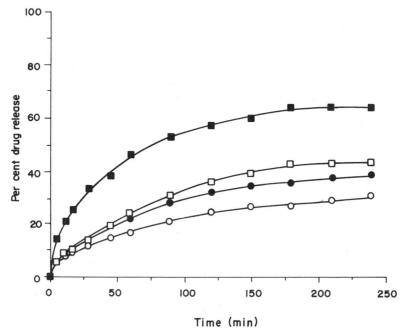

Figure 5. The effect of the polymer-plasticizer (i.e. ethyl-cellulose - diethyl phthalate) ratio on the release of 17β-estradiol from microcapsules containing a 17β-estradiol - 2-hydroxypropyl-β-cyclo-dextrin complex: 1:3, ■ ; 2:1, □ ; 9:1, ● ; only ethylcellulose, ○.The total amount of the polymer and plasticizer used was kept constant at 2.00 g per 1.0g 17β-estradiol-2-hydroxypropyl-β-cyclodextrin complex.

total amount of polymer in the microcapsules apparently opens up the polymer network which results in a faster drug release. Also, the plasticizer used forms a rather labile inclusion complex with 2-hydroxypropyl-β-cyclodextrin and, thus, can accelerate the drug release by competing with the drug for the cyclodextrin cavity.

Fig. 6 shows the effect of surfactant, i.e. sorbitan monooleate, on the release of hydrocortisone from polylactic acid microcapsules containing a hydrocortisone - 2-hydroxypropyl-β-cyclodextrin complex. Increasing the amount of surfactant results in more dense polymer network which results in a slower drug release.

Fig. 7 shows the effect of the dissolution medium on the release rate of hydrocortisone from ethylcellulose microcapsules and the results are compared to the dissolution of free (uncoated and uncomplexed) hydrocortisone under the same conditions. The results show that while the dissolution of free hydrocortisone was accelerated by addition of a surfactant or the water-soluble complexing agent a 2-hydroxypropyl-β-cyclodextrin complex the release rate of hydrocortisone from the microcapsules is virtually unaffected. The dissolution of uncoated hydrocortisone - 2-hydroxypropyl-β-cyclo-dextrin complex was almost instantaneous under these same conditions.

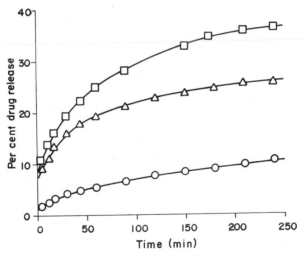

Figure 6. The effect of amount of surfactant, sorbitan monooleate, added to the continuous phase on the release of hydrocortisone from poly(DL-lactic acid) microcapsules containing a hydrocortisone - 2-hydroxypropyl-β-cyclo-dextrin complex. The per cent amount of surfactant in the disperse phase: □ , 1%; △ , 2%; ○, 3%.

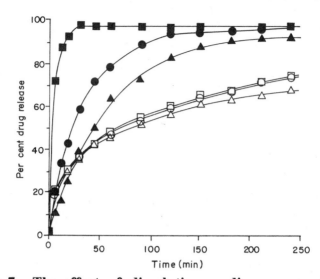

Figure 7. The effect of dissolution medium composition on the release of hydrocortisone from ethylcellulose microcapsules (polymer-plasticizer ratio 2:1) containing water-soluble hydrocortisone - 2-hydroxypropyl-β-cyclodextrin complex and on the dissolution of free (i.e. uncoated and uncomplexed) hydrocortisone. Free hydrocortisone, closed symbols; microcapsules containing hydrocortisone - 2-hydroxypropyl-β-cyclodextrin complex, open symbols: □ and ■ , 0.02% polysorbate 80 in simulated gastric fluid; ○ and ● , 1% 2-hydroxypropyl-β-cyclodextrin in simulated gastric fluid; △ and ▲ , simulated gastric fluid. (Reproduced with permission from ref. 69. Copyright 1992 Taylor & Francis.)

Conclusion

A hydrophile/lipophile version of the emulsion-solvent evaporation method can be used to prepare microcapsules of water-soluble drug - cyclodextrin complexes of lipophilic water-insoluble drugs. Lipophilic water-insoluble drugs possess inherent slow release properties which can limit their release rate from slow-release microcapsule systems. The physical characteristics of the microcapsules prepared could result in better control of the release rate of this type of drugs, i.e. make the release independent of the aqueous solubility of the drug.

Acknowledgement

This work was supported by a grant from the Icelandic Science Foundation. Technical support was provided by K. Ingvarsdóttir and J. Baldvinsdóttir.

Literature Cited

1. Luzzi, L.; Palmieri, A. In *Biomedical Applications of Microencapsulations;* F. Lim, Ed., CRC Press, Boca Raton, Florida, 1984, pp 1-17.
2. Szejtli, J.; Zsadon, B.; Cserhati, T. In *Ordered Media in Chemical Separations;* W. L. Hinze; D. W. Armstrong, Eds.; ACS Symposium Series 342, American Chemical Society, Washington, DC, 1987; pp 200-217.
3. Bergeron, R. J. In *Inclusion Compounds*; Atwood, J.L.; Davies, J.E.D.; MacNicol, D.D., Eds.; Chapter 12, Academic Press, London, 1984, pp 391-443.
4. *Cyclodextrins and their Industrial Uses*; Duchêne, D, Ed.; Editions de Santé, Paris, 1987.
5. Szejtli,J., *Pharm. Techn. Int.,* **1991**, *3(2),* 15-22.
6. Szejtli,J., *Pharm. Techn. Int.,* **1991**, *3(3),* 16-24.
7. Loftsson, L.; Brewster, M.E.; Derendorf, H.; Bodor, N. *Pharm. Ztg. Wiss.,* **1991**, *4/136,* 5-10.
8. Villiers, A.*C. R. Acad. Sci., Paris,* **1891**, *112,* 536-538.
9. Schardinger, F. *Z. Unters. Nahf. u. Genußm.,* **1903**, *6,* 865-880.
10. Schardinger, F. *Wein. Klin. Wschr.,* **1903**, *16,* 486-474.
11. Sicard, P. J.; Saniez, M.-H. In *Cyclodextrins and their Industrial Uses*; Duchêne, D, Ed.; Editions de Santé, Paris, 1987, pp 77-103.
12. *Inclusion Compounds of Physiologically Active Organic Compounds;* Freudenberg, K.; Cramer, F.; Plieninger, H.; Ger. Pat. No. 895,769; 1953.
13. Szejtli, J. *Drug Invest.,* **1990**, *2 (Suppl. 4),* 11-21.
14. Brewster, M.E.; Simpkins, J.W.; Hora, M.S.; Stern, W.C.; Bodor, N. *J. Parenter. Sci. Technol.,* **1989**, *43,* 231-239.
15. Pitha, J; Pitha, J. *J. Pharm. Sci.,* **1985**, *74,* 987-990.
16. Frank, D.; Gray, J.; Weaver, R. *Am. J. Pathol.,* **1976**, *83,* 367-382.
17. Pitha,J.; Milecki, J.; Fales, H.; Panell, L.; Uekama, K. *Int. J. Pharm.,* **1986**, *29,* 73-82.
18. Brewster, M. E.; Loftsson, T.; Baldvinsdóttir, J.; Bodor, N. *Int. J. Pharm.,* **1991**, *75,* R5-R8.
19. Hirayama, F.; Uekama, K. In *Cyclodextrins and their Industrial Uses*, Duchêne, D, Ed.; Editions de Santé, Paris, 1987; pp 131-172.
20. Higuchi, T.; Connors, K. A. In *Advances in Analytical Chemistry and Instrumentation* Reilly, C. N. Ed.; Wiley-Interscience, New York, 1965, Vol. 4, pp 117-212.

21. Loftsson, T.; Björnsdóttir, S.; Pálsdóttir, G.; Bodor, N. *Int. J. Pharm.*, **1989**, *57*, 63-72.
22. Bekers, O.; Bejinen, J. H.; Vis, B. J.; Suenaga, A.; Otagiri, M.; Bult ,A.; Underberg, W. J. M. *Int. J. Pharm.*, **1991**, *72*, 123-130.
23. Brewster, M. E.; Loftsson, T.; Estes, K. S.; Lin, J.-L.; Friðriksdóttir, H. Bodor, N. *Int. J. Pharm.*, **1992**, *79*, 289-299.
24. Loftsson, T.; Ólafsdóttir, B. J.; Baldvinsdóttir, J. *Int. J. Pharm.*, **1992**, *79*, 107-112 (1992).
25. Loftsson, T.; Friðriksdóttir, H. *Int. J. Pharm.*, **1990**, *62*, 243-247.
26. Rainsford, K. D. *Drug Invest.*, **1990**, *2 (Suppl. 4)*, 3-10.
27. Santucci, L.; Fiorucci, S.; Patoia, L.; Farroni, F.; Sicilia, A.; Chiucchiu, S.; Bufalino, L.; Morelli, A. *Drug Invest.*, **1990**, *2 (Suppl. 4)*, 56-60.
28. Backensfeld, T.; Müller, B. W.; Kolter, K. *Int. J. Pharm.*, **1991**, *74*, 85-93.
29. Backensfeld, T.; Müller, B. W.; Wiese, M.; Seydel, K. *Pharm. Res.*, **1990**, *7*, 484-490 .
30. Bettinetti, G. P.; Mura, P.; Liguori, A.; Bramanti, G. *Farmaco*, **1989**, *44*, 195-213.
31. Loftsson, T.; Bodor, N. *Acta Pharm. Nord.*, **1989**,*1*, 185-194.
32. Donbrow, D. In *Topics in Pharmaceutical Sciences 1987*, D.D. Breimer, D.D. and Speiser, P., Eds.; Elsevier Science Publishers B.V. 1987.
33. Bakan, J.A. In *The Theory and Practice of Industrial Pharmacy 2nd ed;* Lachman, L., Lieberman, H.A., Kanig, J.L. Eds.; Marcel Dekker, New York, 1986.
34. Li, S.P.; Kowarski C.R.; Feld, K.M.; Grim. W.A. *Drug Dev. Ind. Pharm.* **1988**, *14*, 353-376 .
35. Lin, S.-Y. *J. Pharm. Sci.*, **1988**, *77*, 229-232.
36. Prapaitrakul, W.; Whitworth, C.W. *Drug Dev.Ind.Pharm.*, **1989**, *15*, 2049-2053.
37. Chemtob, C.; Chaumeil, J.C.; N'Dongo, M. *Int. J. Pharm.*, **1986**, *29*, 83-92.
38. Jalsenjak, I.; Nixon, J.R.; Senjkovic, R; Stivic, I., *J. Pharm. Pharmacol.*, **1980**, *32*, 678-680.
39. Pathak, Y.V.; Dorle, A.K.*J. Microencapsulation*, **1989**, *6*, 199-204.
40. Jalsenjak, I.; Nicolaidou, C.F.; Nixon, J.R. *J. Pharm. Pharmacol.*, **1976**, *28*, 912-914.
41. Prapaitrakul, W.; Whitworth, C.W. *Drug Dev.Ind.Pharm.*, **1990**, *16*, 1427-1434.
42. Nixon, J.R.; Hassan, M. *J. Pharm. Pharm.*, **1980**, *32*, 857-859.

43. Nixon J.R.; Harris, M.S., *Acta Pharm. Tech.*, **1983**, *29*, 41-45.
44. Takeda, Y.; Nambu, N.; Nagai, T. *Chem. Pharm. Bull.*, **1981**, *29*, 264.
45. Dubernet, C.; Rouland, J.C.; Benoit, J.P. *J. Pharm. Sci.*, **1991**, *80*, 1029-1033.
46. Dalal, P.S.; Narurkar, M.M. *Int. J. Pharm.*, **1991**, *73*, 157-162.
47. Sveinsson, S.J.; Kristmundsdóttir, T. *Int. J. Pharm.*, **1992**, *82*, 129-133.
48. Pongpaibul, Y.; Whitworth C.W. *Int. J. Pharm.*, **1986**, *33*, 243-248.
49. Benita, S.; Donbrow, M. *J. Pharm. Pharmacol.*, **1982**, *34*, 77-82.
50. Lin, S.-Y.; Yang, J.-C. *J. Pharm. Sci.*, **1987**, *76*, 219-223.
51. Jalsenjak, I.; Nixon, J.R.; Senjkovic, R.; Stivic, I. *J. Pharm. Pharmacol.*, **1980**, *32*, 678-680.
52. Harris, M.S. *J. Pharm. Sci.*, **1981**, *70*, 391.
53. Beck, L.R.; Cowsar, D.R.; Lewis, D.H.; Cosgrave, R.J.; Riddle, C.T.; Lowry, S.L.; Epperly, T., *Fert. Steril.*, **1979**, *31*, 545-551.
54. Mason, N. S., Thies, C.,and Cicero, T.J. *J. Pharm. Sci.*, **1976**, *65*, 847-850 .
55. Benita, S.; Benoit, J.P.; Puisieux, F. ; Thies, C. *J. Pharm. Sci.*, **1984**, *73*, 1721-1724.
56. Maulding, H.V. *J. Controlled Release*, **1987**, *6*, 167-176
57. Lin, S.-Y. *J. Microencapsulation*, **1985**, *2*, 91-101.
58. Kato, T.; Nemoto, R.; Mori, H.; Kumagai, I. *Cancer*, **1980**, *46*, 14-21.
59. Rowe, R. C. *Acta pharm. Suec.*, **1982**, *19*, 157-169.
60. Rowe, R.C. *Pharm. Int.*, **1985**, *6*, 14-17.
61. Cavalier, M.; Benoit, J.P.; Thies, C. *J. Pharm. Pharmacol.*, **1986**, *38*, 249-253.
62. Jalil, R., Nixon, J.R. *J.Microencapsulation*, **1990**, *7*, 297-325.
63. Benoit, J.P.; Courteille, F.; Thies, C. *Int. J. Pharm.*, **1986**, *29*, 95-102.
64. Jalil, R.; Nixon, J.R. *J.Microencapsulation*, **1990**, *7*, 41-52.
65. Bodmeier, R.; Oh, K.H.; Chen, H. *Int. J. Pharm.*, **1989**, *51*, 1-8.
66. Bodmeier, R.; McGinity, J.W. *J. Microencapsulation*, **1987**, *4*, 279-288.
67. Tsai, D.C.; Howard, S.A.; Hogan, T.F.; Malanga, C.J.; Kandzari, S.J.; Ma, J.K.H. *J. Microencapsulation*, **1986**, *3*, 181-193.
68. Huang, H.-P.; Ghebre-Sellassie, I. *J. Microencapsulation*, **1989**, *6*, 219-225.
69. Loftsson, T.; Kristmundsdóttir, T.; Ingvarsdóttir, K., Ólafsdóttir, B.J.; Baldvinsdóttir, J. *J. Microencapsulation*, **1992**, *9*, 375-382.

RECEIVED October 5, 1992

Chapter 12

Polymeric Microspheres for Controlled-Release Herbicide Formulations

J. Tefft and D. R. Friend

Controlled Release and Biomedical Polymers Department,
SRI International, Menlo Park, CA 94025

A series of polymeric microsphere formulations were prepared and tested for their ability to control the release of an herbicide, Dicamba (DA; 3,6-dichloro-2-methoxybenzoic acid), over an extended period of time. Microspheres (MSs) were produced from both ethyl cellulose (EC) and polyarylsulfone (PS) in the desired size range (20-40 μm) by solvent evaporation or spray drying. Generally, PS MSs could be loaded with more DA (up to nearly 50 wt%) than could the EC MSs (maximum loading obtained was 23 wt%). EC MSs released DA at a greater rate than did comparable PS MSs under sonication and in soil column irrigation tests. The release rate of DA from the EC MSs could be controlled by using different viscosity grades (i.e., molecular weights) of EC: higher viscosity EC led to a lower DA release rate. The release rate of DA was also controlled by using a combination of EC with PS. Overall, MSs prepared from EC, PS, or EC/PS show promise as controlled release formulations for herbicides.

Controlled release formulations can be used to maintain effective local concentrations of herbicides in the soil and to reduce run-off (*1,2*). The number of applications required in the growing season can be reduced through the use of controlled release technology. A microparticulate herbicide formulation is desirable (*3*) because such chemicals are applied to fields by spraying. The formulation should be dispersible in water so that standard spraying equipment can be used. The particles should be under 50 μm in diameter and preferably 20-30 μm in diameter.

In this study, a model herbicide (Dicamba, DA; 3,6-dichloro-2-methoxybenzoic acid) was used to evaluate several microencapsulation procedures. Two polymer systems were investigated in this work: ethyl cellulose (EC) and polyarylsulfone (PS). Both polymers were formulated into DA-loaded microspheres (MSs) using either solvent evaporation or spray drying. This paper presents the conditions used to prepare and evaluate the MSs and measurement of release rates under several conditions.

0097–6156/93/0520–0190$06.00/0
© 1993 American Chemical Society

Materials and Methods

Materials. Various grades of ethyl cellulose (EC) were obtained as gifts from Hercules Incorporated (Wilmington, DE) and Dow Chemical Company (Midland, MI). Polyarylsulfone (PS; Radel A-200) was a gift of Amoco Performance Products, Ridgefield, CT. Dicamba (practical grade, 88%) was a gift of Sandoz Agro. Inc. DA was used in the free-acid form. Polyvinyl alcohol (Airvol, 99.3% hydrolyzed, medium viscosity) was a gift of Air Products and Chemicals, Allentown, PA). All other chemicals were either reagent grade or the highest grade available; all chemicals were used as received.

Solvent Evaporation. DA was encapsulated in the two polymers (EC and PS) using the solvent evaporation technique. The polymers (2.0 g) were dissolved in methylene chloride (10 mL). DA was then dissolved in the polymer solution at 10-25 wt% for EC and 25-50 wt% for PS relative to the amount of polymer used. This solution was then emulsified into an DA-saturated aqueous solution containing polyvinyl alcohol (5 wt%), and 0.01 wt% sodium dodecyl sulfate (SDS). After stirring for 15 min at 500 rpm, vacuum was applied for 15 min to 12.5 cm Hg, then increased to 25 cm Hg for 30 min followed by 37.5 cm Hg for another 30 min. The MSs were collected by filtration, resuspended in water containing 0.5 wt% SDS and dried using a Yamato Pulvis GB-21 spray drying unit.

Spray Drying. Microcapsules containing DA were also prepared by spray drying using a Niro E-4300 spray drying unit. EC or PS (2.5 wt%) were dissolved in methylene chloride and DA was then added at 20 to 40 wt%, depending on the polymer, relative to the total polymer solids dissolved. Microparticles were prepared in the spray dryer using a solvent flow rate of 40 mL/min with an atomizing head air pressure of 2.5 kg/cm^2. No drying heat was required. MSs were collected as a dry, free flowing powder in the hopper.

Characterization of Dicamba-loaded Microspheres. Median size and size distributions were determined on a Horiba CAPA-700 particle size analyzer. MS loading was determined by dissolving a weighed amount of MS (approximately 10 mg) in methylene chloride (25 mL). The organic phase was extracted with aqueous sodium carbonate (0.01 M; 15 mL). The organic phase was washed two more times with aqueous sodium carbonate. The aqueous phase was then adjusted to pH 3.5 using 1 N HCl. The concentration of DA in the extract was measured by HPLC (see below). Scanning electron micrographs (SEMs) of the MSs were obtained using a Cambridge 250 Scanning Electron Microscope. No special procedures were required to obtain the micrographs.

Release Rate Measurements. Release of DA from MSs and unencapsulated DA was measured using two techniques. The first technique was an accelerated release rate determination in a sonicated water bath. In these studies, a Branson 3200 Branisonic (50/60 Hz) was used. A small amount of MSs or DA was weighed in duplicate (ca. 3-4 mg of active agent) and placed into 4-dram vials containing 10 mL distilled water. The vials were placed in a test tube rack which was placed in the sonicating water bath. The samples were sonicated for 90 min (temperature was uncontrolled). The temperature rose from room temperature to approximately 50°C in all the sonication studies. At predetermined time points, the vials were removed and the release medium was collected by filtration. The amount of DA released was measured by HPLC (see below). Without sonication, the release of DA from MSs occurs over the period of days rather than minutes.

The second technique used to evaluated release of DA was a soil column leaching test. The soil leaching studies were performed in glass columns (5 cm deep and 9 cm in diameter). The soil used was sandy loam (mixture of sand, silt, and clay). MSs were placed on top of the column and varying amounts of water applied to the column. The MSs were dispersed evenly over the surface of the soil. Three different amounts or irrigations of water were placed on the column: 1.25 cm, 2.5 cm and 7.5 cm. The water passing through the column was collected and analyzed for released DA by HPLC (see below). Each irrigation was applied three consecutive times to the column. The total amount of water was applied over a 5 to 10 min period. All the water passing through the column was collected before the next irrigation was applied.

Analytical. The measurement of DA in the loading and release rate experiments was performed by HPLC. The column used was a Brownlee Spheri-5 RP-18 (3.9 mm x 15 cm). The mobile phase was 60% 0.2 M triflouroacetic acid, pH 3.2 and 40% acetonitrile/methanol (1:1); flow rate was 1.0 ml/min. The column was used at room temperature. The HPLC analyses were performed on a Waters 840 system consisting of two Model 510 pumps, a Model 481 UV detector, a Model 710B WISP (automatic sampler), and a Digital Computer Model 350. DA was detected at 240 nm. The retention time of DA was 5 min under the conditions used.

Results and Discussion

The long term goal of this research was to develop a long-term (ca. 30 d) particulate-based delivery system for herbicides. The formulation was designed to release the herbicide over this time during periods of rain or irrigation. In order to control the release of the herbicide, two polymers were selected for microencapsulation: EC and PS. In the case of EC, several different viscosities (molecular weights) were examined to determine if release rates were dependent upon molecular weight of the polymer. The MSs were prepared by either solvent evaporation or spray drying.

Microspheres Prepared by Solvent Evaporation. Scanning electron micrographs of MSs prepared by spray drying are shown in Figures 1 (EC) and 2 (PS). As shown, the MSs were within the acceptable size range of under 50 µm in diameter. Figure 3 shows an example of the size range distribution of EC MS prepared by solvent evaporation. All the MSs produced by solvent evaporation were composed of low viscosity EC (N-10; Hercules Incorp.). The average molecular number of the N-10 EC is approximately 10,000. Likewise, grades V7 and V100 have number average molecular weights of 7,000 and 100,000, respectively. As shown in Figure 3, the MSs fell into the desired 10-40 µm size range. Loading of MS with DA was dependent on the polymer used to prepare the MS. In general, EC MSs could be loaded up to a maximum of about 20 wt%; higher loadings led to a plasticizing effect and hence a tacky formulation. PS MSs could be loaded up to about 50 wt% before significant plasticizing effects were noted. Generally, the total amount of MSs recovered at the end of the microencapsulation procedure was 80-90%. Loading of the MSs was almost always theoretical. However, the external aqueous phase was saturated with excess solid DA in order to keep loading level at theoretical.

Figure 1. SEMs of EC MSs containing DA. MSs were prepared by solvent evaporation.

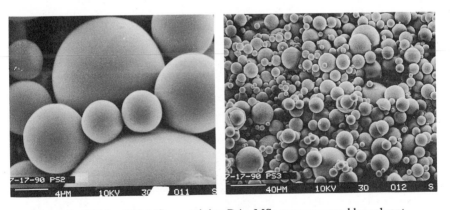

Figure 2. SEMs of PS MSs containing DA. MSs were prepared by solvent evaporation.

The release of DA from the various MS formulations was evaluated by sonicating an aqueous suspension of MSs. Sonication leads to an increase in temperature of the release medium: the temperature increases from about 22-23°C to 50°C over the 90 min release experiment. The increase in temperature, plus vibrational energy accelerates the release and dissolution of DA. In the absence of sonication, the release of DA from the MS formulations occurs over several days rather than 90 min. The sonication technique was used only to accelerate the release of DA; therefore, results from these experiments are used *only* to rank the formulations according to release kinetics rather than to obtain specific information about release rate kinetics under conditions expected in the field (the latter was addressed by the column soil leaching experiments—see below). The mechanism of release from the MS formulations was not addressed in this work. The release of DA from EC MS prepared by solvent evaporation is shown in Figure 4. MSs contained approximately 3-4 mg active agent. Initial loading of the formulations is indicated in the figure. M_t is the mass of DA released at time t; M_0 is the total mass of encapsulated DA. All the EC formulations retarded the release of DA compared with dissolution of unencapsulated DA. As expected, increased loading of the MSs with DA led to an increase in release rate (4).

The fractional release of DA (M_t/M_0) from a PS MS formulation prepared by solvent evaporation is shown in Figure 5. As before, the MSs contained approximately 3-4 mg active agent. The MSs in this case were loaded at 49 wt% with DA. Despite the high loading, the MS released DA at a relatively slow rate after an immediate release of about 10% of the total encapsulated DA. The fractional release of DA from another PS MS formulation is shown in Figure 6. In this case, the total percent released was less than observed in previous formulation. The MSs tested in Figure 6 were loaded at 30 wt% while those in Figure 5 were loaded at 49 wt% DA. The lower loading probably accounts for the lower rate of herbicide release observed in Figure 6 (4). Compared with the EC MS, the PS MS were considerably more effective at controlling the rate of DA released over the 90 min release experiments. In addition, the PS MS could be loaded at much higher levels than was possible with EC.

Microspheres Prepared by Spray Drying. Spray drying is an alternative technique for preparation of microcapsules and microspheres. An example of EC MSs containing DA prepared by spray drying is shown in Figure 7. These MSs were prepared with the N-10 grade of EC (Hercules). The MSs have a porous appearance which is clearly visible in the high magnification SEM in Figure 7. In contrast, MSs prepared by spray drying from PS had a solid, convoluted surface morphology (Figure 8). Morphologic characteristics of spray dried particles are a function of drying conditions (5).

The fractional release of DA from spray dried formulations of EC and PS are shown in Figures 9 and 10, respectively. Despite the marked difference in surface morphology of the MSs prepared by solvent evaporation and spray drying, there was little difference between the release rate kinetics. EC MSs consistently gave about 30-50% released DA during 90 min of sonication while the PS MSs released between about 5 to 25% DA, depending on the loading of the MSs.

In addition to testing of the N-10 grade of EC, other grades were tested for their ability to control release of DA from MSs. In this case, low viscosity EC (7V) and high viscosity EC (100V), both of Dow Chemical Co., were used to prepare MSs using the spray drying technique. The loading of these MSs was the same for each formulation (23 wt% DA). Fractional release of DA from these two formulations under sonication is shown in Figure 11. The higher viscosity EC

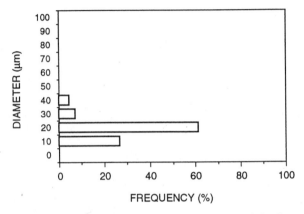

Figure 3. Frequency distribution of EC MS. Median particle diameter was 13.8 μm with a standard deviation of 7.2 μm and a calculated surface area of 0.508 m^2/g.

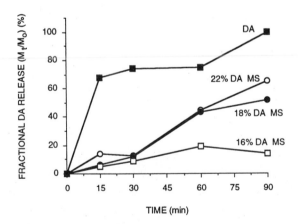

Figure 4. Fractional release (M_t/M_o) of DA during sonication into 10 mL H$_2$O from two EC MS formulations; also shown is the dissolution of unencapsulated DA.

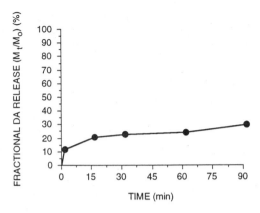

Figure 5. Fractional release (M_t/M_0) of DA during sonication into 10 mL H_2O from a PS MS formulation. MSs were loaded at 49 wt% with DA.

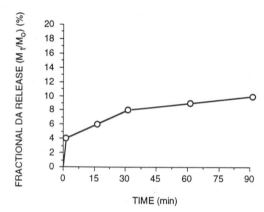

Figure 6. Fractional release (M_t/M_0) of DA during sonication into 10 mL H_2O from a PS MS formulation. MSs were loaded at 30 wt% with DA.

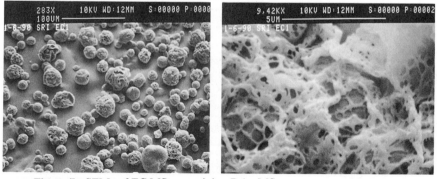

Figure 7. SEMs of EC MSs containing DA. MSs were prepared by spray drying in a Niro E-4300 spray drying unit.

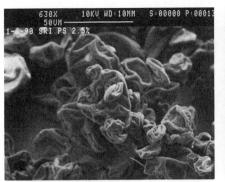
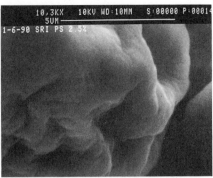

Figure 8. SEMs of PS MSs containing DA. MSs were prepared by spray drying in a Niro E-4300 spray drying unit.

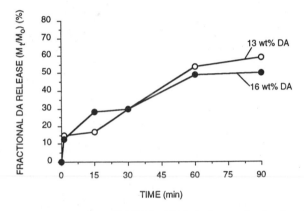

Figure 9. Fractional release (M_t/M_o) of DA during sonication into 10 mL H_2O from several EC MS formulations prepared by spray drying.

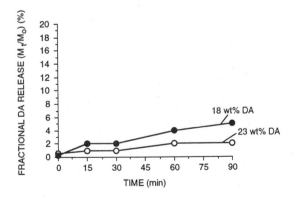

Figure 10. Fractional release (M_t/M_o) of DA during sonication into 10 mL H_2O from a PS MS formulation.

reduced the release rate of DA during the 90 min sonication experiment. In addition, the low viscosity EC MSs prepared from the Dow Chemical EC gave a release rate profile similar to that obtain from MSs prepared from low viscosity EC of Hercules.

We investigated another way to control the release of DA from the MSs: mixing the two polymers (EC and PS). Both EC and PS have very similar solubility parameters (6) suggesting that these polymers are compatible when blended together as described herein. These MSs were prepared by spray drying in the ratio of 1:9, 1:4, and 2:3 (EC/PS). The Hercules N-10 grade EC was used in the preparation of these MSs. The fractional release of DA from the EC/PS formulations during sonication is shown in Figure 12. The MSs were loaded with DA as follows: 1:9 EC/PS were 27 wt% DA, 1:4 EC/PS were 29 wt% DA; and 2:3 EC/PS were 27 wt% DA. The small amount of EC added to PS (1:9 EC/PS) increased the release rate of DA slightly (see Figure 10 for comparison). Increasing the amount of EC in the formulation to 2:3 EC/PS greatly increased the release rate. After 90 min, nearly all the DA was released from the MSs. The 1:4 EC/PS formulation gave a rapid release of DA over the early portion of the experiment followed by a slower release rate (see Figure 12). The reason for this aberrant release behavior is unknown.

Soil Column Leaching Studies. Several of the formulations were examined for their ability to release DA in a soil column leaching study. The column contained sandy loam (mixture of sand, silt, and clay). MSs or unencapsulated DA were placed evenly over the top of the soil. A measured amount of water was then added to column to simulate irrigation. The amount of DA in the water collected at the bottom of the column was measured by HPLC. Two PS MS formulations and one EC MS formulation were tested in the soil column leaching test. Water was placed on top of column at one of three depths (1.25 cm, panel A; 2.5 cm, panel B; 7.5 cm, panel C). Each irrigation was placed on the column three consecutive times. The effluent collected at the bottom of the column was analyzed for DA by HPLC. The total amount of DA (free or encapsulated) added to the column was 2.5 mg. All MSs tested in the irrigation study were prepared by spray drying. The amount of DA recovered following the three irrigation protocols is shown in Figure 13 (1.25 cm, panel A; 2.5 cm, panel B; and 7.5 cm, panel C). Line A indicates the local phytotoxic level of DA and line B indicates the minimum effect amount required for weed control. In this case, the phytotoxic level is an average DA amount in the effluent known to be phytotoxic to crops commonly treated with DA.

The DA-only formulation provided adequate amounts of DA in the soil over the first two irrigations with 1.25 cm of water while the third 1.25 cm irrigation contained an insufficient amount of DA to provide adequate weed control (Figure 13, panel A). The 2.5 and 7.5 cm irrigations of the DA-only formulation (Figure 13, panels B and C) led to phytotoxic levels of DA present in the effluent in the first of the three irrigations while the last two irrigations contained insufficient amounts of DA to control weed growth.

The three MS formulations evaluated in the soil column leaching tests are also shown in Figure 13. Of the three formulations, the EC MS, loaded at 16 wt% DA, released DA in amounts equal to or greater than the minimum effective level (line B) yet no greater than the phytotoxic level (line A). This release profile was observed at the 1.25, 2.5, and 7.5 cm irrigation levels. The two PS MS formulations released less DA than did the EC MS formulations. This finding is generally consistent with the relative release rate kinetics of the EC and PS MS formulations under sonication. Of the two PS MS formulations, the 30 wt% loaded MSs gave

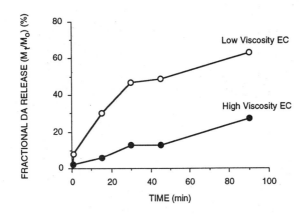

Figure 11. Fractional release (M_t/M_o) of DA during sonication into 10 mL H_2O from several EC MS formulations prepared by spray drying.

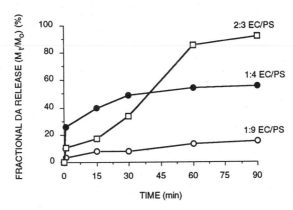

Figure 12. Fractional release (M_t/M_o) of DA during sonication into 10 mL H_2O from MS prepared from EC/PS.

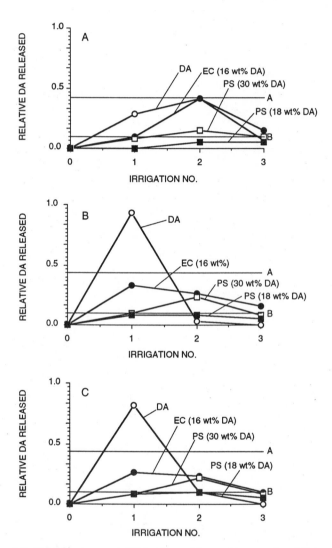

Figure 13. Relative amount of DA released from unencapsulated DA and three MS formlations (EC MSs, 16 wt% DA; PS MSs, 18 wt% DA; and PS MSs, 30 wt% DA).

greater release of DA in all the irrigation experiments. The two PS MS formulations released only about 50% of the total DA applied to the column, regardless of the amount of water used in the irrigations. In contrast, about 90% of the DA was released from the EC formulations.

Conclusions

The conclusions of this work are summarized as follows: 1) MSs could be produced in the desired size range (20-40 μm) from both EC and PS; 2) PS MSs could be loaded at a higher percent DA than could the EC MSs; 3) EC MSs released DA at a greater rate than did PS MSs under sonication and in a soil column irrigation test; 4) release of DA from the EC MSs could be controlled by using different grades of EC; and 5) release of DA could also be controlled by using a combination of EC and PS. While the results indicated that the release of DA could be extended compared with that from unencapsulated DA, field tests must be performed to determine the actual duration of activity of the formulations.

References

1. *Controlled Release Pesticides* , H. B., Ed.; ACS Symposium Series 53, American Chemical Society, Washington, DC; 1977.
2. Schrieber, M. M.; Shasha, B. S.; Trimnell, D.; White, M. D. Controlled release herbicides; In *Methods of Applying Herbicides,* McWhorter C. G.; Gebhardt, M. R., Eds.; Monograph Series, No. 4; WSSA: Champaign, IL, 1987; pp. 177-191.
3. Trimnell, D.; Shasha, B. S. *J. Controlled Release* **1988**, *7*, 263-268.
4. Baker, R. W. *Controlled Release of Biologically Active Agents*; J. Wiley, New York, NY 1987; pp. 68-73.
5. Masters, K. S*pray Drying Handbook,* 4th Ed., Longman Scientific and Technical, Harlow, 1985; pp. 298-342.
6. Burrell, H., Solubility parameter values; In *Polymer Handbook, Sec. Ed.,* Brandrup, J.; Immergut, E. H., Eds.; J. Wiley, New York, NY, 1975; p. IV-358, IV-353.

RECEIVED October 5, 1992

Chapter 13

Encapsulated Systems for Controlled Release and Pest Management

R. Levy, M. A. Nichols, and T. W. Miller, Jr.

Lee County Mosquito Control District, P.O. Box 06005,
Ft. Myers, FL 33906

Controlled release of mosquitocides from Culigel acrylamide or acrylate superabsorbent polymer granules in
aquatic environments is regulated by inert ingredients
that are admixed in the pesticide formulation during
granule manufacturing procedures. Release profiles of
organophosphate larvicides, growth regulators, and
larvicidal bacilli can be tailored to specific mosquito control requirements by altering the type and/or
concentration of dispersants comprising a dispersant
complex.

Culigel (Registered Trademark®, Lee County Mosquito Control
District) systems represent patented pest-control technology
(1-4). Manufacturing procedures and novel applications for non-
toxic Culigel acrylamide and acrylate superabsorbent polymers as
water-management aids and as matrices for controlled release of
bioactive agents used to control aquatic and terrestrial pests are
presented in a comprehensive review article by Levy et al. *(5)*.
Controlled release of pesticides from insoluble Culigel superabsorbent polymer granules was due to a swelling-controlled diffusion
or leaching process that was primarily triggered by water or aqueous
fluids. Water quality was shown to affect the amount of polymer
swelling, and subsequent diffusion rate of an encapsulated pesticide. The rate of pesticide diffusion from Culigel granules could be
regulated or modified by altering the type and concentration of inert
ingredients used in the pesticide formulation. Slow release kinetics
were shown to be first-order or square-root-of-time; however, "modified" zero-order release profiles could be obtained by varying the
pesticide and inert ingredient admixture sequences in the stages
during the manufacturing of the pesticide-loaded Culigel granules.
Secondary release of pesticides from Culigel granules was related to
the rate of matrix biodegradation or decomposition, and was a was a
function of monomer composition, crosslinking density, or the

0097–6156/93/0520–0202$06.00/0
© 1993 American Chemical Society

polymerization process used to manufacture the superabsorbent polymers.

Technology related to the controlled release of a variety of bioactive agents for pest control in aquatic and terrestrial environments has been reviewed by Kydonieus *(6)*, Baker *(7)*, Duncan and Seymour *(8)*, and Wilkins *(9)*. Culigel superabsorbent polymers can also be used to physically control mosquitoes that breed in small collections of water in natural or artificial habitats *(5)*.

Components of the Granular Formulations

Initial laboratory and field research on controlled release of conventional and biological pesticides was conducted with Type I and II superabsorbent polymers; however, recent research on the pesticide entrapment efficiency of Type III and Type IV superabsorbent polymers suggested that one type or several types of these water-activated, commercially available polymers could be used as "generic" pesticide formulation matrices for development of controlled-release products (Table I). Culigel Type III granules were selected to demonstrate

Table I. Types of Granular Culigel Superabsorbent Polymers Evaluated as Controlled-Release Matrices

Type I =	Crosslinked Copolymer of Acrylamide and Sodium Acrylate
Type II =	Lightly Crosslinked Potassium Polyacrylate
Type III =	Partial Sodium Salt of a Lightly Crosslinked Polypropenoic Acid
Type IV =	Crosslinked Potassium Polyacrylate/ Polyacrylamide Copolymer

the broad spectrum controlled-release applications of an acrylate superabsorbent polymer, since previous research had highlighted the controlled-release potential of an acrylamide-base Type I superabsorbent polymer *(5)*. Bioactive agents utilized in the present study were the larvicidal bacilli Vectobac TP and *Bacillus sphaericus* coded ABG-6184 (Abbott Laboratories, North Chicago, IL) and the insect growth regulators Altosid Liquid Larvicide (Zoecon Corporation, Dallas, TX 75234), Nylar Emulsifiable Concentrate (McLaughlin Gormley King Company, Minneapolis, MN 55427-4372), and Dimilin 25W (Uniroyal Chemical Company, Middlebury, CT 06749); however, the organophosphate larvicide temephos (Abate 4-E) has been the main pesticide under evaluation *(5)*. Prolonged mosquito-controlling efficacy was demonstrated with a Culigel Type I formulation of Abate 4-E (Figure 1). Dispersant Complex C (Table III) was used as the pesticide release-rate regulator in this formulation. In general, previous studies utilized a single inert ingredient as a dispersant in the granular compositions (Table II), while in the present study, a combination of inert ingredients were utilized as dispersant complexes in all pesti-

204 POLYMERIC DELIVERY SYSTEMS

cide formulations to regulate or control the rate of release of a
bioactive agent from a Culigel superabsorbent polymer granule (Table
III). A two-part dispersant complex formulation was admixed with a
bacterial or growth regulator formulation at the time of granule
loading at rates that ranged from 0.1-10% and 1-25% by weight, respectively.

Table II. Types of Inert Ingredients used in Dispersant
Complexes to Regulate Pesticide Release from Culigel
Granules

Alcohols	Surface-Active Agents
Surfactants	Binding Agents
Emulsifiers	Suspending Agents
Solvents	Compatibility Agents
Salts	Wetting Agents
Diluents	Oils

Table III. Inert Dispersant Complexes used to Regulate
Pesticide Delivery from Culigel Type III Granules

Dispersant Complex A = 2-Ethyl Hexanol + Magnesium
Chloride

Dispersant Complex B = 2-Ethyl Hexanol + Acrylic Acid,
Copolymer

Dispersant Complex C = Sulfated Alkyl Carboxylate and
Sulfonated Alkyl Naphthalene, Sodium
Salt + Acrylic Acid, Copolymer

Dispersant Complex D = Sodium Alkyl Aryl Sulfonate +
Acrylic Acid, Copolymer

The chemical make-up, concentration, solubility, and ionic characteristic of inert ingredients comprising a dispersant complex were
dependent on the type of Culigel superabsorbent polymers and the type
of bioactive agent formulation. For example, some pesticide formulations required a binder to hold the product within the granular matrix to prevent premature or rapid leaching in storage or in an
aquatic environment, while other pesticide products became entrapped
or bound within microporous channels and required a propellant to
drive the product out of the granule when introduced into an aquatic
habitat. Culigel insecticidal granules were prepared according to
our microsponging manufacturing procedure (5). In general, controlled-release insecticidal granules were prepared for bioassays by
mixing 5-10 g of 2-3 mm irregularly shaped Culigel superabsorbent
polymer granules in 1000 ml Nalgene bottles containing 750 g distilled water solution/suspension of 10%(w/w) insecticide formulation
and 10%(w/w) dispersant complex for ca. 1-3 hr, depending on the
absorbency of the granule type. Although a range of formulation
ratios were evaluated, a 1:1 pesticide formulation to dispersant

complex ratio was used as a standard during aqueous granule preparation in the present study, however; it should be noted that pesticide
release profiles from Culigel Type I-IV granules could be drasti-
cally modified by altering the pesticide to inert formulation ratio
(5) and/or the type, concentration, and number of inert ingredients
used in a dispersant complex (Levy, R., Lee County Mosquito Control
District, unpublished data). During the admixing procedure, the
Culigel granules swell in the aqueous insecticide/dispersant complex
formulations, and in the process absorb many times their weight in
water and active and inert formulation components. A paint shaker
was used to vigorously/homogeneously mix or suspend the formulation components in water to assure that the Culigel granules
would absorb the desired ratio of active and inert ingredients.
Therefore, a 1:1 pesticide formulation to dispersant complex granule
loading ratio is assumed. The hydrated, insecticide-loaded granules
were poured through a series of sieves, washed to remove excess surface coating, and air-dried in a room maintained at ca. 43% RH and
25°C for 1-4 days to remove the entrapped water. Dehydrated granules
were removed from the screens and stored in glass bottles until testing.

Protocol for Mosquito Bioassays

Bioassays were conducted against multiple broods of larvae according
to our standard granule transfer protocol (5). Culigel granules
(Table I) containing a growth regulator or larvicidal bacilli and
dispersant complex (Table III) were evaluated in 1000 ml of brackish
water against ten 2nd instar larvae of *Culex quinquefasciatus*
(=larval brood) at a rate of one, 2-3 mm insecticidal granule per
0.068 m² test habitat (glass pan). Tests were replicated 3-4 times.
Percent mortality was recorded at 24 hr posttreatment intervals. At
the termination of a test series, 24 hr mortality data for each replicate in a series was averaged and graphically plotted against time
to determine the controlled-release profile for a specific growth
regulator or bacterial formulation. Culigel insecticidal granules in
a test series were transferred to new simulated aquatic habitats with
new larval broods if average mortality was 90% or greater. A series
was terminated if test adult emergence or control mortality exceeded
10%. Granule transfers were sequentially continued to challenge new
larval broods until average percent control was ineffective. It
should be noted that bioassays utilized a 10-day interval between
granule transfers. During this 10-day interval, granules remained
submerged in the simulated permanent water habitats with dead larvae
before being transferred to a new test set-up. A transfer interval
is indicated on each graph (Figures 1-6) by a "10-day gap" on the
x-axis between the introduction of a new larval brood. Bioassay data
were analyzed using "z" and "t" tests. Promising granular formulations for each insect growth regulator and larvicidal bacilli were
selected on the basis of the duration of 90-100% control of immature
mosquitoes at the application rates indicated and on the number of
broods controlled over this time period. Culigel bacterial and insect growth regulator formulations showing the most effective mosquito control for the longest duration were scheduled for small-plot
field trials against natural populations of mosquitoes.

Controlled Release of Larvicidal Bacilli

Evaluations against larvae of *Cx. quinquefasciatus* with Culigel
Type III granular formulations of *Bacillus thuringiensis* var.
israelensis (B.t.i.) and Dispersant Complex A have indicated that an
effective acrylate superabsorbent polymer-base granular product can
be developed for prolonged control of multiple broods of mosquitoes
with a single application (Figure 2). Culigel Type III acrylate
granules containing Vectobac Technical Powder and a dispersant complex composed of an aliphatic alcohol and salt blend produced satisfactory controlled-release profiles and mosquito-controlling efficacy
for ca. 4 1/2 months at an application rate of 2.14 lbs/acre. Vectobac TP potency is rated at 5000 ITU/mg. Culigel Type III granules
contained ca. 36% Vectobac TP and Dispersant Complex A formulation
by weight, and were 33 days old at the time of testing.
   A 4 1/2-month controlled-release duration was also obtained with
Culigel Type III superabsorbent polymer granules containing an experimental technical powder formulation of *Bacillus sphaericus* that
was designated as ABG-6184, and a dispersant complex composed of an
aliphatic alcohol and acrylic acid, copolymer blend (Figure 3). *B.
sphaericus* is rated at 2500 ITU/mg. Culigel Type III superabsorbent polymer granules contained ca. 22% *B. sphaericus* and Dispersant
Complex B formulation, and were 49 days old when evaluations were
initiated. Results of bioassays against 2nd instar larvae of *Cx.
quinquefasciatus*at an application rate of 2.03 lbs/acre were
similar to results obtained with *B.t.i.*, and suggested that an
effective acrylate-base controlled-release granular product can be
developed for this larvicidal bacillus.

Controlled Release of Growth Regulators

Culigel Type III granules were also evaluated for the controlled
release of commercially available and experimental insect growth
regulators. Results of bioassays against 2nd instar larve of *Cx.
quinquefasciatus* with Culigel Type III granules loaded with a
microencapsulated formulation of 5% S-methoprene called Altosid Liquid Larvicide and a dispersant complex composed of a naphthalene
sulfonate surfactant and an acrylic acid, copolymer blend indicated
that effective mosquito-controlling levels of methoprene can be
slow-released from the Type III granules for ca. 2 1/2 months at an
application rate of 2.17 lbs/acre (Figure 4). Culigel Type III granules contained ca. 27% Altosid Liquid Larvicide and Dispersant Complex D formulation, and were 30 days old at the time of testing.
   Research was also conducted with an experimental 10% emulsifiable concentrate of pyriproxyfen called Nylar Insect Growth Regulator (Figure 5). A combination of this product and a dispersant complex composed of a naphthalene sulfonate surfactant and acrylic acid,
copolymer blend was loaded into Culigel Type III superabsorbent polymer granules for controlled-release evaluations. Culigel granules
contained ca. 20% Nylar emulsifiable concentrate and Dispersant Complex C formulation, and were 30 days old when evaluated. Results of
our bioassays against 2nd instar larvae of *Cx. quinquefasciatus*
indicated that mosquitocidal levels of pyriproxyfen can be effect-

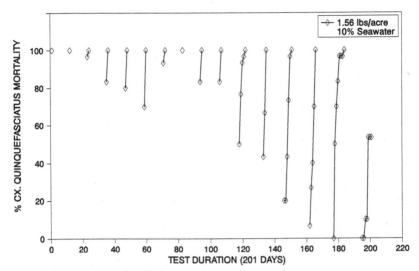

Inert Ingredients: Dispersant Complex C

Figure 1. Controlled release of Abate 4-E from Culigel Type I granules (granule composition = 64% Culigel + 18% Abate 4-E + 18% Dispersant Complex C).

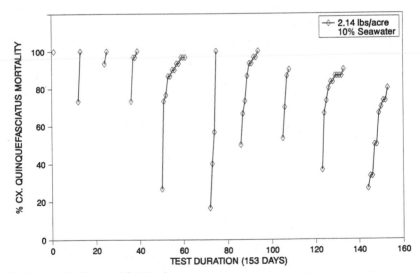

Inert Ingredients: Dispersant Complex A

Figure 2. Controlled release of Vectobac Technical Powder from Culigel Type III granules (granule composition = 64% Culigel + 18% Vectobac TP + 18% Dispersant Complex A).

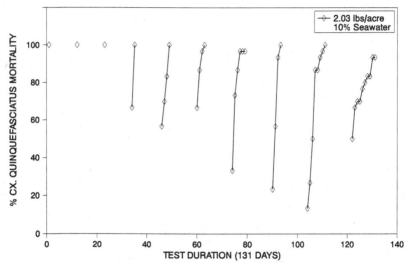

Inert Ingredients: Dispersant Complex B

Figure 3. Controlled release of *Bacillus sphaericus* (ABG-6184)
from Culigel Type III granules (granule composition = 78% Culigel
+ 11% ABG-6184 + 11% Dispersant Complex B).

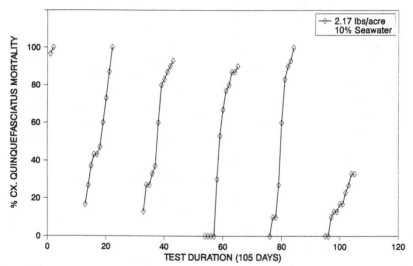

Inert Ingredients: Dispersant Complex D

Figure 4. Controlled release of Altosid Liquid Larvicide from
Culigel Type III granules (granule composition = 73% Culigel +
13.5% Altosid LL + 13.5% Dispersant Complex D).

ively released from Culigel Type III acrylate granules for ca. 5
months at an application rate of 2.04 lbs/acre.
     Culigel superabsorbent polymer granules were also evaluated as
potential matrices for the controlled release of an insect growth
regulator formulation composed of 25% diflubenzuron called Dimilin
25W (Figure 6). An experimental formulation consisting of Dimilin
25W and a dispersant complex composed of a naphthalene sulfonate
surfactant and acrylic acid, copolymer blend was encapsulated within
Culigel Type III granules. Culigel granules contained ca. 26%
Dimilin 25W and Dispersant Complex C formulation, and were 29 days
old at the time of testing. Bioassays against multiple broods of 2nd
instar larvae of *Cx. quinquefasciatus* indicated that larvicidal
levels of Dimilin can be effectively slow released from Culigel Type
III acrylate granules for ca. 4 1/2 months.

Controlled Release of Multiple Products

Aqueous formulations of Altosid Liquid Larvicide and Vectobac TP or
Vectobac TP and Arosurf MSF have been shown to produce significantly
better control of multiple stages of immature mosquitoes at lower
than label application rates for each bioactive component. Simultaneous controlled release of larvicidal bacilli and insect growth
regulators from Culigel Type I and II superabsorbent polymer granules
have been demonstrated *(5)*. This joint-action formulation concept
for the simultaneous controlled-release of two or more pesticide
products from the same Culigel superabsorbent polymer granule is
currently under evaluation with a variety of pesticides and Culigel
superabsorbent polymer types (Table IV). Preliminary results (Levy,
R., Lee County Mosquito Control District, unpublished data) indicate
that the multi-product/multi-spectrum control method can be significantly more cost-effective than single-product applications.

Table IV. Joint-Action Pesticide Formulations Evaluated against
Mosquitoes in the Culigel Controlled-Release System

Larvicidal Bacilli + Insect Growth Regulators

Larvicidal Bacilli + Monomolecular Surface Film

Organophosphate Larvicide + Monomolecular Surface Film

Organophosphate Larvicide, Larvicidal Bacilli or
Insect Growth Regulators + Herbicides (Aquathol;
Sonar; Diquat)

Summary of Results. In general, data obtained from bioassays on
larvae of *Cx. quinquefasciatus* with three insect growth regulators
and two larvicidal bacilli indicated that acrylate Culigel Type III
superabsorbent polymer granules can be formulated with a properly
matched dispersant complex to provide effective slow release of
larvicidal levels of a variety of bioactive agents in shallow brack-
ish-water mosquito habitats for prolonged periods. To date, our
research on the potential use of acrylamide and acrylate Culigel Type

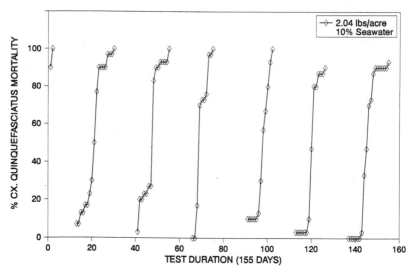

Inert Ingredients: Dispersant Complex C

Figure 5.   Controlled release of Nylar Emulsifiable Concentrate
from Culigel Type III granules (granule composition = 80% Culigel
+ 10% Nylar EC + 10% Dispersant Complex C).

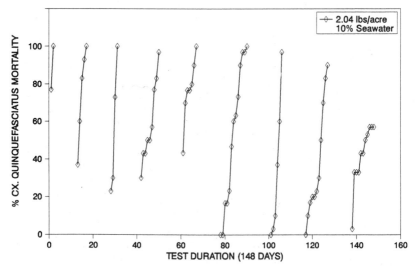

Inert Ingredients: Dispersant Complex C

Figure 6.   Controlled release of 'Dimilin 25W from Culigel Type
III granules (granule composition = 74% Culigel + 13% Dimilin 25W
+ 13% Dispersant Complex C).

I - IV superabsorbent polymer granules as controlled-release matrices
for mosquito larvicides has indicated that the duration of release
was mainly a function of the type and concentration of inert ingredients used as a dispersant or dispersant complex in the granular formulations (Figure 7). Controlled-release profiles were adjusted for
a specific water quality or a range of water qualities by manipulat-
ing the inert ingredients comprising a pesticide formulation. For
the most part, satisfactory controlled release of any insect growth
regulator or larvicidal bacilli from any Culigel granule type is
formulation dependent.
     It should be noted that a wide range of commercial formula-
lations of insect growth regulators, larvicidal bacilli, organo-
phosphates, and monomolecular surface films have been evaluated in
the Culigel controlled-release system for control of mosquito larvae
and pupae (Table V). A combination of products have also been used.
Several types of acrylamide and acrylate Culigel superabsorbent polymers are currently being evaluated for controlled release of one or
more bioactive agents for control of a variety of adult and immature
pests in aquatic and terrestrial environments. Granule, pellet,

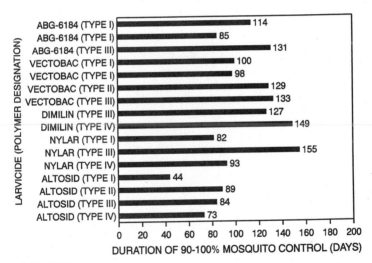

1.56 - 2.17 lbs/acre
10% Seawater

Figure 7. Controlled release of mosquito larvicides from Culigel
Type I-IV superabsorbent polymer granules (granules contained a
dispersant or dispersant complex at varying pesticide to inert
ratios).

Table V. Pesticides Evaluated against Mosquitoes in
the Culigel Controlled-Release System

Larvicidal	Growth
Bacilli	Regulators
Vectobac TP	Altosid Liquid Larvicide
Vectobac 12AS	Altosid Technical
Teknar HP-D	Dimilin 25W
Bactimos	Dimilin 4L
Acrobe	Nylar 10% EC
ABG-6184	Nylar Technical
BSP-1	Fenoxycarb Technical
Organophosphates	Monomolecular Surface
Abate 4-E	Films
Abate Technical	Arosurf MSF

briquet, powder, and oil- or water-base variable-viscosity Culigel
formulations are being developed for use in a variety of habitats.
Patents are pending in the United States, Europe, Japan, Canada, and
Australia.

Literature Cited

1. Levy, R., U.S. Patent No. 4,818,534, April 4, 1989.
2. Levy, R., U.S. Patent No. 4,983,389, Jan. 8, 1991.
3. Levy, R., U.S. Patent No. 9,983,390, Jan. 8, 1991.
4. Levy, R., U.S. Patent No. 4,985,251, Jan. 15, 1991.
5. Levy, R., Nichols, M.A., and Miller, Jr., T.W. In *Pesticide Formulations and Application Systems;* Devisetty, B.N. and Berger, P.D., Eds.; ASTM STP 1146; American Society For Testing and Materials: Philadelphia, PA, 1992, Vol. 12.
6. *Controlled Release Technologies: Methods, Theory, and Applications;* Kydonieus, A.F., Ed.; CRC Press, Inc.: Boca Raton, FL, 1980; Vol. 1 and 2.
7. Baker, R. *Controlled Release of Biologically Active Agents* ; John Wiley & Sons: New York, 1987.
8. Duncan R.; Seymour, L.W. *Controlled Release Technologies, A Survey of Research and Commercial Applications;* Elsevier Advanced Technology: Oxford, UK, 1989.
9. *Controlled Delivery of Crop-Protection Agents;* Wilkins, R.M., Ed.; Taylor & Francis Inc.: Bristol, PA, 1991.

RECEIVED September 24, 1992

Chapter 14

Controlled Release of Herbicide from an Unmodified Starch Matrix

R. E. Wing[1], M. E. Carr[1], W. M. Doane[1], and M. M. Schreiber[2]

[1]Plant Polymer Research, National Center for Agricultural Utilization Research, Agricultural Research Service, U.S. Department of Agriculture, Peoria, IL 61604
[2]Insect and Weed Control Research, Department of Botany and Plant Pathology, Agricultural Research Service, U.S. Department of Agriculture, Purdue University, West Lafayette, IN 47907

Unmodified starch is an effective matrix for encapsulating solid and/or liquid active agents such as herbicides. Gelatinization of starch in the presence of water and herbicide via continuous twin-screw extrusion processing followed by particularization to desirable mesh sizes yields slow release herbicide products. Rate of release can be controlled by varying extruder conditions, particle size and the addition of other chemical additives to the starch matrix. Data will be presented to show efficacy of weed control and reduction of ground water contamination by various herbicide formulations.

Studies were presented earlier emphasizing improving worker safety in handling herbicides and providing a release profile for soil incorporated herbicides such that the release rate of active ingredient was sufficient to control the target weed but slow enough to avoid phytotoxicity to the crop. Encapsulation of herbicides was achieved by dispersing the herbicides into an aqueous dispersion of gelatinized starch and then crosslinking the starch by (a) xanthide (1); (b) calcium chloride (2); (c) borate (3); or (d) calcium-borate (4) methods.
Several researchers have shown that starch encapsulation: a) significantly decreases percutaneous permeability of parathion as compared with clay formulated samples (5); b) extends the duration of weed control by extending the period of release (6-7); c) obviates the need for soil incorporation of certain herbicides (8); and d) reduces amounts of herbicide required (9).
Starch exhibits many properties required of a polymer to function as an encapsulation matrix. Broader acceptance of starch

This chapter not subject to U.S. copyright
Published 1993 American Chemical Society

as an encapsulation matrix depends on improving the economics of encapsulation and/or eliminating the use of chemicals needed to form the crosslinked matrix. Recent articles (10-13) and a patent (14) showed herbicides are effectively encapsulated in starch without the use of crosslinking agents. Cornstarch is first solubilized or highly dispersed by steam injection cooking and then crosslinked through the natural process of retrogradation after the herbicide is added. Retrogradation is the formation of aggregates resulting from hydrogen bonding between hydroxyl groups of adjacent starch chains. Factors such as rate of cooling, pH, amylose content, moisture content, and dispersion temperature affect retrogradation of solubilized starch. This property is beneficial for enhancing encapsulation and controlling the release rate of pesticides from the formed starch matrix. Mixtures of various starches (waxy, pearl, and high amylose) can be selected to control the amylose content and thus the degree of retrogradation. Several researchers (15-16) have reported excellent results in controlling weeds and reducing potential groundwater contamination with products prepared by this encapsulation technique (17-22).

More recently, we produced starch-encapsulated products in a continuous, efficient, and effective process by using a twin-screw extruder to gelatinize starch and incorporate the herbicide (23-25). A preblend of starch and atrazine was fed into the twin-screw extruder while water was injected to gelatinize the starch and allow the atrazine to enter the matrix. Other herbicides (alachlor or metolachlor) were metered into other starch formulations as liquids. The extrudates were dried and ground to provide products with 94-99% atrazine, 72-95% alachlor, and 87-92% metolachlor encapsulated. We now report a scale-up of the extrusion process to produce thousand-kilogram quantities of these starch encapsulated herbicides. Preliminary data of their evaluation in greenhouse bioassay and soil columns are presented. Preliminary results of starch-clay blends with metolachlor are also discussed.

EXPERIMENTAL METHODS

Chemicals. Starch from CPC International, Englewood Cliffs, NJ and Agsorb 30/60 LVM-GA clay, Oil Dri, Chicago, IL was used. Technical grade atrazine [2-chloro-4-ethyl-amino-6-isopropyl-amino-s-triazine (97%)], technical grade, metolachlor [2-chloro-2'-ethyl-6'-methyl- N-(1-methyl- 2-methoxyethyl) acetanilide (95%)] and Dual 8E (86% metolachlor) were supplied by Ciba Geigy Corp., Greensboro, NC. Technical grade alachlor [2-chloro-2',6'-diethyl-N-(methoxy- methyl)-acetanilide (94.3%] was supplied by Monsanto Co., St. Louis, MO. Alachlor was melted at 50°C before use.

Extrusion and Encapsulation. The extruder used was a ZSK 57 corotating, fully intermeshing twin-screw extruder (Werner and Pfleiderer). The barrel length/screw diameter (L/D) ratio was 30. Starch or a starch-clay blend was fed into barrel section (BS) 1 at the rate of 150 lb/hr (135 lb/hr dry basis). Atrazine as a solid was also fed into BS-1 at the 10% addition level based on starch, while in other formulations Dual or melted alachlor were metered into BS-3. Barrel temperatures were 25°C at BS-1, 75°C at

BS-2 to BS-4, and 95°C at BS-5 to BS-10. The screw at 200 rpm was composed of alternating conveying and kneading block elements in the starch gelatinization zones. The die head assembly was equipped with a die having twenty 5-mm diameter holes. Production rate was 100 kg/hr product with 30% moisture. Products were cut with a die-face cutter at 70% solids at the extruder exit. The extrudate was dried and then ground to the desired particle sizes in a Bauer disc mill. The milled samples (2 to 25%) moisture were sieved to obtain 14-20 and 20-40 mesh products. A ZSK 30 extruder was used for the starch-clay blends.

Percent herbicide encapsulated. The amount of herbicide entrapped in the sieved products was determined from samples (100 g) washed three times with chloroform (300 ml total) to remove absorbed herbicide. Products were analyzed for percent active ingredient (a.i.) via nitrogen analysis for atrazine and by chlorine analysis for metolachlor and alachlor as previously described (23-25).

Swellability. Samples (0.20 g, 14-20 and 20-40 mesh) were placed in a 10-ml graduated cylinder with water (4.0 ml) at 30°C and gently stirred several times during the first three hours to prevent clumping. After 24 h, the height of the swollen product in the cylinder was used to calculate the percent increase in volume.

Soil column leaching studies. Dry-screened (1/2 cm hardware cloth) Miami silt loam top soil was packed into 7.5 cm dia. aluminum tubes. Each tube had a 2.5-cm wide slot, 40-cm long down one side. This slot was covered and sealed with floral clay to produce a water-tight seal. The dry soil was added into the columns in portions. After each portion the column was dropped 4 times onto a rubber stopper from a height of 5 cm. This procedure was repeated until the soil reached 2 cm above the top of the slot. The columns were then saturated with ~750 ml water and allowed to drain. Starch-encapsulated products (20-40 mesh) and commercial emulsifiable concentrate (EC) samples were applied to the top of the columns at the rate of 3.36 kg/ha. Sand (1 cm) was placed over the samples and 375 ml water (equivalent to 7.5 cm rainfall) was passed through the column at a rate of 4 ml/min. A 1-cm head of water was always maintained during the leaching period to prevent channeling. After leaching, the column was placed on its side and the slot was opened. Foxtail seeds (~200) for alachlor and metolachlor and bentgrass for atrazine were sown on the exposed soil and a 0.5-cm layer of sand was added to cover the seeds. The sand was kept slightly moist during the 14-day growing period. The distance from the top of the slot to where growth started determined the depth of leaching.

Herbicide release studies into water. Samples (14-20 mesh) of encapsulated atrazine(25 mg), alachlor (100 mg), or Dual (100 mg) with water (100 ml) were placed in 125-ml Erlenmeyer flasks. The mixtures were agitated in an orbital shaker (Lab-Line, Melrose Park, IL) at 250 shakes per min and sampled at 1, 2, 3, 24 and 48 h. This water volume is in large excess of that required to dissolve all of the herbicides (atrazine solubility is 33 mg/l water at 27°C, alachlor solubility is 240 mg/l at 23°C, and metolachlor solubility is 530 mg/l at 20°C - Herbicide Handbook.   Herbicide concentrations in the water phase were determined spectrophoto- metrically (Beckman DU-50, Irvine, CA) at their ultraviolet absorption maxima (262-264 nm). Corrections were made at these wavelengths using starch controls without herbicide.

RESULTS AND DISCUSSION

Large quantities (750 kg) of starch encapsulated atrazine, alachlor, and metolachlor were prepared on a ZSK 57 extruder. Table I shows that these herbicides were encapsulated effectively

Table I. Scale-up Encapsulation of Atrazine, Alachlor,
and Metolachlor[a]

Active agent	% Water at grinding	Drying method	14-20 Mesh		20-40 Mesh		% swellability in water[c]
			% total active agent	% encapsulation[b]	% total active agent	% encapsulation[b]	
atrazine	4	air	11.3	94	11.1	95	340
alachlor	16.5	air	9.6	80	9.6	66	220
alachlor	7.8	air	9.4.	93	9.5	91	180
alachlor	2.7	oven[d]	10.0	90	9.7	90	180
alachlor	2.3	oven	9.7	95	9.8	89	220
metolachlor	25	air	9.2	50	9.4	34	200
metolachlor	18	air	9.2	41	9.2	29	300
metolachlor	10	oven	9.0	70	9.0	55	180
metolachlor	7	air	8.7	77	8.9	62	180

[a] Starch at 70% concentration, ~ 10% herbicide.
[b] Herbicide remaining after extracting surface material.
[c] Sample (0.2 g) in water (4 ml).
[d] Dried at 50°C, 24 hr.

in the scale-up at a production rate of 70 kg/h. The swellabilities listed in Table I correlate well on how fast the herbicides will be released. Portions of the products were ground at different moisture contents after either air or oven drying. Drying the samples prior to grinding seemed to improve the encapsulation process. Therefore it may be important during continuous, commercial scale processing that the products are dried to about 10% moisture if grinding is necessary.

Metolachlor was encapsulated in various starch-clay blends using a ZSK 30 extruder. Table II shows a slight decrease in encapsulation efficiency as the percent clay increases. However, the product processing improves, surface quality improves, density increases, and the cost decreases as the amount of clay increases.

Table II. Encapsulation of Metolachlor in Starch-Clay Blends[a]

Percent Clay	14-20 Mesh			20-40 Mesh		
	% total active agent	% encapsulation[b]	% swellability in water[c]	% total active agent	% encapsulation[b]	% swellability in water[c]
0	10.8	79	240	10.0	81	280
20	10.6	71	240	10.1	61	280
50	11.1	72	200	11.2	63	200
60	9.9	68	160	9.8	55	200
80	9.7	67	100	11.0	51	100

[a] Starch-clay blend at 65% concentration, ~10% metolachlor.
[b] Metolachlor remaining after extracting surface material.
[c] Sample (0.2g) in water (4 ml).

In Table III the release data in a laboratory assay are reported. The data shows controlled release in a very harsh test. As the water solubility of the active agent increases, the rate of release increases. Initial rate of metolachlor release from the starch-clay matrix increases slightly up to a 50% clay level.

Table III. Rate of Release of Encapsulated Herbicides into Water[a]

Product encapsulated	% Released				
	1 hr	2 hr	3 hr	24 hr	48hr
atrazine	9	17	22	56	67
alachlor	31	46	56	80	91
metolachlor	37	49	61	91	97
0% Clay	29	39	50	80	85
20% Clay	32	42	55	91	98
50% Clay	40	51	58	93	100
60% Clay	40	48	58	93	100
80% Clay	27	40	51	100	

[a] 14-20 mesh unwashed samples (25 or 100 mg) in 100 ml of water swirled at 250 rpm.

The starch encapsulated herbicides were compared with the commercial EC formulations in soil column tests using 7.5 cm simulated rainfall. The results in Table IV show a significantly reduced movement of active agent from encapsulated products. All the encapsulated samples were unwashed so the surface herbicide was still present.

Table IV. Soil Column Leaching of Herbicides from Starch
and Commercial EC Formulations[a]

Herbicide formulation	Depth, cm
atrazine	
starch	6
EC	33
alachlor	
starch	5
EC	24
metolachlor	
starch	5
EC	27

[a] Starch (20-40 mesh) and EC formulations at 3.36 kg/ha and
7.5 cm rainfall.

CONCLUSIONS

Using a twin-screw extruder for encapsulating herbicides in starch
and starch–clay matricies is efficient, effective and continuous.
Laboratory evaluation of the products shows them to be slow
release, effective in controlling weeds, and able to reduce
leaching under controlled conditions.

*The mention of firm names or trade products does not imply that
they are endorsed or recommended by the U.S. Department of
Agriculture over other firms or similar products not mentioned.

LITERATURE CITED

1. Wing, R. E.; Otey, F. H. *J. Polym. Sci., Polym. Lett. Ed.*; **1983**; *21*(1):121-140.
2. Shasha, B. S.; Trimnell, D.; Otey, F. H. *J. Polym. Sci., Polym. Chem. Ed.*, **1981**; *19*(11):1891-1899.
3. Trimnell, D.; Shasha, B. S.; Wing, R. E.; Otey, F. H. *J. Appl. Polym. Sci.*, **1982**; *27*(11):3919-3928.
4. Wing, R. E.; Maiti, S.; Doane, W. M. *J. Contr. Rel.*, **1987**; *5*(7):79-89.
5. Reily, R. T. *J. Ag. Food Chem.*, **1983**; *31*(2):202-206.
6. Schreiber, M. M.; Shasha, B. S.; Ross, M. A.; Orwick, P. L.; Edgecomb, Jr., D. W. *Weed Sci.*, **1978**; *26*(6):679-686.
7. Devisetty, B. N.; McCormick, C. L.; Shasha, B. S. *Proc. 7th Inter. Sym. Contr. Rel. Bioact. Mat.*, 1980; 187-188.
8. White, M. D.; Schreiber, M. M. *Weed Sci.*, **1984**; *32*(4):387-392.
9. Coffman, C. B.; Genter, W. A. *Ind. J. Ag. Sci.*, **1984**; *54*(2):117-122.
10. Wing, R. E.; Maiti, S.; Doane, W. M. *Starch*, **1987**; *39*(12):422-425.
11. Wing, R. E.; Maiti, S.; Doane, W. M. *J. Contr. Rel.*, **1988**; *7*(1):33-37.

12. Wing, R. E. *Proc. Corn Util. Conf.*, 2nd., 1988; 1-17.
13. Wing, R. E. *Proc. 16th Inter. Symp. Contr. Rel. Bioact. Mat.*, 1989; 430-431.
14. Doane, W. M.; Maiti, S.; Wing, R. E. U. S. Patent 4,911,952 1990.
15. Schreiber, M. M.; Wing, R. E.; Shasha, B. S.; White, M. D. *Proc. 15th Inter. Symp. Contr. Rel. Bioact. Mat.*, 1988; 223-224.
16. Schreiber, M. M.; White, M. D.; Wing, R. E.; Trimnell, D.; Shasha, B. S. *J. Contr. Rel.*, **1988**; *7*(3):237-242.
17. Schoppet, M. J.; Gish, T. J.; Helling, C. S. *Proc. Am. Soc. Ag. Eng.*, 1989; 1-13.
18. Mills, M. S.; Thurman, E. M.; Wing, R. E.; Barnes, P. L. *Trans. Am. Geophy. Un.*, EOS, **1990**; *71*(43):1331.
19. Mills M. S.; Thurman, E. M. *Environ. Sci. Technol.* (In Press).
20. Gish, T. J.; Shoppet, M. J.; Helling, C. S.; Shirmohammadi, A.; Schreiber, M. M.; Wing, R. E. *Trans. Am. Soc. Agric. Engin.*, **1991**; *34*(4):1738-1744.
21. Gish, T. J.; Schoppet, M. J.; Weinhold, B. J.; Helling, C. S.; Wing, R. E.; Schreiber, M. M. *Weed Sci.* (In Press).
22. Fleming, G. F.; Simmons, F. W.; Wax, L. M.; Wing, R. E.; Carr, M.E. *Weed. Sci.* (In Press).
23. Carr, M. E.; Wing, R. E.; Doane, W. M. *Cereal Chem.*, **1991**; *68*(3):262-266.
24. Wing, R. E.; Carr, M. E.; Trimnell, D.; Doane, W. M. *J. Contr. Rel.*, **1991**; *16*(30):267-278.
25. Trimnell, D.; Wing, R. E.; Carr, M. E.; Doane, W. M. *Starch*, **1991**; *43*(4):146-151.

RECEIVED October 1, 1992

Chapter 15

Transport Studies of Oil-Soluble Polymers

Brian A. Harvey[1], Thelma M. Herrington[1], and Rodney Bee[2]

[1]Department of Chemistry, The University of Reading RG6 2AD,
United Kingdom
[2]Unilever Research, Colworth Laboratory, Sharnbrook, Bedford MK44 1LQ,
United Kingdom

In connection with our current studies of the factors affecting
the stability of water/oil/water multiple emulsions for the food
industry, we have carried out some fundamental studies of
solute transfer across an oil membrane by polymeric molecules.
The overall rates of transfer of a series of carboxylic acids
across a polar oil (glycerol triester) was determined by using a
rotating diffusion cell. The polymers used were a polyglycerol
fatty acid ester and an ABA block copolymer of poly(12-
hydroxy)stearic acid with ethylene oxide. The effect of other oil
soluble surfactants and of electrolyte in the aqueous phase on
the rates of transfer was also studied. By measuring the
diffusion coefficients and partition coefficient of the solute it is
possible to differentiate between the contributions to the
transport from the transfer at the oil/water interface and that
from diffusion in the oil phase.

In a multiple emulsion two liquid phases are separated by another immiscible
liquid layer. This type of system was first discovered by Seifriz (1) in studies
of the phase inversion of ordinary emulsions. Since the sixties many potential
applications have been suggested for water-in-oil-in-water (W/O/W) multiple
emulsions. In the pharmaceutical industry, Engel (2) tried to improve the
efficiency of intestinal adsorption by immobilizing insulin in the inner aqueous
phase of a multiple emulsion and several workers have formulated recipes for
prolonged drug delivery systems (3-5), or suggested usage as solvent reservoirs
in the treatment of drug overdose cases (6, 7). At Exxon progress has been
made in the application of multiple emulsions for liquid membrane extraction
processes (8). In the food industry it has potential use in the manufacture of
low fat food products, such as mayonnaise, and for the controlled release of
nutrients for special dietary requirements. However, more effort is required

0097–6156/93/0520–0220$06.00/0
© 1993 American Chemical Society

to obtain fundamental information on the factors affecting the dispersion state and stability of multiple emulsions. Matsumoto et al (9) attempted to clarify the efficiency of the hydrophobic and hydrophillic emulsifiers used in a 2 step procedure of emulsifying a water-in-oil emulsion in an aqueous surfactant solution. Florence and Whitehill (10) studied the kinetics of breakdown and suggested possible mechanisms. Tomita et al (11) reported that the changes in phase volume of the aqueous compartments in W/O/W emulsions are not only brought about by the migration of water under the osmotic pressure gradient, but are partially caused by the permeation of solutes, such as glucose, across the oil layer.

In order to determine how solute transport affects the stability of W/O/W multiple emulsions, a model system, with planar interfaces, was chosen. In earlier experiments in our laboratory (Harvey, B., Reading University, UK, unpublished data), using a bulk W/O/W emulsion made from food components, studies of solute transport were not reproducible. This multiple emulsion consisted of three phases: an internal aqueous phase with droplet diameters of 1 μm; an oil phase of groundnut oil containing polyglycerol polyricinoleate (5 wt%); a continuous outer aqueous phase containing sodium caseinate (2 wt%). The oil phase is predominantly a mixture of triglycerides containing the oil-soluble surfactant. The multiple emulsion was stable for 2 to 3 weeks, but the stability could be enhanced by adding electrolyte to the aqueous phase. The properties were affected by the relative volumes of the 3 phases, the size and distribution of the internal water and oil droplets, the concentration of the secondary surfactant, the osmotic properties of the aqueous phase and the viscosity. Studies of solute transfer from the inner to outer aqueous phases were carried out, but these met with little success, as the properties are interdependent and altering one parameter changed another. It is also difficult to separate the phases for analysis.

The W/O/W system with planar interfaces was modelled using a Rotating Diffusion Cell in which the planar oil-water interfaces are supported on a membrane filter. In order to determine the resistance to interfacial transfer for a solute, it is necessary to measure diffusion coefficients in both the oil and aqueous phases and the oil/water partition coefficients.

Experimental

Materials. Analytical Grade reagents were used throughout unless otherwise stated. The triglyceride esters used were glycerol trioleate (99%) and a commercial triglyceride ester (groundnut oil) The polymeric surfactants used were: polyglycerol polyricinoleate; an ABA block copolymer of poly(12-hydroxystearic) acid/ethylene oxide - number average molar mass 3543 $\pm$ 30 g mol^{-1} (obtained by vapour pressure osmometry) (polyethylene oxide chain 1500 g mol^{-1}) (12). The monomeric surfactants used were sucrose tristearate, sorbitan trioleate, sorbitan monooleate and glycerol monooleate. Double-distilled-deionised water was used throughout and carbon dioxide free nitrogen.

The Rotating Diffusion Cell. This is designed hydrodynamically so that stationary diffusion layers of known thickness are created each side of the oil layer *(13)*. The cell is shown in Figure 1. The oil layer is supported on a porous membrane filter which divides the apparatus into 2 compartments, separating the inner and outer aqueous phases. The cell is mounted in a thermostatted glass jacket. The central assembly is rotated by a stepper motor at constant known speeds up to 6 Hz. The fixed slotted baffle, positioned a short distance above the filter, ensures a stationary diffusion layer. The 60 μm teflon filter, of average pore size 0.2 μm, is located by screwing the stainless steel filter holder against the steel plate attached to the perspex cylinder, leaving an exposed circular area 2 cm in diameter. To measure the flux of an organic acid, the inner compartment, within the perspex cylinder, was filled with the acid and the outer compartment was filled to the same level with 5 x 10^{-3} mol dm^{-3} NaCl solution. Care must be taken to ensure that no air bubbles are trapped beneath the filter and precautions were taken to remove and exclude CO_2 from the solutions by working under nitrogen.

The flux of acid passing through the oil membrane was monitored by maintaining the pH in the outer compartment at 7.0 by means of a combined electrode and pH meter attached to a Dosimat automatic titrator which injects carbon dioxide-free 0.1 mol dm^{-3} NaOH solution.

The Stokes Diffusion Cell. A modified Stokes Diffusion Cell was used to determine the diffusion coefficients, D_{aq} and D_o, of the aqueous and organic phases respectively *(14)*. This is shown in Figure 2. The stirrer and follower closely sweep both sides of a porous glass sinter at a fixed rotation speed of 2 Hz to prevent stagnant diffusion layers forming. The nylon lid has holes drilled for a pH-electrode, nitrogen supply and alkali delivery tube. To determine D_{aq}, the lower compartment and the sinter are filled with the acid, whose diffusion coefficient is to be measured, the upper compartment is filled with 5 x 10^{-3} mol dm^{-3} KCl and the flux is measured with the pH-stat. For D_o, the lower compartment and the sinter are filled with a solution of the acid in oil and the upper compartment is filled with the oil. The upper compartment is sampled periodically and the concentration of the acid found by partitioning against water; the partition coefficient/concentration relationships required were determined using oscillated flasks mounted in an air thermostat.

Theory

Rotating disc hydrodynamics are created within the cell *(15,16)*. A stagnant layer is created on each side of the filter, whose thickness Z is given by the Levich equation:

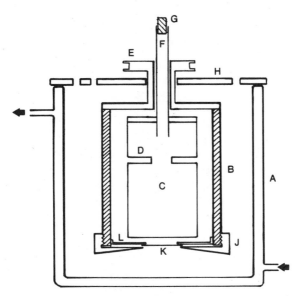

Figure 1. The Rotating Diffusion Cell.

A	Thermostatted outer jacket	G	Rubber bung
B	Perspex cylinder	H	Lid
C	Teflon baffle	J	Stainless steel filter holder
D	Slots	K	Membrane filter
E	Pulley	L	Stainless steel plate
F	Stainless steel filling tube		

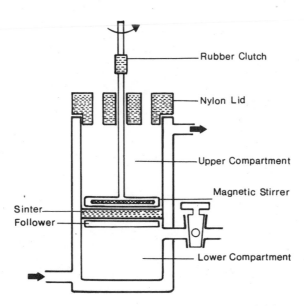

Figure 2. Cell for measurement of diffusion coefficients.

$$Z = \beta\eta^{1/6} D_{aq}^{1/3} \omega^{-1/2} \tag{1}$$

where η is the kinematic viscosity, D_{aq} is the diffusion coefficient in aqueous solution, ω is the rotation speed, β is a constant. The rate of transfer of the diffusing solute species from the inner to the outer compartment is given by

$$J = kAc \tag{2}$$

where A is the area of the filter, c is the bulk concentration of the solute in the inner compartment, k is given by

$$\frac{1}{k} = \frac{2Z}{D_{aq}} + \frac{2}{\alpha k_{-1}} + \frac{Kl}{\alpha D_o} \tag{3}$$

The significance of the 3 terms on the right hand side of equation 3 are:

1    $2Z/D_{aq}$ describes diffusion through the 2 stagnant aqueous diffusion layers of thickness Z;

2    $2/\alpha k_{-1}$ is the contribution from interfacial transfer at the 2 oil-water boundaries; α is the cross-sectional area of the pores divided by the filter area and k_{-1} is the rate constant for the species M defined by

$$M(org) \underset{k_{-1}}{\overset{k_1}{\rightleftharpoons}} M(aq) \qquad K = \frac{k_1}{k_{-1}} \tag{4}$$

3    $Kl/\alpha D_o$ is the contribution from the diffusion of the solute through the oil phase in the filter; K is the partition coefficient, l is the filter thickness and D_o the diffusion coefficient in the oil phase.
    By substituting equation 1 in equation 3, equation 5 is obtained

$$\frac{1}{k} = 2\beta\eta^{1/6} D_{aq}^{-2/3} \omega^{-1/2} + \frac{2}{\alpha k_{-1}} + \frac{Kl}{\alpha D_o} \tag{5}$$

and it follows that a plot of $1/k$ against $\omega^{-1/2}$, a Levich plot, has an intercept of:

$$2/\alpha k_{-1} + Kl/\alpha D_o.$$
Interfacial transfer + Diffusion in the oil phase

A typical value for Z, the thickness of the stagnant layer, is 30 to 40 μm. The results were fitted to a Levich plot with a theoretical slope given by equation 5. To obtain the rate constant for transfer at the oil/water interface, k_{-1}, it is necessary to determine the diffusion coefficient, D_o in the oil phase and the partition coefficient, K. Also the first term of the intercept must be significant compared with the second.

Results

There has been no systematic literature study of the interfacial transfer of small molecules. The effect of the length of the alkyl chain of the alkanoic acids and of the temperature on the rate of transfer were studied. The rotating diffusion cell was set running for 1 hour at a low speed to equilibrate. The speed was then successively increased in 4 steps from 1.2 to 4 Hz, maintaining each speed for 30 minutes, to obtain the Levich plot for each system. Kinetic data were obtained at 25.00 $\pm$ 0.01 °C. The effect of various additives, polymers and surfactants, dissolved in the membrane bound oil phase on the rate of transfer of different solutes, of varying degrees of polarity, was investigated.

For the oil phase of triglyceride ester containing 5% by weight of polyglycerol polyricinoleate the Levich plot is shown in Figure 3. The results show that there is a progressive increase in transfer rate with increasing chain length of the acid. The results are reproducible: for example for 4 runs with valeric acid the error in the intercept is $\pm$ 2%. Using the pure triglyceride ester, glycerol trioleate, as the oil phase increased the rate of transport by 5%.

The effect of changing the ionic strength of the aqueous phase was studied. The Levich plots for the transport of valeric acid through the same oil phase but increasing the concentration of electrolyte in the aqueous phase from zero to 0.4 mol dm^{-3} is shown in Figure 4. The effect of temperature on the transport of valeric acid is shown in Figure 5 over the temperature range 25 to 45 °C.

The effect of the presence of dissolved polymer in the oil phase was investigated. It was found that a 5 wt% concentration of polyglycerol polyricinoleate slowed down the transport of valeric acid (Figure 6), whereas a 5 wt% concentration of the block copolymer increased the rate. The effect of several nonionic surfactants, dissolved in the oil phase, was determined. Glycerol monooleate, 5 wt%, decreases the rate of transport of valeric acid through the triglyceride ester (Figure 6). A temperature of 45 °C was used to enhance the solubility of the surfactants; it was found that sucrose tristearate (1 wt%) decreased the rate of transport of valeric acid, whereas sorbitan trioleate (1 wt%) and sorbitan monooleate (1 wt%) increased it, the latter quite significantly (Figure 7).

The transport of acetic and formic acids through the triglyceride oil membrane was very slight and not reproducible, also the Levich plots were not straight lines. The triglyceride ester oil phase was found to be impermeable to the following polar solutes: sorbitol; glucose; raffinose; glycine; leucine; sodium chloride; potassium chloride; barium chloride; calcium chloride and caesium chloride. It was also impermeable to the following acids which are allowed in foodstuffs: lactic; malic; tartaric; citric; fumaric; ascorbic; sorbic and benzoic.

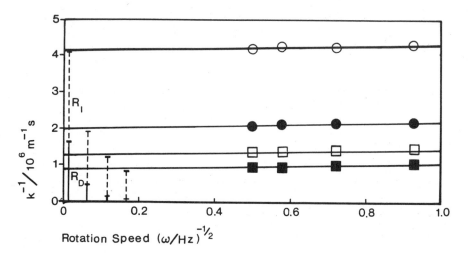

Figure 3. Levich Plot of n-alkyl carboxylic acids through triglyceride ester containing 5 wt% polyglycerol polyricinoleate at 25 °C: ○ Propanoic acid; ● Butyric acid; □ Valeric acid; ■ Caproic acid.

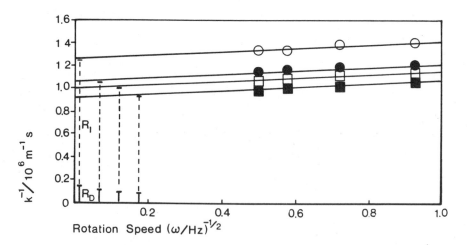

Figure 4. Levich Plot of valeric acid through triglyceride ester containing 5 wt% polyglycerol polyricinoleate, as a function of the NaCl concentration, at 25 °C: ○ no NaCl; ● 0.1 M NaCl; □ 0.2 M NaCl; ■ 0.4 M NaCl.

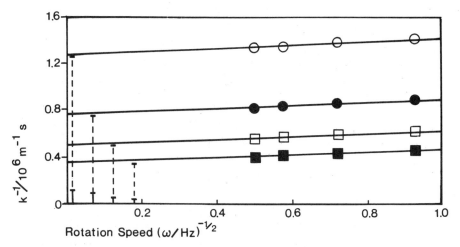

Figure 5. Levich Plot of valeric acid through triglyceride ester containing 5 wt% polyglycerol polyricinoleate, as a function of the temperature: ○ 25.0 °C ; ● 30.0 °C; □ 37.0 °C; ■ 45.0 °C.

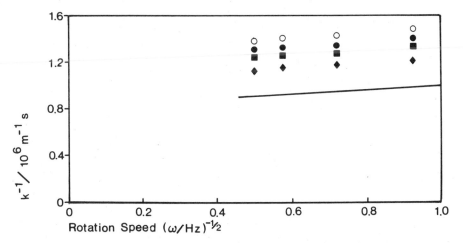

Figure 6. Levich Plot of valeric acid through triglyceride ester containing different surfactants at 25 °C: ○ 5 wt% polyglyceryl polyricinoleate; ● 5 wt% glycerol monooleate; ■ no surfactant; ♦ 5 wt% block copolymer; ——— Levich slope.

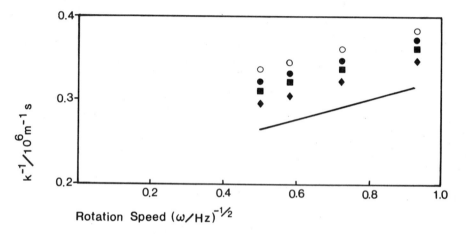

Figure 7. Levich Plot of valeric acid through triglyceride ester containing different surfactants at 45 °C: ○ 1% sucrose tristearate; ● no surfactant; ■ 1 wt% sorbitan trioleate; ♦ 1 wt% sorbitan monooleate; ——— Levich slope.

Discussion

The intercept of the Levich plot, $2/\alpha k_{-1} + Kl/\alpha D_o$, gives the contribution of the transport at the oil-aqueous interfacial barrier (first term) and the transport by diffusion through the oil phase (second term). It is convenient to consider these terms as the resistance to transport of the acid through the cell. Thus the first term is R_I, the resistance due to interfacial transfer, and the second term is R_D, the resistance due to diffusion. The diffusional resistance term can be calculated from our experimental measurements of the Distribution Coefficient, K, and the diffusion coefficient in the oil phase, D_o. The resistance caused by transfer at the oil-aqueous interface, R_I, is then obtained as the difference between the experimental Levich intercept and R_D.

The overall resistance to transport decreases with increasing chain length, for the triglyceride oil with polyglycerol polyricinoleate (5 wt%). The diffusional resistance decreases with increasing length of the alkyl chain, so that the interfacial resistance also decreases from butyric to caproic acid (Table I). Both these trends are consistent with the increasing nonpolar nature of the acids.

Table I. The contributions of transfer at the oil/water interface and of diffusion through the oil phase, to the rate of transport of the n-alkyl carboxylic acids through the oil phase of triglyceride ester containing 5 wt% of polyglycerol polyricinoleate

Acid	$R_D/10^5$ m^{-1} s	$R_I/10^5$ m^{-1} s
Propionic	15.1	26.6
Butyric	4.6	17.4
Valeric	1.4	11.4
Caproic	0.6	8.4
Valeric Acid		
No NaCl	1.42	11.4
0.1 M NaCl	1.39	9.3
0.2 M NaCl	1.36	8.8
0.4 M NaCl	1.30	8.1
Valeric Acid		
25 °C	1.42	11.4
30 °C	1.02	6.9
37 °C	0.70	4.5
45 °C	0.46	3.2

The effect of increasing the ionic strength of the aqueous phase (Figure 4) shows that the resistance to overall transport, given by the intercept of the Levich plot, decreases with increasing sodium chloride concentration. From Table I it can be seen that the diffusional contribution to the transport only changes very slightly with electrolyte concentration. Thus it is the interfacial contribution to the transport which is increased by increasing the concentration of electrolyte. As shown by the intercepts of the Levich plot of Figure 5, increasing the temperature has a large effect on the transport of valeric acid. Both the overall resistance to transport and the diffusional resistance decrease with increasing temperature (Table I).

That transport of valeric acid is slowed down by the addition of polyglycerol polyricinoleate to the oil phase (Figure 6) is probably caused by the increased viscosity of the oil. The transport of the acid is facilitated by the block copolymer (Figure 6), possibly by reverse micelles acting as carriers although the effect is very slight. Of the monomeric surfactants (Figure 6 and 7), only sorbitan trioleate markedly facilitated the transport of the valeric acid. Again reverse micelles may be acting as carriers.

Conclusions

The rotating diffusion cell provides an exact method for determining the rates of transport of molecules from one aqueous phase to another through an oil membrane. The rate constant for transport increased, with increasing number of carbon atoms in the alkyl chain, for the 4 straight chain alkyl carboxylic acids, propanoic to caproic. Solute molecules that were highly polar did not pass through the membrane. The effect on the rate of transfer of several oil soluble polymeric and monomeric surfactants was studied but no evidence for facilitated transport was found.

Acknowledgments

We thank both SERC and Unilever for financial assistance towards this research.

Literature Cited

1   Seifriz, W. *J. Phys. Chem.*, **1925**, *29*, 738.
2   Engel, R. H.; Riggi, S. J.; Fahrenbach, M. J. *Nature*, **1968**, *219*, 856.
3   Taylor, P. J.; Miller, C. L.; Pollock, T. M.; Perkins F. T.;
     Westwood, M. A. *J. Hygienics Cambridge*, **1969**, *67*, 485.
4   Benoy, C. H.; Elson, L. A.; Schneider, R. *Br. J. Pharmacol.*, **1972**, *45*, 135.
5   Whitehill, D. *Chemist Druggist*, **1980**, *213*, 130.
6   Frankenfeld, J. W.; Fuller, G. C.; Rhodes, C. T. *Drug Dev. Commun.*,
     1976, *2*, 405.
7   Chiang, C.; Fuller, G. C.; Frankenfeld, J. W. *J. Pharm. Sci.*, **1978**, *67*, 63.

8   Li, N. N.; Shrier, A. L. In *Recent Developments in Separation Science;*
     Li, N. N., Ed.; Chemical Rubber Co: Cleveland, Ohio, 1972;
     Vol.1; pp 163-181.

9   Matsumoto, S.; Kita, Y.; Yonezawa, D. *J. Coll. Interface Sci.,*
     1976, *57,* 353.

10   Florence, A. T.; Whitehill, D. *J. Coll. Interface Sci.,* **1981,** *79,* 243.

11   Tomita, M.; Abe, Y.; Kondo, T. *J. Pharm. Sci.,* **1982,** *71,* 268.

12   Aston, M. S.; Bowden, C. J.; Herrington, T. M.; Sahi, S. S.
     J. Am. Oil Chem. Soc., **1985,** *62,* 1705.

13   Riddiford, A. C. *Adv. Electrochem and Electrochem Eng.,* **1966,** *4,* 47.

14   Stokes, R. H. *J. Am. Chem. Soc.,* **1950,** *72,* 763.

15   Levich, V. G. *Physicochemical Hydrodynamics;*
     Prentice Hall: Englewood Cliffs, New Jersey, 1962; pp 60-72.

16   Albery, W. J.; Couper, A. M.; Hadgraft, J. and Ryan, C. *J. Chem. Soc.,*
     Farad. Trans. 1, **1974,** *70,* 1124.

RECEIVED October 1, 1992

Chapter 16

Temperature-Compensating Films for Modified Atmosphere Packaging of Fresh Produce

Ray F. Stewart[1], Judy M. Mohr[1], Elizabeth A. Budd[1], Loc X. Phan[1], and Joseph Arul[2]

[1]Landec Corporation, 3603 Haven Avenue, Menlo Park, CA 94025
[2]Department of Food Science, Laval University, University City, Quebec G1K 7P4, Canada

Temperature responsive gas permeable films based on side chain crystallizable polymers have been developed that can match or exceed the increasing respiration rates of fresh produce subjected to temperature fluctuations. The development of the films is discussed and experimental results of test produce packages are compared with computer simulations. There is potential to use this film to compensate for temperature fluctuations in the cold chain of storage and distribution.

Fruits and vegetables are perishable produce which actively metabolize during the postharvest phase. The use of low temperature is probably the most important means of extending the storage life of postharvest produce. Modifying the gas atmosphere inside an enclosure containing the produce can also decrease the respiration rate and thus further extend the storage life. Altered gas atmospheres can be passively created and maintained by flexible film packaging, a process known as Modified Atmosphere Packaging (MAP).

In a sealed package containing a produce with a permeable film, a modified atmosphere is created naturally as a result of dynamic molecular balance between the respiratory process (O_2 uptake and CO_2 external production) and the exchange of gases with the atmosphere through the film. Eventually, steady-state concentrations of O_2, CO_2 and the third gas N_2 are established passively at a given temperature when the rates of O_2 consumption and CO_2 production of the produce equal the rates of permeation of these gases through the package. The magnitude of the CO_2 increase and O_2 decrease at steady-state is dependent on the gas flux through the film and its CO_2/O_2 selectivity.

In order to obtain the maximum benefit from MAP, the steady-state gas concentrations should correspond to the storage optima for a given crop. On the other hand, when CO_2 accumulates above the tolerance limit of the crop, it can cause injury to the crops. Lowering O_2 levels below the critical levels

0097–6156/93/0520–0232$06.00/0
© 1993 American Chemical Society

may result in anaerobic conditions, with the development of off-flavors and undesirable texture changes. At extremely low O_2 levels, toxin production by anaerobic pathogenic organisms can also occur (*1,2,3*). A number of researchers have developed design methodologies to aid the construction of MA packages suitable for a specific crop assuming that the packages would be constantly maintained at the optimal storage temperature of the produce (*4,5,6,7,8, Exama, A., Laval University, personal communication, 1992*).

The practical application and the commercial impact of MAP has thus far been quite limited, in spite of its beneficial affects on the quality and the longevity of fresh crops. This is because some of the serious practical limitations of MAP to ensure the safety of the produce remain unresolved. The primary hurdle relates to the low gas flux and/or improper selectivity of the commercially available films to create and maintain optimal MA for many crops with a few exceptions such as apple and green pepper (*Exama, A., Laval University, personal communication, 1992*). The second problem arises from the condensation of water on the packaged produce which favors fungal growth, as well as on the films which could add resistance to gases through the film.

Even if a films could be developed to meet the required package permeabilities at the optimum storage temperature, as well as prevent the condensation of water vapor prevented, MAP should still prove safe against the temperature fluctuations encountered in the cold chain of storage and distribution. The temperature may fluctuate from the expected storage temperature up to ambient temperatures at which produce packages are often displayed. This has two implications: (1) creation of anoxic atmospheres inside the package at the higher temperature; (2) aggravation of the condensation of the saturated vapor.

Both respiration of produce and the film permeability are dependent on temperature. While the values of respiration rate coefficient (R_{10} value) of most crops are generally between 2.0 to 3.0, the permeability coefficient (P_{10}) of most films ranges between 1.0 and 2.0 for the same temperature range (*Exama, A., Laval University, personal communication, 1992*). This disparity could lead to excessive accumulation of CO_2 and/or depletion of O_2 at higher temperatures, a situation which must be avoided. This problem can be avoided by using films possessing appropriate P_{10} values matching the R_{10} value of the produce. With regard to this possibility, it would seem to be a daunting task to develop a film that would possess the required O_2 and CO_2 permeabilities, in addition to appropriate P_{10} values. The second possibility could involve a temperature sensitive safety value - a value of a membrane - that would open at elevated temperatures and reclose when the desired temperature is re-established (*Exama, A., Laval University, personal communication, 1992*).

Work conducted at Landec Corporation in the area of Intelimer (a registered trademark of Landec Corporation) temperature responsive polymers has led to the development of semipermeable membranes that exhibit large changes in gas permeation in response to small temperature changes. Intelimers belong to a family of materials known as Side-Chain-Crystallizable (SCC) polymers which exhibit abrupt thermal transitions associated with the side chain rather than the polymer backbone. This unique feature allows membrane materials to be designed that exhibit tailored permeation properties over a range of temperatures useful in food packaging.

The general structure of Intelimer polymer is as follow:

```
        R
        |            Where:   R = H, CH₃
   -(CH₂-CH)-                 Z = oxygen, carbonyl, ester, amide
        |                     x = 11 to 21
       (Z)
        |
      (CH₂)x
        |
       CH₃
```

The change in the permeation property mentioned above is associated with the melting point (Tm) of the side chains. That is, at temperatures below the Tm the polymer is not very permeable, but at temperatures above the Tm, the permeability increases dramatically. The melt temperature of the side chains can be varied by systematically changing their chain lengths. The melting point of the polymer changes directly with respect to the change in the chain lengths. Specific compositions and uses of this class of polymer is the subject of issued and pending patents (U.S. Patent No. 4,830,855; 5,129,180 and 5,129,349) assigned to Landec Corporation.

A development effort was made at Landec Corporation in collaboration with Laval University to determine the feasibility of using these materials for compensation strategy against temperature abuse for MAP.

Experimental Procedures and Results

Polymer Preparation. Side chain crystallizable polymers exhibiting melting transitions in the range of 4°C to 20°C were prepared by radical polymerization of n-alkyl acrylates having average side chain lengths of 12 to 14 carbon atoms. The polymers were purified by precipitation into cold ethanol and isolated. Reported thermal transitions were measured via DSC at 10°C/min.

Film Preparation and Permeation Measurement. Intelimer films were prepared as 0.001" thick films and laminated to a highly permeable microporous polyolefin for permeation testing. Pure gas O_2 and CO_2 permeability values were measured on an automated system constructed in our laboratories. All films were tested in quadruplicate, first for O_2 permeation as a function of temperature and followed by CO_2 permeability measurements. The P_{10} values of each film for O_2 and CO_2 were calculated by comparing the permeabilities over a 10°C range below and above the side chain transition. The CO_2/O_2 selectivities were calculated above and below the transition temperature.

A summary of the permeability behavior for a selected series of Intelimer films is given in Table I. It was found that the permeability could be varied over a range of approximately 3,000 to 21,000 cc $\cdot$ mil/m^2 $\cdot$ day $\cdot$ atm. Films with permeabilities in the upper end of this range offer intriguing possibilities for the packaging of rapidly respiring produce such as mushrooms and broccoli. The gas selectivity of the films ranged from approximately 3 to 7, comparable with existing films.

Table I. Permeation Characteristics for Intelimer Films

Film	Transition (°C)	CO$_2$/O$_2$ Selectivity below Tt[a]	above Tt[a]	P$_{10}$ Values CO$_2$	O$_2$	O$_2$ Permeability[b] (below Tt[a])
LL47-183	4	5	6	6	4	5,500
LL47-181	9	5	6	3	3	8,400
LL47-185	16	4	4	5	7	5,250
LL47-177	16	6	7	5	6	21,000
LL47-165	19	7	5	6	8	3,300

[a]Tt is an abbreviation for transition temperature
[b]Units: cc $\cdot$ mil/m^2 $\cdot$ day $\cdot$ atm

Evaluation of Film Suitability for Produce Packaging. The ideal packaging film would have a P$_{10}$ value that closely matches the R$_{10}$ value of the produce to be packaged so that the gas concentrations in the package would remain in balance even under variable temperature conditions. By comparing the selectivities and P$_{10}$ values of the films with reported values of optimum storage conditions and R$_{10}$ values for various produce, one can estimate the suitability of these films for use in specific packaging applications.

Figure 1 shows the effect of temperature on the respiration rate of cabbage, the carbon dioxide permeability of one of the films tested and comparative data for a film used commercially, Cryovac SSD-310 (Cryovac). The Cryovac film is a multi-ply, co-extruded polyolefin film, often used in food packaging. Based on the permeation rate data for these films it would appear that the Intelimer film would be an excellent fit with cabbage.

Computer Modeling of Food Packages. Cabbage was selected for initial evaluation and testing because it is an important commodity, would benefit from MAP, is not chilling sensitive, and is available year round. There are reliable data in the literature on the respiration response of cabbage to different temperatures and atmospheres.

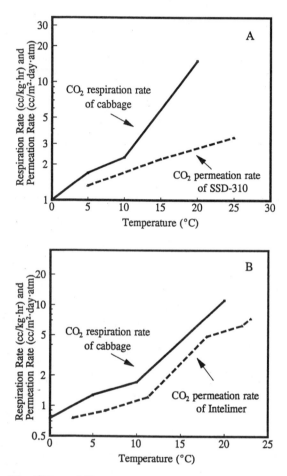

Figure 1. The Effect of Temperature on the Respiration Rate of Shredded Cabbage on (A) the Commercial Film Cryovac SSD-310 and on (B) the Intelimer Film LL47-173.

An iterative computer model developed for purposes of package simulation by a team at the University of California, Davis was used. The model takes into account the produce respiration properties as a function of temperature and atmosphere conditions, film permeability and headspace in the package, and calculates the O_2 and CO_2 concentrations within the package as a function of time until equilibrium is established.

Simulations were preformed for three different films in order to compare their performance as packaging materials for shredded cabbage. Two films were temperature responsive Intelimer films and a third was a commercial Cryovac film, SSD-310. Simulations were done at 2.5, 5, 7.5, 10 and 20°C, a range of temperatures typically encountered in the distribution chain for fresh produce. Results of the simulations for the Intelimer film LL47-177 and for the Cryovac film are presented in Figures 2 and 3. The commercial film is suitable at temperatures up to approximately 10°C, but at 20°C anaerobic conditions result after a short period of time. An identical package constructed with the temperature responsive Intelimer film, in contrast, maintains suitable conditions at all temperatures tested clearly showing an ability to respond to significant temperature changes.

Construction and Evaluation of Prototype Packages. Two types of tests were performed on packages that were prepared utilizing the Intelimer film and the Cryovac film: an isothermal test and a temperature step experiment.

Isothermal Experiment. In the isothermal test, four identical packages were constructed: two utilizing Cryovac and two utilizing the Intelimer film. Each package had 324 cm^2 of film surface area. The amount of shredded cabbage, W to be loaded into the packages was determined from the following equation:

$$\frac{Po_2 \times A}{l} = \frac{RR \times W}{(O_2 - O_{2_{PKG}})} \tag{1}$$

Where PO_2 is the oxygen permeability of the package film, A is the film surface area, L is the film thickness, RR is the respiration rate of the cabbage, and the quantity $(O_2 - O_{2pkg})$ is the difference between atmospheric oxygen concentration and the oxygen concentration inside the package. In this manner each package was balanced for the relative permeability of the films at the design temperature of 10°C. Comparative tests were then conducted isothermally at 10°C and 20°C and the gas concentrations inside the packages were monitored for up to 200 hours.

The gas concentrations inside the packages versus time are shown in Figures 4 and 5. As expected, when maintained at either 10°C or 20°C both packages show an initial rapid drop in O_2 and an increase in CO_2. This response is expected because freshly cut cabbage respires rapidly. As the

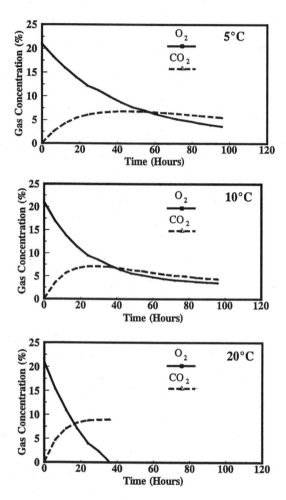

Figure 2. Computer Simulated Changes at Three Temperatures in a Package Constructed with Cryovac Film SSD-310 and Filled with Shredded Cabbage.

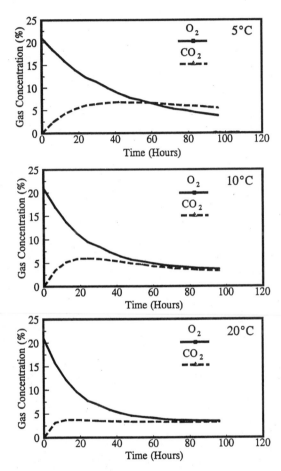

Figure 3. Computer Simulated Changes at Three Temperatures in a Package Constructed with the Intelimer Film LL47-177 and Filled with Shredded Cabbage.

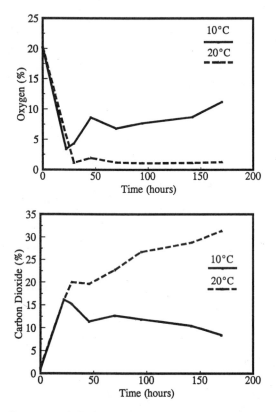

Figure 4. Oxygen and Carbon Dioxide Composition Inside Packages Constructed from Cryovac Film SSD-310 and Filled with Shredded Cabbage and Stored at 10°C and 20°C.

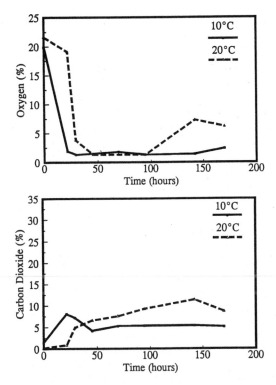

Figure 5. Oxygen and Carbon Dioxide Composition Inside Packages Constructed from the Intelimer Film LL47-177 and Filled with Shredded Cabbage and stored at 10°C and 20°C.

cabbage recovers from the cutting injury and the atmosphere is modified, respiration rates decrease and the packages begin to reach equilibrium conditions.

In the case of the Cryovac film package, it can be seen that at 10°C the package would equilibrate at about 8% O_2 and 10% CO_2. At 20°C however, a very different result is obtained. The O_2 concentration drops to 1% while the CO_2 concentration rises initially to 20% and then continues to increase throughout the experiment. These conditions are not healthy for cabbage and clearly demonstrate the practical problem of designing a package which can both generate a suitable modified atmosphere at the low design temperature and prevent rapid generation of anaerobic conditions at higher temperatures.

In the case of the Intelimer film package maintained at 10°C, the system equilibrates at about 1.5% O_2 and 5% CO_2 which is a suitable environment. When maintained at 20°C the package again equilibrates with a suitable atmosphere not much different from that obtained at the lower temperature. It should be pointed out that these packages do not represent optimized systems and the ideal film for cabbage would exhibit a CO_2/O_2 selectivity of about 3 whereas the tested film exhibit a selectivity of about 6-7.

Temperature Step Experiment. A more realistic test of the performance of the temperature sensitive food package is a temperature step experiment. Test packages were prepared as before, loaded with the appropriate amount of shredded cabbage and placed for 3 days at 3°C. As expected, the gas composition inside the packages is similar, in both cases the O_2 drops to approximately 1.5% and the CO_2 levels rise to about 13%. When the packages are moved to 20°C the CO_2 level in the Cryovac film package continues to increase even faster up to 20 % while the CO_2 level in the Intelimer film package drops to an equilibrium value of about 6% which is maintained even after the temperature is again lowered to 3°C.

While this is an encouraging result it should be noted that during the initial 3°C period of these experiments true equilibrium conditions were not yet attained. Experiments probing additional iterations of film surface area or cabbage loading levels would be needed to find the specific conditions that give comparable equilibrium gas concentrations at the target temperature of 3°C. The experimental results obtained in conjunction with the computer simulations clearly indicate that Intelimer films can be used to design temperature compensating modified atmosphere produce packages.

Conclusions

Work conducted to date shows that Intelimer membranes may be utilized to prepare temperature compensating food packages. A variety of gas selectivities, P_{10} values and permeabilities can be designed into these films to meet the specific needs of various commodity and high value crops. Because actively respiring produce is a dynamic material that changes with time and conditions, computer simulations can serve as a valuable aid in the design of a package construction optimized for specific produce subjected to a variety of probable time-temperature conditions. Temperature abuse compensating food packages may play an important role in the commercial growth of modified atmosphere produce packaging.

Acknowledgments

This material is based upon work supported in part by the U.S. Department of Agriculture under Grant 91-33610-5932. The authors would like to acknowledge helpful discussions with Dr. Devon Zagory of Zagory & Associates.

Literature Cited

1. Kader, A.; Zagory, D. and Kerbel, E. *Critical Reviews in Food Sci.* **1989**, *28*, pp 1.
2. Sugiyama, H. and Yang, K. H. *Applied Microbiology*, **1975**, *30*(30), pp 964.
3. Aylsworth, J. *Fruit Grower*, **1989**, 2, pp 10.
4. Cameron, A. C.; Boylanpett, W. and Lee, J., *J. Food Sci.*, **1989**, *54*(6), pp 1413.
5. Massignan, L.; in *Computerized Selection of Plastic Films for the Packaging of Fresh Fruits and Vegetables*; in Third Subproject: Conservation and Processing of Foods. A Research Report 1982-1986, National Research Council of Italy, Milano, 1987.
6. Zagory, D.; Mannapperuma, J. D.; Kader, A. A. and Singh, R. P. In *Use of a Computer Model in the Design of Modified Atmosphere Packages for Fresh Fruits and Vegetables;* in Proceedings of the 5th International Controlled Atmosphere Research Conference; Feuman, John F. (Ed.); Wenatchee, Washington, 1989, Vol. 1; pp 479.
7. Deily, R. F. and Rizvi, S. S. H. *J. Food Process Eng.* **1981**, 5, pp 23.
8. Henig, Y. S. and Gilbert, S. G. *J. Food Sci.* **1975**, 40, pp 1033.

RECEIVED October 1, 1992

Chapter 17

Side-Chain Crystallizable Polymers for Temperature-Activated Controlled Release

Larry Greene, Loc X. Phan, Ed E. Schmitt, and Judy M. Mohr

Landec Corporation, 3603 Haven Avenue, Menlo Park, CA 94025

A novel controlled release technology based on side chain crystallizable polymers has been developed. The rationale for the use of these polymers in agricultural and medical applications is discussed. The polymer structure and characteristics are described. Laboratory and field data are presented confirming that the release of active ingredients from microcapsule formulations of pesticides can be triggered by increasing soil temperature. This can result in a reduction in pesticide application rates. Temperature activated membranes have been developed to control the release of nicotine in a model transdermal delivery system. The fabrication and release rate characteristics of the membrane system are described.

Development of devices that control the release of active ingredients to various biological systems for use in medical, agrichemical and industrial applications has been ongoing for the past two decades. (1,2) Generally, these devices have provided for a constant rate of release over time that is difficult to vary. A unique class of polymers have been developed (Intelimers, a Landec Corporation registered trademark) that exhibit dramatic changes in permeability in response to small temperature changes.(3) This development will allow the design of controlled release systems that can vary the release of active ingredients over time in response to either passive (i.e. atmospheric temperature changes) or active (i.e. heat applied through a resistive film) temperature changes.

The synthesis, structure and physical characteristics of these polymers will be described. Devices and formulations have been developed utilizing these polymers to control the delivery of a drug, nicotine, a soil insecticide, diazinon and a proprietary organophosphate insecticide, LL825. The rationale and advantages for utilizing temperature regulated delivery in these type of applications will be discussed. The product form for each area will be

0097–6156/93/0520–0244$06.00/0
© 1993 American Chemical Society

described and experimental data will be shown that highlights the uniqueness and advantages of using temperature to trigger the delivery of a variety of biologically active molecules.

The polymers have the generic structure shown in Figure 1. The side chain crystallizable polymers (SCC) are composed of a flexible polymeric backbone to which are attached many hydrocarbon chains that are then held in close conformation for rapid crystallization. Acrylic esters are readily prepared from acrylic acid and long chain fatty alcohols. The acrylic ester monomer is polymerized to yield the side chain crystallizable polymer. Figure 1 illustrates the basic synthetic pathway for the production of acrylic polymer. The molecular weight of the polymer is controlled by the amount and type of initiator and chain terminator added to the process.

The SCC polymers have a temperature "switch" built into them. Below a selected switch temperature, the polymer exists in a crystalline state, but when heated above the switch temperature, conversion to an amorphous or molten state occurs. This change of state is freely reversible an infinite number of times. The switch temperature can be set anywhere between 0°C and 65°C by adjusting the side chain length (4,5,6) and can be controlled within 1-2°C. For instance a polymer with C_{16} side chains will have a melt temperature of 36°C while a polymer with C_{12} side chains will have a melt temperature of 2°C. Figure 3 illustrates how we can then mix monomers containing either C_{12} or C_{16} side chains in the appropriate ratio to obtain any switch temperature between 2°C and 36°C.

DSC analysis has shown that this narrow transition is a result of a very sharp endothermic melting transition with only about 3-5°C between the endotherm onset and the peak. (Figure 2 illustrates this crystal to melt transition.) Below the switch temperature the side chains align through weak hydrophobic interactions that are disrupted above the switch.

Agricultural Applications

In agriculture accurate timing of a pesticide application is one of the keys to an efficient pest control strategy. Ideally, application should coincide with arrival of the pest so that the minimum amount of pesticide, and fewest number of applications will be necessary for effective control. In practice, however, optimum timing is seldom achieved.

One way to achieve better timing of pesticide delivery is to utilize the increase in soil temperature that occurs each spring. There is a predictable increase in soil temperature during the growing season which triggers a variety of biological events such as seed germination, egg hatching, and pupation. Figure 4 illustrates a typical soil temperature profile for the Northern Hemisphere through the growing season. As can be seen, the soil temperature gradually increases during the season and has a daily fluctuation of about 10°C. It is possible to take advantage of this known temperature profile by designing a formulation that releases the chemical only when the specific pest is emerging at a specific temperature.

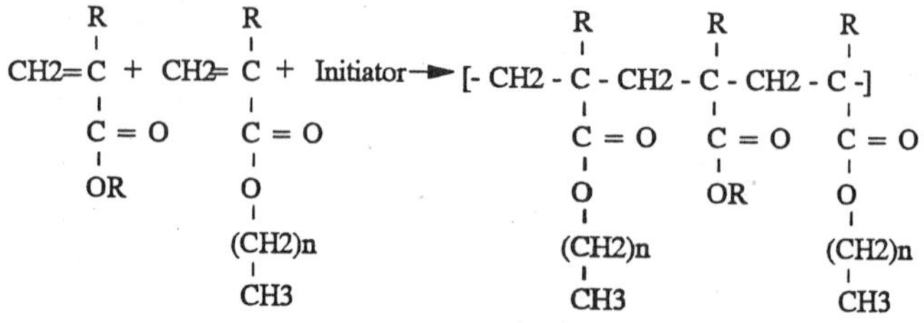

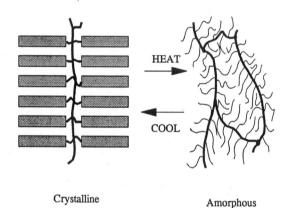

n = Alkyl Acrylate

n = 12 - 22

R = H, CH3

MW = 6,000 to 1,000,000

Figure 1. Side chain crystallizable polymer structure and its synthesis.

Crystalline

Amorphous

Figure 2. The transition of Intelimer polymer from crystalline to amorphous activated by a temperature change.

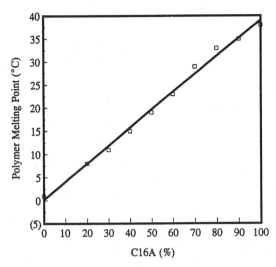

Figure 3. Melting point of side chain crystallizable polymers at different C_{16}/C_{12} ratios.

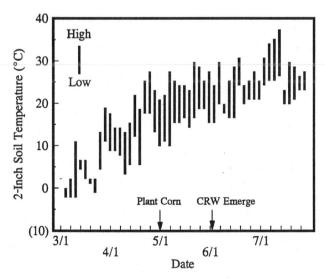

Figure 4. Soil temperature at 2" depth in York, Nebraska from March through July.

Figure 5 shows how delaying the release of an active ingredient is effective in the control of a pest such as the corn rootworm. The graph on the left depicts the current practice of applying a large dose of active ingredient at planting to control an insect whose population will not increase until approximately thirty days after planting. With a pesticide encapsulated in a SCC polymer on the other hand (right graph), release of the chemical is delayed until the desired temperature is reached, allowing for a significant reduction in the application rates.

Improved pesticide formulations designed to deliver the chemical when and where it is needed will eliminate the need for high rates and therefore reduce or eliminate many of the associated problems. If a pesticide could be contained and protected in a microcapsule for extended periods of time and then rapidly released at a preselected temperature, then accurate, automatic delivery would be possible. This can be achieved using new formulations developed by Landec, where delivery is passively controlled by atmospheric or soil temperature.

Medical Applications

In drug delivery, controlled release systems were originally developed to address the problems of fluctuations in plasma concentrations when drugs are administered conventionally. Transdermal therapeutic systems, in particular, were designed to provide a constant rate of drug release to the circulation system for the lifetime of the device. These topical systems have found medical acceptance because they avoid the "first pass effect" and they enable drugs that have short half lives to be administered conveniently. However, constant (or zero-order) release may not always be the best therapeutic regimen due to the development of tolerance for the drug, development of localized irritation, or inconsistency with circadian fluctuations.

An example of the development of tolerance occurs with nitroglycerin transdermal patches. Nitroglycerin is used to reduce ischemia in the heart tissues which results in severe chest pain. This reaction is caused by a pressure differential in the heart. As nitroglycerin is absorbed into the blood system the ischemia is reduced. Figure 6 illustrates this process. A transdermal patch containing nitroglycerine was applied to the skin. This results in an initial high reduction in ischemia(75%), however it is attenuated (reduction in ischemia is not as pronounced) within 12 hours, even with an adequate serum level of nitroglycerin, due to the development of tolerance. The patch was prematurely removed and after the subjects experienced a 12 hour patch free interval, a new patch was applied and the subjects were retested. The reduction in ischemia rose again (indicating effectiveness) and then fell to levels observed during the previous 12 hour post administration period. Once again the effect of nitroglycerine had decreased with time, indicating that the body was building up a tolerance to the drug. These findings suggest that discontinuous transdermal administration of nitroglycerin, in which near steady-state blood concentrations are maintained for several hours followed by a period in which

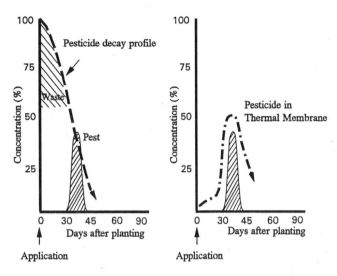

Figure 5. The left figure shows the current practice of pesticide application. The right figure shows that encapsulated pesticide is applied at the same time as planting for convenience, but it is not activated until the pest emerges.

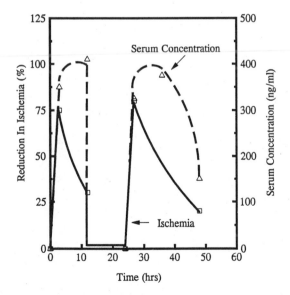

Figure 6. A reduction of the ischemic effect due to the development of nitroglycerine tolerance.

nitroglycerin concentrations are quite low, may enable 24 hour therapeutic coverage without the development of tolerance. While the actual time periods cannot be defined at this time, it is clear that a transdermal device that has the ability to turn on and off would be highly desired to implement this or variations of this regimen.

As a result of this desire for a pulsatile drug delivery device, a thermally responsive Intelimer membrane transdermal device has been developed that will allow the drug to be delivered in a pulsatile fashion. This will enable the patient to receive the drug in a manner that is more closely mimics their normal biological rhythms and reduce the development of tolerance.

Experimental

Atactic side chain crystallizable acrylate polymers were prepared from hexadecyl acrylate either by solution or bulk techniques initiated with AIBN or t-butyl peroctoate. Molecular weight was controlled with dodecyl mercaptan. Solution polymerization products were precipitated into ethanol, filtered and dried under reduced pressure. Bulk polymerization yielded a product of controlled molecular weight suitable for use directly. Polymers were characterized and compared by DSC by initially heating all samples above the melt temperature followed by cooling and a second heating. The second heat endotherm was monitored so that all polymer samples were compared without adulteration by synthetic processing or annealing.

Microcapsules were prepared using a standard emulsion encapsulation process (7). The active ingredient was incorporated into the oil phase, emulsified in water and crosslinked. The microcapsules contained approximately 90 wt% of active ingredient. Microcapsule particles size was monitored using a Hiac Royco particle size analyzer and confirmed through optical and electron microscopy. The microcapsules were supplied in an aqueous medium with no adjuvants, emulsifiers or other additives. Laboratory release rates of diazinon and trifluralin were conducted in water and ethanol:water, 1:1, respectively and monitored by UV spectroscopy.

A laboratory bioassay was conducted using a diazinon microcapsule formulation. A capsule formulation with a melting point of 25°C was compared to a commercial formulation of diazinon, 14 G (14 % active granular). Sufficient formulation was mixed with a dry soil to achieve a 2.5 ppm concentration of diazinon. The soil was placed in a chamber at 20°C. At the end of week 1, 2, and 4, four replicates of 100 g of soil treated with each of the formulations were removed. The soil was placed in cups with a sprouted corn (*Zea mays L.*) seed and 10 larvae of *Diabrotica balteata* (banded cucumber beetle), a corn insect pest. Four replicates of untreated soil were also used for comparison. After four days, the soil was sifted and the number of live larvae were determined. After the four week sample was taken, the temperature was increased to 32°C. Soil was sampled at week 5, 6, and 8 and were processed the same as described above. The data were analyzed by dividing the number of dead larvae/cup by the number of live larvae that had survived in the non-treated soil and multiplying by 100. An analysis of variance was conducted on the data with the means compared by LSD at the 0.05 level of probability.

A field trial using an experimental organophosphate insecticide, LL852, was conducted in York, Nebraska. In experiments conducted in 1989 and 1990, the biological activity of encapsulated LL852 (a Landec formulation) was compared to a LL852 granule and commercially available granule of Counter insecticide.   Both experiments were conducted in a field containing a Sharpsburg silt loam soil (fine, montmorillonitic, mesic Typic Argiudoll) to evaluate the damage to corn roots of each formulation at different concentrations.   Counter was applied at 1.0 lb ai/ac (active ingredients per acre), LL852 10G at 0.5 and 0.25 lb ai/ac, and LL852 Capsule at 0.5 and 0.125 lb ai/ac. The capsule treatments were applied in 165 L of water/ha with a hand held spray boom that was powered by CO_2. The granule formulations were applied using a standard granule applicator. Plots were 3 by 9 m and the treatments were replicated three times in a randomized complete block design. All formulations were lightly incorporated after the corn seed had been planted and covered. Eight weeks after planting, three corn plants in each replicate were pulled from the ground, their roots cleaned and a root rating conducted. On the root rating scale, a 5.0 is maximum damage and a 0 is no damage.

Release of nicotine through SCC polymer membranes was accomplished by preparing either a free standing film of polymer or a porous polypropylene supported film. Nicotine was allowed to permeate the film from an aqueous solution to water that was continually monitored by an automatic sampling UV spectrophotometer. The cell was thermostated and designed to increase the temperature in one degree increments from 5°C to 40°C.

Results and discussion

Pesticide Microcapsules.   Diazinon was microencapsulated with Intelimer polymer that has a melt temperature of 30°C. The formulation had a particle size of 90% less than 10 microns and contained 25% active ingredient. The release profile in water compared to a standard polyurethane microcapsule is shown in Figure 7. At 20°C both formulations had the same release profile; however, at 30°C the release rate of the Intelimer increased dramatically, while the standard capsule release rate did not increase with temperature. The release profile shown here is typical of all microcapsule Intelimer formulations.

Figure 8 illustrates the results of the laboratory biological efficacy of the Landec formulation compared to a commercial diazinon granule (14G) when tested against corn rootworm. This test was designed to demonstrate reduced efficacy at low temperature when the microcapsules are below the transition temperature and commercially acceptable efficacy above the transition temperature when the microcapsules are turned on.

The 14G (commercial granule product) exhibited the greatest efficacy at 20°C with a significant decrease in efficacy at 32°C. Because the 14G granule formulation releases all of its active over a short period of time, the

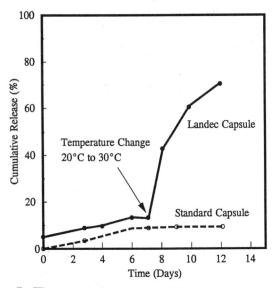

Figure 7. The cumulative release of Diazinon from standard polyurethane microcapsules and Landec microcapsules.

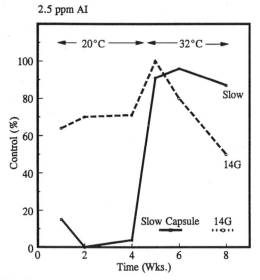

Figure 8. Bioefficacy testing of corn root worm (CRW) using a Landec formulation and a commercial 14G formulation.

chemical is readily available for any environmental degradation. This is seen in the rapid decrease in efficacy during the 32°C test period. After eight weeks, the efficacy of the granule had dropped to 50% control.

The Landec formulation on the other hand gave significantly less control at 20°C, but control increased to 90% when the temperature was increased to 32°C. This higher level of control continued for 4 weeks at 32°C after which time the experiment was terminated. At 2.5 ppm, this formulation was superior to the granule in its duration of control. These data suggest that for the commercial granule to achieve equivalent efficacy to the Landec formulation, amounts greater than 2.5 ppm of active ingredient would have been required.

Corn rootworm field trials were conducted using an experimental organophosphate insecticide, LL852. Table I shows the results of testing in 1989 and 1990.

Table I.   Corn Rootworm 1989 & 1990 Field Trial Results using LL852 Granules and Capsules

Treatment	Rate (lb ai/ac)	Root Rating	
		1989	1990
Control	None	3.9	4.6
Counter 15G	1.0	1.8	2.4
LL852 10G	0.5	1.5	2.9
LL852 10G	0.25	--	2.7
LL852 Capsule	0.5	1.6	2.8
LL852 Capsule	0.125	1.8	2.9

Counter 15G is the commercial standard and the LL852 10G is a standard clay based granular formulation, and the capsule formulation contained the active plus Intelimer polymer. In the root rating system used here, the higher numbers indicate less insect control. Economic control is achieved at a root rating of less than 3. In both 1989 and 1990, the LL852 capsule formulation gave equivalent control at 0.125 lb ai/ac to the standard granule formulation at 0.25 lb ai/ac (the lowest rate tested), and also gave equivalent control to the Counter formulation. This demonstrates that a significant reduction in application rates is possible under field conditions using temperature responsive microcapsule formulations.

Nicotine Transdermal Delivery. There are two components to the Intelimer transdermal system, the temperature sensitive membrane and a mechanism to activate the membrane. The transdermal patch is comprised of the drug reservoir and the temperature sensitive polymer membrane. The patch must also contain a material that warms in response to a stimulus (e.g. electrical impulse) in order to raise the temperature of the polymer membrane, which in turn allows for permeation of the drug. Figure 9 shows a model of such a system. A typical transdermal type system is shown which consists of a drug reservoir and a backing. An Intelimer membrane that can be heated is placed between the drug reservoir and the adhesive. When the membrane is in the crystalline phase no drug is released. As the membrane is heated, the polymer becomes amorphous, and drug is delivered to the adhesive, then to the skin and eventually to the circulatory system. The second component of the transdermal system is the electronics and the power source to provide the stimulus to the receiving material. Both components of the transdermal therapeutic system have been developed.

Drug release was demonstrated using nicotine as the probe molecule. Membrane release through an Intelimer film was effectively controlled in response to temperature where the slope of the release rate vs time exhibits a dramatic increase at the switch temperature as shown in Figure 10. Here, over a 2.5°C temperature change a 1000 fold increase in release rate was observed. In a separate experiment, nicotine diffusion was switched on and off in response to resistive heating of a membrane, yielding greater than two orders of magnitude difference in drug release in the on versus off state. (Figure 11)

The feasibility of constructing a transdermal device that is capable of delivering a drug in response to an external stimulus has been shown. Both a transdermal patch containing Intelimer polymer and a heater that can be programmed and miniaturized have been produced.

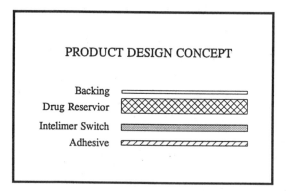

Figure 9. Model of Landec transdermal delivery system.

Conclusions

Side chain crystallizable polymers have been developed that can control the release of a variety of active agents in response to changes in temperature. Heat activated membrane based devices have been constructed that can deliver drugs through a transdermal patch in a pulsatile fashion. Pesticide containing microcapsules have been formulated that can deliver the active to the soil in

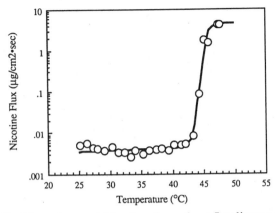

Figure 10. The release of Nicotine through an Intelimer polymer membrane.

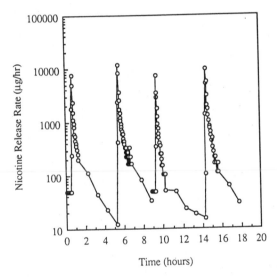

Figure 11. Pusatile release of Nicotine. Each spike of drug release occured after applying 1.8 watts for 5.6 minutes.

response to temperature changes. This will allow for significant reductions in application rates of insecticides and herbicides. New applications are being developed for other medical, agricultural and industrial uses.

Acknowledgments

This work was supported in part by SBIR grant No. 88-39410-4710 from the United States Department of Agriculture and SBIR grant No. 1 R43 GM46156-01 from the United States Department of Health and Human Services. Landec gratefully acknowledges the assistance of Dow-Elanco, Griffin Corp. and Ciba Geigy, Ltd. in providing active ingredients.

Literatures Cited

1. Plate, N. A. and Shibaev, V. P. *Comb Shaped Polymers and Liquid Crystals*; Plenum Press, 1987.
2. Jordan, E. F.; Feldeisen, D. W. and Wrigley, A. N. *J. Polym. SCi.* **1971**, Part A-1, 2, pp 1835.
3. Stewart, R. F.; U.S. Patent 4,830,855
4. Greenberg, S. A. and Alfrey, T. *J.Am. Chem. Soc.* **1954**, 76, pp 6280.
5. Kyodnieus, A. F. *Controlled Release Technologies: Methods, Theory, and Applications*, 1, CRC Press, Inc., 1980.
6. Ruppel, R. F. *J. Econ. Entomol.*, **1984**, 77, pp 1084.
7. Stewart, R. F. Stewart; Greene, C. L. and Bhaskar, R. K. U.S. Patent 5,120,349.

RECEIVED September 24, 1992

Chapter 18

Transdermal Films of Diclophenac Sodium

S. C. Mandal, M. Bhattacharyya, S. C. Chattaraj, and S. K. Ghosal

Division of Pharmaceutics, Department of Pharmaceutical Technology,
Jadavpur University, Calcutta 700032, India

A matrix-dispersion type Transdermal Drug Delivery System
(TDDS) of Diclophenac Sodium (DS) was fabricated, for its
controlled delivery, based on the rate - controlling polymers, namely Eudragit RS100 (RS) and RL100 (RL), with an
objective of studying the effects of a variety of polymer
combinations on the drug - release profile. Effect of
coadministration of varying concentrations of permeation
promoter (Isopropyl myristate), to overcome the diffusional
resistance in course of in vitro release and in vitro skin
permeation, was quantitatively evaluated. In-depth in vitro
release and in vitro skin permeation studies (using excised
pretreated abdominal skin of male albino mice) were conducted with the formulations. Matrix - diffusion type kinetics
for the in vitro release study, and zero order kinetics for
the skin permeation were obtained over 12 hours and 24
hours span of study, for both the cases respectively, with
and without enhancers.

A strong therapeutic rationale and some unique advantages over conventional dosage forms prompted the choice of TDDS as a favourable
mode of delivery for DS (1-3). The drug (advocated for use in rheuma-
toid disorders and degenerative joint diseases), possesses a short
biological half-life (approx. 2 hrs.), low oral bioavailability (54%)
due to extensive first pass metabolism (4,5), gastro-intestinal irritation, and a frequent dosing schedule. The physico-chemical parameters of DS were also suitable for formulation into a TDDS (5). Controlled delivery of DS is deemed to be advantageous in many diseased
states, and the TDDS designed in this study aims at achieving a prolonged delivery of DS, based on permeability characters of the poly-
methyl-methacrylate copolymers (Eudragits) (6,7) which are being extensively used for rate-controlled drug delivery of a multitude of
drugs (8,9). Eudragits RL100 and RS100 are co-polymers of acrylic and
methacrylic acid esters with a low content of quaternary ammonium
groups. The molar ratio of the ammonium groups and the remaining

0097–6156/93/0520–0257$06.00/0
© 1993 American Chemical Society

neutral(meth)acrylic acid esters is 1:20 for Eudragit RL and 1:40 for the RS variety respectively. The letters RL and RS stand for the German equivalents of the words freely permeable and slightly permeable respectively, and refer to the permeability characters of these two polymers. The ammonium groups are present in the salt form in RS and RL and they are responsible for the permeabilities of films of RL100 and RS100. The non-toxic and non-sensitizing properties of Eudragits, make it suitable for use in TDDS.

The suitability of isopropyl myristate (IPM) as permeation-rate enhancer was studied, from the range of 10 to 30% w/w. DS, with a log $P_{octanol/water}$ value of 1.441, has an intermediate lipophilic-hydrophilic character (since $P < 1$ signifies hydrophilic nature, and $P > 3$, lipophilic nature) (10). From the theory of Hildebrand solubility (δ) we can postulate that enhancers with a δ value of less than 12 are suitable for use with lipophilic drug entities, and those with a value greater than 12, for those of hydrophilic nature. The value of IPM (8.5) is quite near to that of the δ value of skin (10.5) and so, can be considered suitable for use with drugs having an intermediate hydrophile - lipophile character (11).

Preformulation Studies

Determination of n-octanol/H_2O Apparent Partition Coefficient (P). The partition coefficient of DS was determined (by the technique of Diez et al.) (12), at $37\pm1°C$ over 72 hours. Calculations were done by taking readings of DS in aqueous phase at 276 n.m. The DS concentration in octanol was calculated by computing the difference between the total concentration, and that of the aqueous phase. Measurements were performed in triplicate.

Drug Permeation Rate Calculation. This was determined by a modified Franz diffusion apparatus. Four sq.cm of freshly excised, full-thickness abdominal skin of male albino mice (5-7 weeks old) was taken after sacrificing the animal by cervical dislocation and clipping off the hair. The adherent fatty material was trimmed out, the skin sample washed with distilled water, blotted dry, and fixed with its stratum corneum (SC) facing the donor compartment. A saturated solution of drug in 20% Polyethylene glycol 400 - saline solution was placed on the donor compartment in contact with the SC, while the receptor compartment was filled with the blank PEG - saline solution. The temperature of the elution medium was maintained at $37\pm1°C$ and it was magnetically stirred at 600 rpm. Aliquots withdrawn over 12 hours were assayed spectrophotometrically at 276 nm. Blanks were run simultaneously.

Calculation of Required Drug Release Rate. For dose designing, the required absorption rate of DS through skin, which is required to achieve an effective plasma concentration of 2 mcg/ml was calculated to be 0.192 mg/hr (vide equation of Sanvordekar et al.) (13).

Fabrication of TD Films. The films were fabricated by the procedure of Mandal et al. (14). In short, a backing membrane of PVA was first cast, followed by casting of a solution of polymers, drug, plasticizer (Dibutyl phthalate) in acetone. The drug load was maintained constant at 20% w/w in all formulation. The films were dried at 40°C, and cut into strips of required dimensions by a die cutter. Use of uniform dies and equal volumes of solution yielded films of uniform thickness. The films were stored under low humidity conditions.

Characterization of TD Films

Content Uniformity. One square cm of samples were cut out from each film from four different points, 5 ml chloroform was added to disrupt the polymer matrix. 100 ml of 0.1 (N) NaOH was added to dissolve the DS, the solution stirred for 20 minutes magnetically, followed by complete evaporation of chloroform in water bath. After suitable dilutions of the alkali solution, DS was estimated spectrophotometrically at 276 nm (Hitachi 200-20 model).

Thickness Determination. Film thickness was measured using a Comparator (Doall Company, Illinois, USA).

In Vitro Release Studies of TD Films. These studies were performed using a hydrodynamically well calibrated, modified Franz diffusion cell, whose receptor compartment was filled with triple glass distilled water (elution medium), maintained at 32±1°C by a surrounding thermostated water jacket. The elution medium was agitated at a constant speed of 600 rpm by a Teflon coated magnetic bar. A four square cm of the film sample was fixed with its drug releasing surface facing the receptor compartment (*15*). Five ml of withdrawn aliquots (replenished by equal volumes of blank solution) were estimated spectrophotometrically, at 276 nm.

In Vitro Skin-Permeation Profiles. A verticle-type, flow-pattern controlled, thermodynamically calibrated modified Franz - apparatus was used, with an elution medium of 20% PEG 400 - normal saline, maintained at 32±1°C by a surrounding thermostated water jacket. 5-7 week old male albino mice were sacrificed by cervical dislocation just prior to experiment initiation, and the hair from abdominal skin were electrically clipped off. Full thickness of the abdominal skin was excised and the adherent fatty material trimmed off, and the skin soaked for 20 mins. in normal saline, blotted dry gently. The skin sample was then taken and the film was fixed on to the skin by pressing the adhesive film located around the edge of the drug releasing area, in such a way that the drug releasing surface was in contact of the SC. Next the skin was evenly mounted on the cell in between the donor and the receptor compartments with the dermal surface in contact with elution media, which was magnetically stirred at 600 rpm (*16*). Aliquots withdrawn over 24 hrs were analyzed spectrophotometrically at 276 nm. Blanks were run simultaneously.

Results and Discussion

The determined n-octanol/H_2O partition coefficient (P) of Diclophenac Sodium was 27.62±0.868, and log P value was 1.441. This indicates that the drug has an optimum hydrophilic-lipophilic character, (since P $\leqslant$ 1 signifies hydrophilic, and P $\geqslant$ 3 lipophilic character) thus showing its suitability for administration through a TDDS (*10*).

The skin permeation rate of drug through mice skin was calculated as 0.4285 ± 0.0517 mg/cm^2/hr which is too high for maintaining therapeutic level and so needs modulation to achieve an optimum permeation rate.

The preparation process of TD films was validated by casting each formulation in triplicate and measuring the content and film thicknesses. Films showed uniform thickness of 85-90 microns, with inter and intrabatch variations being 4-5%. Content determinations showed consistent results, with variations of only 4-6 per cent.

Data from in vitro release studies showed a linear relationship, when plotted as cumulative percent released versus square root of time, thereby following the classical Higuchi equation for matrix diffusion controlled release (17), represented by the equation.

$$Q = [(2A - C_p) C_p Dpt]^{\frac{1}{2}}$$

Higuchi kinetics is generally followed by matrix systems where the drug load exceeds 1% w/w (18). In this TDD system, the drug load is very high, to the extent of 20% w/w. So it can be assumed that diffusion controlled kinetics has every possibility of manifesting itself from the TDDS matrix, and this can be confirmed from the Figure 1.

Increased release rate was observed with increase of proportion of RL in films (Figure 1). Films of 100 percent RL gave a maximum total release of 4.039 ± 0.156 mg/cm^2 (70 percent) followed by 3.016 ± 0.103 mg/cm^2 (61 percent) release from RS-RL combination and 2.924 ± 0.124 mg/cm^2 (56 percent) from RS preparation over a span of 24 hrs (Table I).

Table I. Film Compositions and Permeation Profiles

FN. code	Polymeric composition % RS	RL	Amount of Drug/cm^2 (mg)	Amount of Drug released in 12 hours (mg/cm^2)[a]	Amount of drug permeated in 24 hours (mcg/cm^2)[b]
RS	100	–	5.222	2.924 (±0.124)	187.500 (±17.500)
RL	–	100	5.770	4.039 (±0.156)	450.000 (±19.851)
RS-RL	70	30	4.945	3.016 (±0.103)	225.000 (±20.325)

( ) = Standard deviation
(a) n = 6
(b) n = 4

This release can be attributed to the greater permeable nature of RL, due to a higher content of quaternary ammonium groups than RS (19). With increasing RL proportions in the RS matrix, drug-leaching by the elution medium is prominent, since the medium gains facilitated entry via open channels created by RL, in the slightly permeable RS matrix.

In vitro skin permeation studies showed an apparent zero order release when cumulative amount permeated (Q) was plotted against time (T) (Figure 2). Results depict a gradual increase in permeation rate (J) with increase of RL proportion in the polymeric composition, maintaining the drug load constant at 20 percent w/w for all films. The

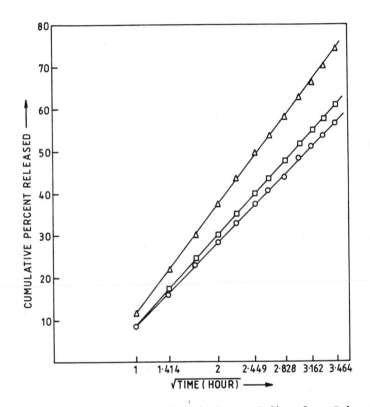

Figure 1. Release Profiles of Diclophenac Sodium from Polymeric Films. O———O, RS100; △——△ , RL100; ☐——☐ , RS:RL::7:3

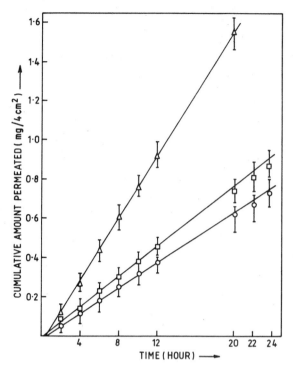

Figure 2. Permeation Profiles of Diclophenac Sodium in 20% PEG 400 in normal saline. O——O, RS100; △——△ , RL100; □——□ , RS:RL::7:3

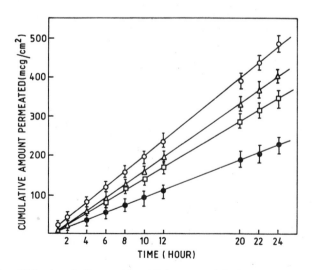

Figure 3. Effect of Permeation Enhancer (IPM) on Permeation of DS. ●——●, RS:RL::7:3, without IPM; □——□, 10% IPM; △——△ , 20% IPM; O——O, 30% IPM.

maximum J value of 0.01937 mg/cm^2/hr was shown by films of RL, followed by those of RS-RL (J = 0.00937 mg/cm^2/hr) and RS (J = 0.00833 mg/cm^2/hr) respectively.

The desired release rate of DS is 0.192 mg/hr and the RL films give a rate of 0.01937 mg/cm^2/hr. Thus, a TD film of area 9.909 cm^2 would result in a release rate of 0.19199 mg/hr. which is appreciably equal to the required release rate needed to maintain a plasma concentration of 2 mcg/ml of DS.

Films of RS and RS-RL showed a permeation rate much lower than that desired to obtain a therapeutic plasma level of 2 mcg/ml. Therefore, films of RS-RL with IPM were chosen for further studies.

Results of permeation studies conducted with films of RS-RL, containing increasing percentages of IPM, depict an increased rate and amount of permeation in comparison to films without IPM. The rate of permeation was calculated from the slope of the graph (Figure 3). The permeation rates and enhancement factors are represented in Table II.

Table II. Effect of Permeation Enhancer (IPM) on Drug Permeation from RS:RL :: 70:30 Film

Permeation rate (J) without IPM[a] (mcg/cm^2/hr)	IPM content % w/w	Permeation rate (J) with IPM[a] (mcg/cm^2/hr)	Enhancement factor
9.375	10	14.50	1.5466
9.375	20	18.33	1.9552
9.375	30	19.64	2.0950

(a) n = 4

The enhancement values of the films with 10, 20 and 30 percent w/w IPM, prove the apparent suitability of IPM in improving drug availability from the fabricated TDDS.

IPM possibly interferes with the lipid components of the skin, increases the solubility parameter of the lipid phase, and blocks the polar routes of absorption. Another hypothesis is that permeation enhancement may have been due to greater drug partitioning of DS into the skin from the TDDS in presence of IPM. In order to alter the permeation properties of stratum corneum lipid components, the IPM possibly fluidizes the non-polar moieties of skin components, causing lipid barrier disruption.

Extended studies on this TDDS need to be continued to develop it into a suitable form for human trials.

Acknowledgment

We would like to thank Win-Medicare Limited, (India) for the gift sample of Diclophenac Sodium. Rohm Pharma (Germany) for Eudragit samples, CSIR for the financial grant. The authors wish to thank Mr. P.K. Khatry for his assistance.

Literature Cited

1. Chien, Y.W. In *Transdermal Controlled Systemic Medication*; Chien, Y.W., Marcel Dekker, New York and Basel, 1987, pp. 1-21.

2. Chang, S.F.; Moore, L.; Chien, Y.W. *Pharm. Res.* 1988, 5, 718.
3. Chien, Y.W. In *Controlled Drug Delivery Fundamentals and Applications*; Robinson, J.R.; Lee, V.H.L., Ed. 2; Drugs and the Pharmaceutical Sciences; Marcel Dekker, Inc. New York and Basel, 1987, Vol. 29; pp. 524-549.
4. Brogden, R.N.; Heel, R.C.; Pakes, G.E.; Speight, T.M.; Avery, G.S. *Drugs.* 1980, 20, 24.
5. Martindale - *The Extra Pharmacopoeia*; ; Reynolds, J.E.F., Ed. 28 : The Pharmaceutical Press : London, 1982, pp. 250.
6. Barkai, A.; Pathak, Y.V.; Benita, S. *Drug Dev. Ind. Pharm.* 1990, 16(3), 2057.
7. Cameron, C.G.; Mcginity, J.W. *Drug Dev. Ind. Pharm.* 1987, 13(8), 1409.
8. Chang, R.K.; Price, J.C.; Whitworth, C.W. *Drug Dev. Ind. Pharm.* 1987, 13(6), 1119.
9. Benita, S.; Babay, P.; Hoffmann, A.; Donbrow, H. *Pharm. Res.* 1988, 5(3), 178.
10. Hori, M.; Satoh, S.; Maibach, H.I.; Grey, R.H. *J. Pharm. Sci.* 1991, 80(1), 32.
11. Pfister, W.R.; Hsieh, D.S.T. *Pharm. Technol.* 1990, 14, 134.
12. Diez, I.; Colom, H.; Moreno, J.; Obach, R.; Peraire, C. *J. Pharm. Sci.* 1991, 80(10), 931.
13. Sanvordekar, D.R.; Cooney, J.G.; Westen, R.C. U.S. Patent No. 4, 336, 243, 1982.
14. Mandal, S.C.; Bhattacharyya, M.; Chattaraj, S.C.; Ghosal, S.K. *Indian Drugs* 1991, 28(19), 478.
15. Toddywallah, R.; Chien, Y.W. *Drug Dev. Ind. Pharm.* 1991, 17(12), 245.
16. Keshary, P.R.; Chien, Y.W. *Drug Dev. Ind. Pharm.* 1984, 10(10), 1663.
17. Higuchi, T. *J. Pharm. Sci.* 1961, 50, 874.
18. *Microcapsulation and related Drug processes*; Deasy, P.B., Drugs And Pharmaceutical Sciences; Marcel Dekker Inc., New York and Basel, 1984; Vol. 20.
19. Akbuga, J. *Drug Dev. Ind. Pharm.*, 1991, 17(4), 593.

RECEIVED October 5, 1992

Chapter 19

Hydronium Ion Diffusion into Microcapsules and Its Effect on the pH of Encapsulated Aqueous Solutions

Theoretical Analysis

G. D. Svoboda, C. Thies, P. S. Cheng, J. Zhou, M. Asif, and
D. L. Distelrath

Department of Chemical Engineering, Washington University,
One Brookings Drive, St. Louis, MO 63130

Fundamental microcapsule mass balance equations are used to describe the mass transport behavior of hydronium ions into a microcapsule with a uniform spherical core/shell structure. The analysis is applied to CAP microcapsules which are used for enteric drug release. Unbuffered CAP solutions appear unable to protect pH sensitive agents for the time required to reach the small intestine. Buffered aqueous solutions are found to offer much greater protection than their unbuffered counterparts and appear capable of delivering pH sensitive agents to the small intestine.

Recently, Svoboda and co-workers *(1)* demonstrated the utility of applying fundamental transport models to describe the mass transport behavior of a diffusing species into or out of a microcapsule. The analysis yielded some insight into the effect of system parameters, such as diffusivity and capsule wall thickness, on the mass transport behavior of the system. One unanticipated result established that a capsule wall thickness equal to one-sixth the outer capsule diameter maximizes the barrier properties of a spherical microcapsule. This corollary is an important design criterion if one wishes to protect or retain an encapsulated material. Since microcapsules are candidate enteric drug delivery devices *(2-5)*, it is of value to apply a similar analysis to encapsulated aqueous solutions suspended in acidic media. The analysis is used here to characterize the effect of hydronium ion diffusion into a microcapsule on the pH of the encapsulated solution. Both unbuffered and buffered solutions are considered to determine the benefit of a buffer solution to an encapsulated pH sensitive agent.

Mass Transport Equations

Unbuffered Encapsulated Aqueous Solution. The assumptions and mass transfer analysis for unbuffered solutions are identical to those described in Svoboda et al. such that only the final equation is presented here. The concentration, C_i, of a single diffusing species inside a microcapsule with an unbuffered aqueous inner core region at time t is given by:

0097–6156/93/0520–0265$06.00/0
© 1993 American Chemical Society

$$C_i = (C_i^o - C_b)e^{-Kt} + C_b \tag{1}$$

where C_i^o is the initial concentration of diffusing species inside the capsule, and C_b is the bulk concentration outside the microcapsule. K is a constant defined by the following relationship for spherical microcapsules:

$$K = \frac{3D}{R_i^2\left(1 - \frac{R_i}{R_o}\right)} = \frac{12D}{d_o^2}\left\{\left[\left(\frac{1}{\theta} - 1\right)\left(\frac{\rho_{IP}}{\rho_W}\right) + 1\right]^{-\frac{2}{3}} - \left[\left(\frac{1}{\theta} - 1\right)\left(\frac{\rho_{IP}}{\rho_W}\right) + 1\right]^{-1}\right\}^{-1} \tag{2}$$

where D is the diffusivity of the diffusing species, d_o is the outer capsule diameter, R_i and R_o are the inner and outer capsule radii, respectively, ρ_{IP} and ρ_W are the inner core and capsule wall densities, respectively, and θ is the inner core weight fraction.

When the diffusing species is a hydronium ion, equation 1 can be written in terms of pH:

$$pH_i = -\log\left[(10^{-pH_i^o} - 10^{-pH_b})e^{-Kt} + 10^{-pH_b}\right] \tag{3}$$

where pH_i is the pH inside the capsule, pH_i^o is the initial pH inside the capsule, and pH_b is the bulk pH outside the microcapsule, which is assumed constant. Equation 3 defines how the pH of an unbuffered aqueous solution inside a microcapsule varies with time of immersion in an acidic medium of constant pH. When $pH_i^o - pH_b > 2$, equation 3 can be simplified to give:

$$pH_i = pH_b - \log\left[1 - e^{-Kt}\right] \tag{4}$$

Equation 4 is a special case of equation 3 and is valid for a large initial hydronium ion concentration gradient where such ions diffuse into a capsule.

Buffered Encapsulated Aqueous Solution. Equation 1 describes the mass transfer behavior of microcapsules with a single diffusing species that does not undergo chemical reaction. This is the situation that exists when a capsule is loaded with an unbuffered aqueous phase. In cases where the capsule carries a buffered aqueous solution, it is necessary to incorporate a reaction term, R_{C_i}, into the mass transfer equation:

$$\frac{dC_i}{dt} + K(C_i - C_b) - R_{C_i} = 0 \tag{5}$$

For diffusion of hydronium ions into a buffered aqueous encapsulated solution, it is appropriate to define some new variables. The dissociated hydronium ion concentration inside the microcapsule is designated C_H, the dissociated salt ion concentration C_A, the undissociated weak acid concentration C_{HA}, and the bulk concentration C_b. Equation 5 then becomes:

$$\frac{dC_H}{dt} + K(C_H - C_b) - \frac{dC_A}{dt} = 0 \tag{6}$$

The dissociation of a weak acid, HA, in water is given by :

$$HA \Leftrightarrow H^+ + A^- \qquad (7)$$

The equilibrium dissociation constant, K_a, for equation 7 is the following:

$$K_a = \frac{C_H C_A}{C_{HA}} \qquad (8)$$

In order to greatly simplify the mathematical analysis, the buffer is assumed to be unable to diffuse out of the microcapsule. This assumption makes the total buffer concentration inside the capsule, C_T, constant. Thus, $C_T = C_{HA} + C_A$, and equation 8 can be written as:

$$C_A = \frac{K_a C_T}{C_H + K_a} \qquad (9)$$

Differentiation of equation 9 with respect to time gives:

$$\frac{dC_A}{dt} = -\frac{dC_H}{dt}\left[\frac{K_a C_T}{(C_H + K_a)^2}\right] \qquad (10)$$

Substitution of equation 10 into equation 6 produces:

$$\frac{dC_H}{dt} = \frac{K(C_b - C_H)}{1 + \dfrac{K_a C_T}{(C_H + K_a)^2}} \qquad (11)$$

The solution of equation 11 yields the time-dependent hydronium ion concentration within the microcapsule. Equation 11 can be written in integral form as follows:

$$t = \int_{C_H^o}^{C_H^f} \frac{(C_H + K_a)^2 + K_a C_T}{K(C_b - C_H)(C_H + K_a)^2}\, dC_H \qquad (12)$$

where C_H^o and C_H^f are the initial and final hydronium ion concentrations, respectively. Differentiation of equation 12 with respect to R_i, and subsequently setting dt/dR_i equal to zero gives the required optimum shell thickness of one-sixth the outer diameter.

Results and Discussion

The mass transfer equations presented above provide a means of defining how hydronium ion diffusion through a capsule shell affects the pH of encapsulated unbuffered and buffered aqueous solutions. They provide insight into the influence of capsule size and shell thickness on the change in pH of unbuffered and buffered aqueous solutions inside a microcapsule with time of immersion in an acidic medium. The equations are applicable to any capsule that has a uniform spherical core/shell structure. However, they are of particular importance for enteric capsules designed to protect pH-sensitive active agents.

Unbuffered Encapsulated Aqueous Solution. As a specific example, consider a hypothetical 100 micron diameter microcapsule fabricated from cellulose acetate phthalate (CAP), a well-established enteric polymer. The microcapsule has a uniform spherical core/shell structure and is initially filled with 0.9 wt. % saline at pH 7. Figure 1 contains a series of curves calculated from equation 4 for varying active agent loading, or inner core weight fraction θ. The curves were constructed using a CAP density of 1.25 g/cm^3 and a hydronium ion diffusivity through the CAP of 1.4×10^{-11} cm^2/sec.

The density of CAP was unavailable in the literature, so a value representative of similar cellulose ester polymers was used here (6). The diffusivity used was the lowest of several values reported by Spitael and Kinget (7). It was obtained experimentally by placing a CAP film cast from an ethyl acetate/isopropanol (77/23) mixture in a diffusion cell, one side of which contained 0.9 % saline and the other a solution that was 0.9 % saline and 0.1 M HCl. These curves show how the pH of the saline solution carried by these capsules decreases with immersion time in a 37 °C solution that is 0.9 % saline and 0.1 M HCl. The calculated curves in Figure 1 show that pH values decrease from 7 to below 4 after 1-2 minutes immersion time. Figure 1 suggests that a 100 micron diameter capsule with a CAP shell is incapable of providing prolonged protection to unbuffered solutions exposed to a low pH medium like that encountered in the stomach.

Other microcapsule sizes behave similarly. Figure 2 shows the effect of capsule diameter on the rate of decrease of pH of an encapsulated, unbuffered 0.9 % saline solution when θ is 0.255. θ was fixed at 0.255 since this value represents the optimum active agent loading for this particular system, corresponding to a capsule wall thickness of one-sixth the outer capsule diameter. According to Figure 2, none of the capsules considered are able to maintain the pH above 6 for the prolonged period which microcapsules can spend in the stomach, 4-7 hours. The 3000 micron diameter capsule is able to maintain an unbuffered 0.9 % saline solution above pH 5 for only 150 min.

Since the calculated curves in Figures 1 and 2 suggest that a range of CAP capsules immersed in 0.1 M HCl + 0.9 % saline are not capable of providing prolonged protection to pH-sensitive active agents dissolved in unbuffered aqueous media, it is of value to use equation 4 to calculate the hydronium diffusivity that a capsule shell material must possess in order to maintain the pH of an unbuffered solution above a specified pH for a specified time. Assume that the 100 micron capsule considered above has a θ of 0.5 and must be maintained at or above pH 6, 5, and 4 respectively, when it is immersed for 8 hrs in a medium of pH = 1. Equation 4 determines that the shell material for such a capsule must have a hydronium ion diffusivity of 3.5×10^{-16}, 3.4×10^{-15}, and 3.5×10^{-14} cm^2/s in order to maintain the pH inside the capsule > 6, 5, and 4, respectively. Thus, for unbuffered aqueous solutions, capsules designed to maintain the pH of unbuffered solutions inside a capsule at values like 6 or 7 when the capsule is immersed in acidic media for a finite time may require shell materials that have hydronium diffusivity values significantly smaller than that of CAP.

Buffered Encapsulated Aqueous Solution. Significantly, CAP appears to be a suitable enteric shell material if the encapsulated aqueous solution is buffered. Figure 3 contains a series of plots analogous to Figure 1 for a 0.2 M aqueous buffer solution with a pK$_a$ of 7.2. The curves were calculated from equation 12 with $d_o =$ 100 microns and θ varying from 0.10 to 0.95. The addition of the buffer to the 0.9 % saline solution is assumed to have no effect on the diffusivity of the hydronium ions

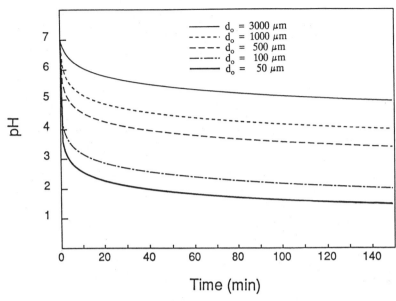

Figure 1. pH-time profiles of unbuffered encapsulated aqueous solutions immersed in acidic media for varying active agent loading. Conditions: $d_o = 100$ microns, $\rho_W = 1.25$ g/cm^3, $\rho_{IP} = 1.0$ g/cm^3, $D = 1.4 \times 10^{-11}$ cm^2/s, $pH_i^o = 7.0$, $pH_b = 1.0$.

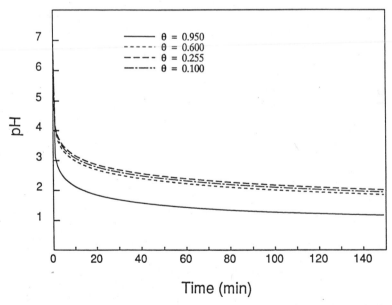

Figure 2. pH-time profiles of unbuffered encapsulated aqueous solutions immersed in acidic media for varying capsule outer diameter. Conditions: $\theta = 0.255$, $\rho_W = 1.25$ g/cm^3, $\rho_{IP} = 1.0$ g/cm^3, $D = 1.4 \times 10^{-11}$ cm^2/s, $pH_i^o = 7.0$, $pH_b = 1.0$.

through the CAP shell, so the hydronium ion diffusivity value reported by Spitael and Kinget for CAP is assumed to be valid. It also is assumed that the buffer ions are unable to diffuse out of the capsule. As shown in Figure 3, the addition of the buffer greatly increases protection to pH sensitive agents, maintaining the pH above 7 for at least 300 min after immersion for $\theta < 0.60$.

Figures 4 and 5 contain a series of calculated plots that illustrate how buffer concentration affects the pH versus time behavior. The plots in Figure 4 are for capsules with $\theta = 0.255$ while those in Figure 5 are for capsules with $\theta = 0.90$. All other parameters are the same as those used to construct Figure 3. Both figures show the major effect that the addition of the buffer has on pH versus time behavior. When the capsule has the optimum θ, a buffer concentration of only 0.02 M maintains the pH of the encapsulated saline solution above 6.2 for at least 120 min. When the buffer concentration is increased to 0.05 M, the pH decreases to approximately 6.9 in 150 min. There is little pH change in 150 min when the buffer concentration is raised to 0.1 M or more. These data suggest that relatively low buffer concentrations are sufficient to protect active agents encapsulated in a CAP shell provided θ is optimum. Figure 5 shows that higher buffer concentrations are required to prevent a significant reduction in pH of an encapsulated 0.9 % saline solution when θ is increased to 0.90. Nevertheless, a 0.2 M buffer solution will maintain the pH inside a 100 micron diameter CAP capsule above 6.5 for over 150 min.

Since CAP is reportedly soluble at pH 7.2, it is reasonable to question the significance of calculations that generate plots like those shown in Figures 3-5. However, it is relevant to note that the CAP hydronium ion diffusivity value used throughout this paper was obtained by exposing one side of a CAP film to an initially 0.9 % saline solution while the other side was exposed to a 0.9 % saline solution that also contained 0.1 M HCl. Spitael and Kinget did not report that the 0.9 % saline solution attacked the CAP film. Since the other side of the CAP film was exposed to 0.1 M HCl, the membrane remained intact. How a buffered pH 7.2 solution inside a microcapsule and an acidic medium outside a microcapsule affect the hydronium diffusivity through the CAP membrane remains to be defined experimentally.

Even if such experiments establish that the hydronium ion diffusivity through a CAP capsule wall differs significantly from that reported by Spitael and Kinget, the plots in Figures 3-5 still contribute to our understanding of the requirements that a microcapsule shell material must meet in order to function as intended in acidic media. Garcia et al. (5) recently described the behavior of 250-300 micron diameter CAP capsules loaded with invertase, a pH sensitive active agent, when exposed to an acidic media. CMC was encapsulated with the invertase because it is a protective agent. Invertase carried by such CAP capsules retained 70 % of its initial value when the capsules were incubated 1 hr in 37 °C pH 1.5 HCl, but only 30 % of this activity after 2 hrs of incubation. Invertase carried by CAP capsules incubated 1 hr in gastric juice retained only 30-40 % of its initial activity. Although the invertase carried by all of the capsule samples evaluated retained activity longer than free or unencapsulated invertase, significant loss of activity still occurred. The CAP capsule shell did not provide total protection. It would be interesting to reexamine the performance of invertase-loaded CAP capsules that also contain an established buffer system that meets the requirements of the mass transfer analysis presented here.

Nomenclature

C_i = inner core concentration of diffusing species, mol/cm^3
C_b = bulk concentration of diffusing species, mol/cm^3
C_i^o = initial inner core concentration of diffusing species, mol/cm^3
C_H = inner core hydronium ion concentration, mol/cm^3
C_H^o = initial inner core hydronium ion concentration, mol/cm^3

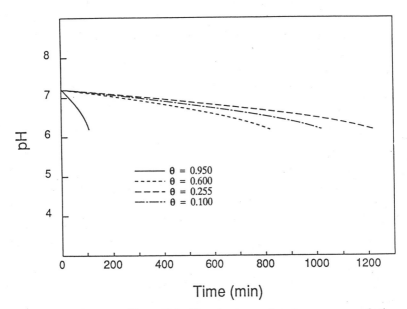

Figure 3. pH-time profiles of buffered encapsulated aqueous solutions immersed in acidic media for varying active agent loading. Conditions: $d_o = 100$ microns, $\rho_W = 1.25$ g/cm^3, $\rho_{IP} = 1.0$ g/cm^3, $D = 1.4 \times 10^{-11}$ cm^2/s, $C_T = 0.2$ M, $pK_a = 7.2$, $pH_b = 1.0$.

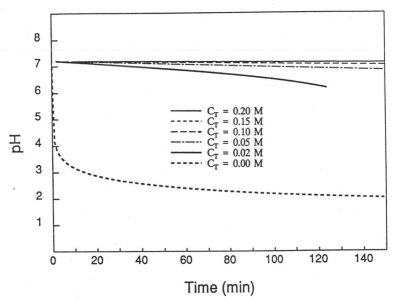

Figure 4. pH-time profiles of buffered encapsulated aqueous solutions immersed in acidic media for varying buffer concentration. Conditions: $d_o = 100$ microns, $\theta = 0.255$, $\rho_W = 1.25$ g/cm^3, $\rho_{IP} = 1.0$ g/cm^3, $D = 1.4 \times 10^{-11}$ cm^2/s, $pK_a = 7.2$, $pH_b = 1.0$.

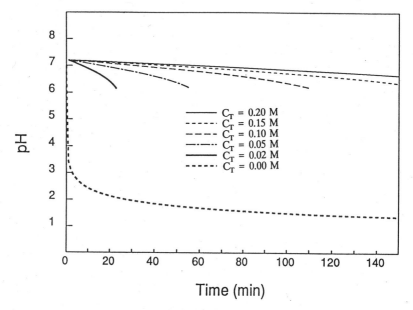

Figure 5. pH-time profiles of buffered encapsulated aqueous solutions immersed in acidic media for varying buffer concentration. Conditions: $d_o = 100$ microns, $\theta = 0.900$, $\rho_W = 1.25$ g/cm³, $\rho_{IP} = 1.0$ g/cm³, $D = 1.4 \times 10^{-11}$ cm²/s, $pK_a = 7.2$, $pH_b = 1.0$.

$C_H{}^f$ = final inner core hydronium ion concentration, mol/cm^3
C_{HA} = inner core undissociated acid concentration, mol/cm^3
C_A = inner core salt ion concentration, mol/cm^3
C_T = inner core total buffer concentration, mol/cm^3
d_o = outer diameter, cm
D = diffusivity, cm^2/s
K = constant, s^{-1}
pH = inner core pH
pH_b = bulk pH
$pH_i{}^o$ = initial inner core pH
R_i = inner radius, cm
R_o = outer radius, cm
R_{C_i} = rate of chemical production of diffusing species, mol/cm^3/s
t = time, s

Greek
ρ_{IP} = inner core density, g/cm^3
ρ_W = capsule wall density, g/cm^3
θ = inner core weight fraction, dimensionless

Literature Cited

1. Svoboda, G.D.; Zhou, J.; Cheng, P.S.; Asif, M.; Distelrath, D.L.; Thies, C. *J. Controlled Release* **1992**, *20*, 195.
2. Merkle, H.P.; Speiser, P. *J. Pharm. Sci.* **1973**, *62*, 1444.
3. Beyger, J.W.; Nairn, J.G. *J. Pharm. Sci.* **1986**, *75*, 573.
4. Maharaj, I.; Nairn, J.G.; Campbell, J.B. *J. Pharm. Sci.* **1984**, *73*, 39.
5. Garcia, I.; Aisina, R.B.; Ancheta, O.; Pascual, C. *Enzyme Microb. Technol.* **1989**, *11*, 247.
6. Eastman Chemical Products *An Introduction to Eastman Cellulose Esters* ; 1974.
7. Spitael, J.; Kinget, R. *Pharm. Acta. Helv.* **1977**, *52*, 47.

RECEIVED October 5, 1992

Chapter 20

Drug Release from Triblock Copolymers of Poly(hydroxyalkyl L-glutamine)–Poly(ethylene oxide)–Poly(hydroxyalkyl L-glutamine)

Chong Su Cho, You Han Bae, and Sung Wan Kim

Center for Controlled Chemical Delivery, Department of Pharmaceutics and Pharmaceutical Chemistry, University of Utah, Salt Lake City, UT 84112

The loading and release behavior of drugs with various aqueous solubilities from potentially biodegradable block copolymer discs were investigated. The block copolymers used were crosslinked poly(hydroxyalkyl L-glutamine)(PHAG)/poly(ethylene oxide)(PEO)/PHAG block copolymers, which were obtained by aminoalcoholysis of poly(γ-benzyl L-glutamate) (PBLG)/PEO/PBLG using various aminoalcohols. The release pattern and rate were influenced by the ratio of the two domains, microstructure morphology, drug solubility, and chemical composition which was determined by alkyl length and degree of aminoalcoholysis. In particular, a combination with low PEO content, low drug solubility, low degree of aminoalcoholysis, and longer alkyl chains led to long term constant release from these matrices. This constant release can be explained by a barrier effect of the continuous PHAG phase.

Drug release from biodegradable or bioerodable polymeric matrices has been intensively investigated for the last decade. One major advantage of using a biodegradable system is to eliminate surgical removal of an implanted delivery device after the delivery system is exhausted. Biodegradable materials used for drug delivery include poly(lactide-co-glycolide)s, polyanhydrides, poly(ortho esters), poly(α-amino acids), and polyphosphazenes. Besides the good biocompatibility and nontoxic by-products of the matrix degradation, one important factor in designing a biodegradable delivery system is to control the release kinetics. Biodegradable delivery systems can also be used to achieve zero-order release kinetics which would be desirable for a long term delivery of drugs having a narrow therapeutic window.

One way to obtain constant release from a biodegradable monolithic device is to use a polymeric prodrug, where drug molecules are bound to a biodegradable polymeric backbone. Poly(α-amino acids) is commonly used as a biodegradable via hydrolytically labile bonds, polymeric backbone (1-4). The drug release rate from a polymeric biodegradable matrix is controlled by several processes, such as penetration of water into the device, hydrolysis of the labile bonds, and diffusion of free drugs out of the device. Zero-order release can be achieved if water penetration or drug diffusion is the rate determining step. Following release, the polymeric backbone would be further degraded.

0097–6156/93/0520–0274$06.00/0
© 1993 American Chemical Society

Another approach to constant release is surface erosion of bioerodable polymer matrices containing dispersed or dissolved drugs (5, 6). The release of the drugs is governed by the erosion of the polymer at the interface between the device surface and the release media. Constant release can be achieved if the device maintains a constant surface area during erosion.

There are few studies on drug release from microphase-separated polymeric matrices (7-10). The factors influencing the drug release pattern and/or the release rate are specific molecular interactions between loaded drugs and polymer microdomains (7), the mode of microdomain structure (8), degradation rate (9, 10), and hydrophilicity of involved segments (10).

We synthesized and characterized potentially biodegradable, microphase-separated block copolymer hydrogels, aiming at a long term implantable drug delivery device. It has been reported that ABA block copolymers, consisting of poly(γ-benzyl L-glutamate)(PBLG) as the A block and poly(ethylene oxide)(PEO) as the B block, showed a microphase-separated structure (11) and that the block copolymer was degraded by proteolytic enzymes both *in vitro* and *in vivo* (12, 13). The PBLG block was derivatized with varying alkyl groups to yield poly(hydroxyalkyl L-glutamine)(PHAG) blocks and then the resulting block copolymers were crosslinked. Hayashi et al.(14) reported that the *in vitro* degradation, by pronase E, of the poly(hydroxyalkyl L-glutamine-co- γ-methyl L-glutamate) fibers was highly dependent on the swelling of the fibers. Pytela et al.(15) also reported that the degradation, by proteolytic enzymes, of the copolymers of hydroxyethyl L-glutamine(HEG) was enhanced by the introduction, or copolymerization, of hydrophobic groups into the poly(hydroxyethyl L-glutamine)(PHEG).

The loading and release behaviors of model solutes with varying aqueous solubilities were investigated using our polymer matrices.

Experimental

Chemicals and Reagents. γ-Benzyl L-glutamate, hydrocortisone (HC), timolol maleate (TM), and verapamil hydrochloride (VP) were purchased from Sigma Chem. Co.(St. Louis, MO). 1,8-Octamethylene diamine (OMDA), 3-amino-1-propanol, 5-amino-1-pentanol, 6-amino-1-hexanol, and triphosgene were purchased from Aldrich Chem. Co., Inc.(Milwaukee, WI). Ruthenium tetroxide was purchased from Polysciences, Inc.(Warrington, PA). Amine-terminated poly(ethylene oxide)(ATPEO, MW:4, 000) was obtained from Texaco Chem. Co.(Houston, TX). Poly(γ-benzyl L-glutamate)(PBLG, MW: 40,000) was obtained from Miles-Yeda LTD.(Israel). All chemicals used were of reagent or spectrometric grade. Tetrahydrofuran, n-hexane, methylene dichloride, dioxane and absolute ethanol were stored with 4 Å molecular sieves and used without further purification.

Synthesis of PBLG/PEO/PBLG Block Copolymer. γ-Benzyl L-glutamate N-carboxyanhydride(BLG-NCA) was prepared by a method described in the literature (16). The method of the block copolymer synthesis was previously reported (11). Briefly, the PBLG/PEO/PBLG block copolymer was obtained by polymerization of BLG-NCA initiated by the amine-terminated PEO in methylene dichloride, at a total concentration of BLG-NCA and ATPEO of 3 % (W/V), at room temperature for 72 hrs. The reaction mixture was poured into a large excess of diethyl ether to precipitate the PBLG/PEO/PBLG copolymer. The resulting copolymer was washed with diethyl ether and then dried *in vacuo*. The unreacted monomer and ATPEO do not precipitate from a mixture of methylene dichloride and diethyl ether. The reaction scheme is shown in Fig. 1.

Preparation of Poly(hydroxyalkyl L-glutamine)(PHAG)/PEO/PHAG Block Copolymer Hydrogel. The PBLG/PEO/PBLG block copolymer was dissolved in dioxane (6 W/V %) at 60°C, casted into a membrane and then the solvent was evaporated for 48 hrs. The prepared membrane was immersed in ethanol solution which contained aminoalcohol (containing 25 x BLG repeat units) and OMDA (0.1 x BLG repeat units) as the crosslinking agent at 65°C for 7 days. The resulting PHAG/PEO/PHAG copolymer hydrogel membranes were washed with ethanol to remove the unreacted aminoalcohols and OMDA. Poly(hydroxyhexyl L-glutamine)(PHHG) homopolymer hydrogel was prepared by a similar method. Elemental analysis of the polymers was performed to determine the degree of aminoalcoholysis of the PBLG homopolymer and PBLG/PEO/PBLG copolymer, using CHN-600 elemental analyzer (LECO Corp., St. Joseph, MI). Once the degree of polymerization of PBLG is known, one can calculate the degree of aminoalcoholysis from the C, H, N values in the elemental analysis after aminoalcoholysis. The reaction scheme is presented in Fig. 2.

[1] H NMR Spectroscopy. ^{1}H NMR spectra of the PBLG/PEO/PBLG block copolymers were measured in a mixed solvent of $CDCl_3$ and trifluoroacetic acid (9/1:V/V) to estimate the copolymer compositions and the molecular weights of PBLG blocks, using a IBM NR/200 FT NMR spectrometer. As the number-average molecular weight (4,000) of PEO is known, one can estimate the number-average molecular weights of the PBLG block and of the copolymer from the copolymer composition calculated from the peak intensities in the spectrum assigned to each polymers (17).

X-ray Diffraction. X-ray diagrams were obtained to detect crystallinity of the copolymer with Philips Electronic Instruments using Ni-filtered CuK_a radiation (35 kV, 15 mA).

Differential Scanning Calorimeter (DSC). The glass transition temperature (Tg) and melting temperature (Tm) of the dried samples, which were weighed (10 to 15 mg) and sealed in stainless steel pans with a rubber O-ring (pan no. : 319-0218), were determined by using DSC (Perkin-Elmer DSC-4) at a heating rate of 20 K/min from 183 K to 623 K. Helium was used as sweep gas (24 ml/min.).

Swelling. Solvent uptake in water and DMF of PHAG/PEO/PHAG discs were obtained by monitoring solvent absorption periodically at 37°C and were determined when no significant weight change was observed. The equilibrium was reached within one day. Solvent uptake was defined as, W_s/W_p, where W_s is the absorbed solvent weight and W_p is the dried polymer weight.

Transmission Electron Microscopy (TEM). The block copolymer discs were exposed to the vapor of ruthenium tetroxide(0.5 wt.-% aqueous solution) at room temperature for 12 hrs and then embedded in epoxy resin by a method described in the literature (18). The embedded specimens were microtomed to obtain ultrathin sections of 50 nm thickness. The ultrathin sectioned specimens were again stained before TEM observation. A JEOL transmission electron microscope (Model 100 CX) was used at an accelerating voltage of 100 kV.

Drug Loading. The dried polymer discs (7 mm in diameter and 0.9 mm in thickness) cut from PHAG/PEO/PHAG copolymer hydrogel membranes were equilibrated in 10 W/V % drug solutions in DMF at 37°C for 24 hrs. After blotting the swollen discs with a paper towel, the discs were vacuum-dried (20-30 mm Hg) at 65°C for at least two days. The weights of dried samples were determined when there

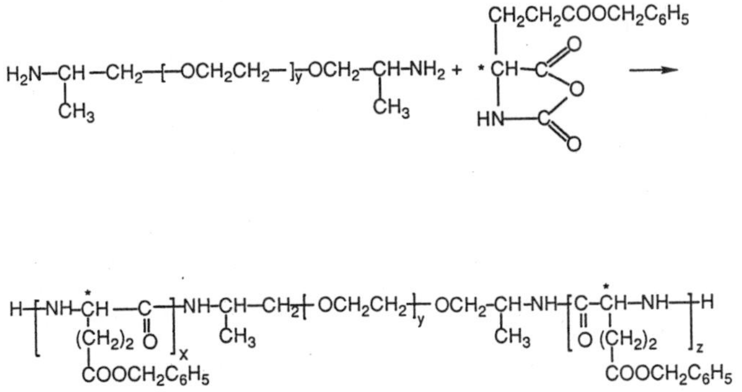

Fig. 1. Synthesis of PBLG/PEO/PBLG Block Copolymer.

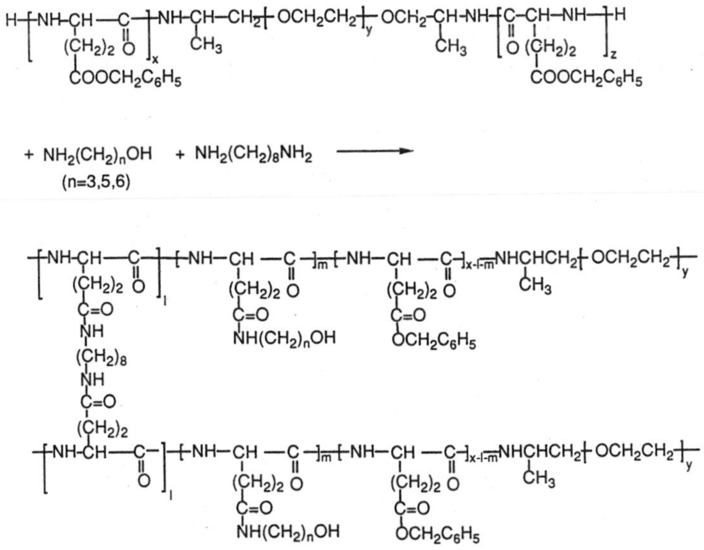

Fig. 2. Synthesis of Crosslinked PHAG/PEO/PHAG Block Coplymer.

was no change in weight over time. The weight of loaded drug (mg) was obtained by substracting the weight of unloaded polymer (mg) from the weight of drug loaded polymer (mg). Drug loading content was calculated as, $[W_{drug}/(W_{polymer} + W_{drug})] \times 100$.

In Vitro Release Studies. The release experiments were carried out in one ml phosphate buffered saline (PBS)(pH=7.4) in a shaking waterbath at 37°C. One ml aliquot was taken and replaced with fresh PBS at specific time points. The concentration of the samples were determined by UV spectrophotometer (Perkin Elmer Lamba 19, Norwalk, Conn.). The released amount of HC, TM and VP in PBS were determined at 242, 294 and 232 nm, respectively.

Results and Discussion

Table I summarizes the results of characterization of the PBLG/PEO/PBLG block copolymers. The copolymer composition was estimated from peak intensities of the benzyl proton signal (5.03 PPM) of the PBLG block and the methylene proton signal (3.67 PPM) of the PEO block in the NMR spectrum. Fig. 3 shows NMR spectrum of the PBLG/PEO/PBLG-3 block copolymer. Also, the number-average molecular weights, Mn, of the copolymers were estimated from the copolymer composition and the molecular weight of PEO chains used.

Table II summarizes the characterization of the PHAG/PEO/PHAG block copolymers and PHHG homopolymer. Aminoalcoholysis of PBLG/PEO/PBLG produces PHAG/PEO/PHAG according to the reaction scheme described in Fig. 2. The membranes used were produced by 20.3 to 85.5 mol-% aminoalcoholysis of the BLG unit in PBLG/PEO/PBLG block copolymer membranes. All the membranes were cross-linked by adding OMDA during aminoalcoholysis. The crosslinking was introduced to prevent the block copolymers from dissolving in DMF during the drug loading process. The thermal properties of the polymers are also given in Table II. The observed glass transition temperature (Tg) for the PEO segments of the PHHG/PEO/PHHG block copolymers was independent of the PEO content except for PHHG/PEO/PHHG-1, where the Tg of the PEO segment was not apparent. The PEO homopolymer was found to have a melting temperature (Tm) of 331 K. No endothermic peak was seen for the block copolymers around the Tm of PEO. This indicates that the PEO segments in these block copolymers may be present in a amorphous state. But the Tm of the PHHG segments was observed at around 615 K for the block copolymers independent of the PEO content, whereas the Tg of the PHHG segments in the block copolymers was not apparent. Thermal degradation for the PHHG segment in the block copolymers occurred near the Tm.

Crystallinities for the PHHG segment in the block copolymers were examined by X-ray diffraction. The wide-angle X-ray diffraction (WAXD) patterns for the PHHG/PEO/PHHG block copolymers and PHHG homopolymer are shown in Fig. 4. The pattern observed for the block copolymer discs depends on the content of PEO and varies from very sharp and intense reflections to diffuse ones with an increase of PEO content in the block copolymers. The first main reflection, corresponding to the intermolecular spacing of α-helical chains (*19*) for the PHHG segment, increases with an increase of PEO content. Also, it was found that PEO forms amorphous domains from the WAXD patterns.

DSC and X-ray results indicate that these PHHG/PEO/PHHG block copolymers exhibit a microphase-separated structure composed of PHHG as the crystalline hard segments and PEO as the amorphous soft segments, although it is not clear which segment is the continuous phase in the block copolymer.

The swelling properties and drug loading content of the crosslinked block copolymers were summarized in Table III. It is apparent from Table III that, as

Table I. Characterization of PBLG/PEO/PBLG Block Copolymers and PBLG Homopolymer

Sample	PEO Content		$\overline{Mn}$ of Polymer[a]
	mol-%	wt.-%	
PBLG	0.0	0	40,000
PBLG/PEO/PBLG-1	18.3	4.3	92,900
PBLG/PEO/PBLG-2	28.6	7.7	52,200
PBLG/PEO/PBLG-3	46.8	15.4	25,900

a) Composition and $\overline{Mn}$ of the copolymers were obtained by NMR measurement.

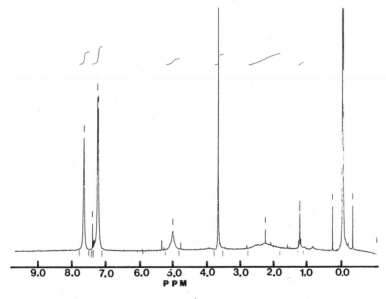

Fig. 3. NMR Spectrum of PBLG/PEO/PBLG-3 Block Copolymer.

Table II. Characterization of PHAG/PEO/PHAG Block Copolymers and
PHHG Homopolymer

Polymer Sample	EO Unit(mol-%)	Degree of Amino-alcoholysis(%)	PEO/Tg (K)	Tm (K)
PHHG	–	54.8	–	615.0
PHPG/PEO/PHPG-1	18.3	85.5	–	–
PHPeG/PEO/PHPeG-1	18.3	83.7	–	–
PHHG/PEO/PHHG-1	18.3	20.3	N.O. a)	N.O. b)
PHHG/PEO/PHHG-2	28.6	22.3	250.6	N.O. b)
PHHG/PEO/PHHG-3	46.8	83.8	251.0	N.O. b)
PEO	100.0	–	231.3	331.0

a) N.O.: not observed

b) Tm of PEO

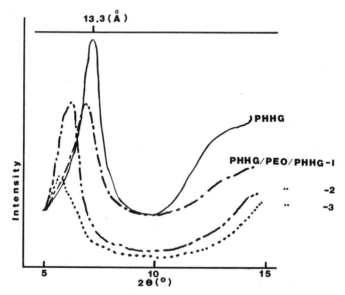

Fig. 4. Wide-angle X-ray Diffraction Patterns of PHHG/PEO/PHHG Block
Copolymers and PHHG Homopolymer.

expected, the degree of swelling of PHHG/PEO/PHHG in water increases with increasing content of PEO and decreasing alkyl chain length of PHAG for the same content of PEO in the block copolymer. With higher PEO content in the composition, the degree of swelling in DMF increased, resulting in higher hydrocortisone loading . As can be seen in the table, the ratio of PHHG to PHHG/PEO/PHHG-1 in solvent sorption uptake is approximately 1.5, while the ratio in the loading content is about 3. Even though there is a considerable deviation in the degree of aminoalcoholysis, the drug loading content in PHHG/PEO/PHHG series is proportional to the PEO content and the degree of swelling in DMF. These observations may indicate that hydrocortisone is preferentialy partitioned in PEO phase in the block copolymers. In fact, the solubility parameter of hydrocortisone [12.4 (cal /ml)$^{1/2}$] is similar to that of the PEO segment (*20*). When the PEO content and degree of aminoalcoholysis remained similar, for example in the case of PHPG/PEO/PHPG-1 and PHPeG/PEO/PHPeG-1, the hydrocortisone loading was affected by alkyl chain length.

To demonstrate the effect of PEO contents on the hydrocortisone release, the cumulative releases from PHHG/PEO/PHHG-1 and PHHG/PEO/PHHG-2, which have similar degrees of loading content (17.8 and 19.3 wt%) and aminoalcoholysis (20.3 and 22.3 %), are compared in Fig. 5. Hydrocortisone released from PHHG/PEO/PHHG-2 in a first-order like pattern during the beginning days, and in this time period about 40 % of loaded drug was released. In case of PHHG/PEO/PHHG-1, the time period (12 hr) and the released amount (less than 5 %) for initial rapid release were significantly reduced in comparision with those of the PHHG/PEO/PHHG-2 matrix. The release rate after initial rapid release fluctuated around 0.1 mg/day/100 mg disc for 120 days and then the rate decreased to about 0.05 mg/day/100 mg disc and continued to keep this rate until 205 days, as demonstrated in Fig. 6. With higher PEO content and degree of aminoalcoholysis (PHHG/PEO/PHHG-3), 80 % of loaded drug was depleted within one week in a first-order like pattern.

These observed phenomena could be explained in relation to either the degree of swelling, microphase-separated structure of the block copolymer matrices or both factors. Fig. 7 shows hydrocortisone release curves from PHPG/PEO/PHPG-1 and PHPeG/PEO/PHPeG-1 matrices. These matrices have equilibrium water uptake (W_s/W_p) of 6.23 and 0.21, respectively (see Table III). The large difference in water uptake may be caused by the side chain packing effects in addition to hydrophobicity of pentyl groups on swelling. Hydrocortisone was released from PHPG/PEO/PHPG-1 matrix up to 95% within one day. With a reduced swelling level, the release rate was reduced as was the initial rapid release. With careful observation from Fig.'s 5 and 7, the release rates and patterns from PHHG/PEO/PHHG-2 and PHPeG/PEO/PHPeG-1 are very close to each other, regardless of the different swelling levels of 0.13 and 0.21. This means that at low swelling levels, the small change in swelling may not be critical in release behavior. Another example can be found with PHPeG/PEO/PHPeG-1 (Fig. 7) and PHHG/PEO/PHHG-3 (Fig. 5). They have exactly the same swelling level, the release from PHHG/PEO/PHHG-3 was much faster than that from PHPeG/PEO/PHPeG-1.

Fig. 8 shows TEM of the crosslinked PHHG/PEO/PHHG block copolymers. By close observation, it was found that bright and dark images were seen in all cases, demonstrating that the block copolymers showed microphase-separated structure consisting of bilayers. It was found that the dark part increases with increasing PEO content in the copolymers. Thus it seems that the bright part should be assigned to the PHHG layer and the dark one to the PEO layer. This darkening is due to ruthenium tetroxide selectively diffusing into PEO domains. It appears that PEO soft segment domains are the isolated phase and are dispersed into the hydrophobic PHHG segment matrices in PHHG/PEO/PHHG-1 (Fig. 8(a)). This result suggests that the PEO domains of the isolated phase can be considered as a drug reservoir, while the

Table III. Swelling and Drug Loading Content of PHAG/PEO/PHAG Block Copolymers and PHHG Homopolymer

Polymer Sample	Swelling(Ws/Wp)		Loading Content(wt.-%)		
	in H_2O	in DMF	HC [1]	TM [2]	VP [3]
PHHG	0.02	0.38	4.9±2.5	–	–
PHPG/PEO/PHPG-1	6.23	–	22.8	–	–
PHPeG/PEO/PHPeG-1	0.21	–	14.2	–	–
PHHG/PEO/PHHG-1	0.05	0.58	17.8±1.2	14.5	11.7
PHHG/PEO/PHHG-2	0.13	0.68	19.3±0.1	–	–
PHHG/PEO/PHHG-3	0.22	1.35	28.6±2.5	–	–

[1] Solubility of Hydrocortisone in water; 0.38 mg/ml

[2] Solubility of Timolol Maleate in water; 130 mg/ml

[3] Solubility of Verapamil Hydrochloride in water; 42 mg/ml

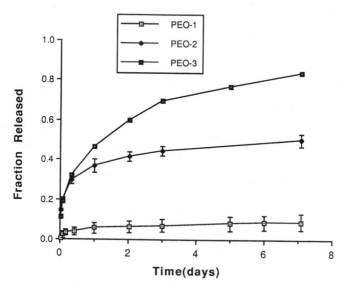

Fig. 5. Fraction of Hydrocortisone Released from the PHHG/PEO/PHHG Block Copolymers with Varying PEO Content.

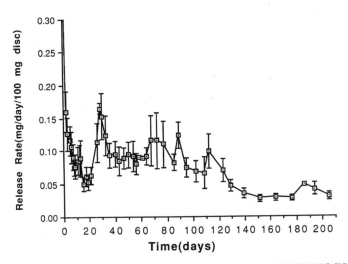

Fig. 6. Long Term Release Rate of Hydrocortisone from the PHHG/PEO/PHHG-1 Block Copolymer Disc(mean: n=3).

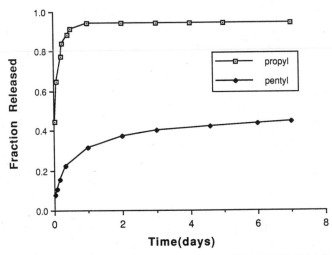

Fig. 7. Fraction of Hydrocortisone Released from the PHAG/PEO/PHAG-1 Block Copolymer with Different Length Alkyl Side Chain.

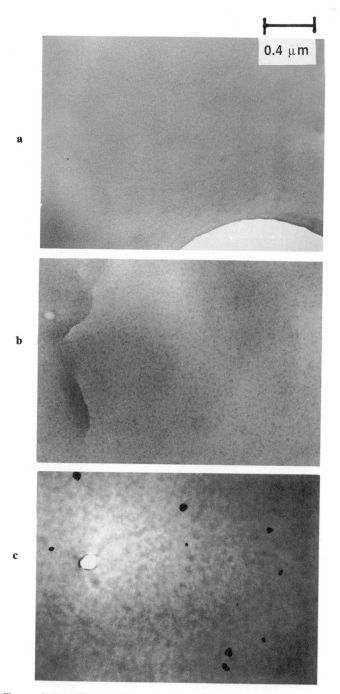

Fig. 8. Transmission Electron Micrographs of PHHG/PEO/PHHG Block
Copolymers. (a): PHHG/PEO/PHHG-1, (b): PHHG/PEO/PHHG-2, and (c):
PHHG/PEO/PHHG-3.

PHHG domains of the surrounding one can be a diffusional barrier for the drug release. Therefore, the long term constant release of hydrocortisone from the PHHG/PEO/PHHG-1 block copolymer device is well correlated with the microstructure of the copolymer. Increased PEO content in the block copolymer seems to promote formation of isolated PHHG segment domains which are dispersed into the PEO segment domains as shown in PHHG/PEO/PHHG-3 (Fig. 8(c)). This result suggests that the continuous PEO segment phase may also act as the transport channel of hydrocortisone, which may lead to the first-order like release pattern of the drug. In case of PHHG/PEO/PHHG-2 (Fig. 8(b)), it seems that the mixing between the PEO and PHHG segment domains occurs. Therefore, some of hydrocortisone loaded into the PEO domains is rapidly released by diffusion and the remaining drug release exhibits constant release kinetics due to the diffusional barrier of PHHG phase.

Assuming similar morphology of PHPeG/PEO/PHPeG -1 to PHHG/PEO/PHHG-1, the role of chemical structure of continuous phase in the hydrocortisone release can be estimated. The major difference between two polymers is the degree of aminoalcoholysis. In PHPeG/PEO/PHPeG-1, 83.7% of BLG units were converted to hydroxy pentyl groups plus crosslinking points, while only 20.3 % of BLG units were reacted to result in hydroxy hexyl groups and crosslinking in PHHG/PEO/PHHG-1. The dominant property of continuous phase of PHHG/PEO/PHHG-1 could result from the population of benzyl glutamate side chains, while it is expected that hydroxy pentyl groups govern the continuous phase in PHPeG/PEO/PHPeG-1. Under this situation, the difference between pentyl and hexyl group should be minor. These effects on release can be observed from Fig.'s 5 and 7. The hydroxy pentyl environment allows for a greatly enhanced release rate relative to the aromatic environment, resulting in rapid depletion of the loaded drug, since the PBLG part in this type of block copolymer constitutes a tight α-helix (*11*). The α-helix in the PBLG part may permit the continuous phase of PHHG/PEO/PHHG-1 to act as a diffusional barrier (*21-23*) for hydrocortisone release.

Another important factor affecting the solute release pattern from a microphase-separated polymer matrix is the hydrophobic/hydrophilic balance of matrix including loaded drug (*24*). This prompted us to examine the nature of loaded drug in the release hehavior. Drugs with increased water solubility (see Table III), verapamil (VM) and timolol maleate (TM), the release curves from the PHHG/PEO/PHHG -1 matrix showed no regions of reduced or constant release, as presented in Fig. 9. This result indicates that even though there is a barrier effect in drug transport from one phase in microstructured polymers, osmotic pressure developed by dissolved drug inside the matrix can disturb the barrier property.

Conclusions

The crosslinked PHAG/PEO/PHAG generated by partial aminoalcoholysis showed microphase-separated structure. The drug release was affected by PEO content, degree of aminoalcoholysis and drug solubility. Constant release kinetics for a drug with low solubility can be obtained from these microphase-separated polymer matrices. The isolated phase can be a drug reservoir, while the surrounding phase can be a diffusional barrier for drug release in a particular situation. The chemical properties of each phase and the nature of loaded drug should be taken into account in the design of an implantable biodegradable polymer system for long term drug delivery.

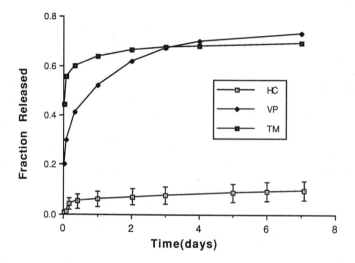

Fig. 9. Fraction of Drug Released from the PHHG/PEO/PHHG-1 Block Copolymers Loaded with Drugs of Varying Solubility.

Acknowledgements

We would like to thank Dr. K. J. Ihn for taking electron micrographs.

This research was supported, in part, by NIH grant HL 44539-05.

Literature Cited

1. Feijen, J.; Gregonis, D.; Anderson, C. G.; Petersen, R. V.; Anderson, J. M. *J. Pharm. Sci.* **1980**, *69*, 871.
2. Negishi, N.; Bennett, D. B.; Cho, C. S.; Jeong, S. Y.; Van Heeswijk, W. A. R.; Feijen, J.; Kim, S. W. *Pharm. Res.* **1987**, *4*, 305.
3. Petersen, R. V.; Anderson, C. G.; Fang, S-M.; Gregonis, D. E.; Kim, S. W.; Feijen, J.; Anderson, J. M.; Mitra, S. In *Controlled Release of Bioactive Materials;* Baker, R., Ed.; Academic Press, New York, NY, **1980**, pp 45-61.
4. Bennett, D. B.; Adams, N. W.; Li, X.; Kim, S. W. *J. Bioact. Compat. Polym.* **1988**, *3*, 44.
5. Heller, J.; Frizinger, B. K.; Ng, S. Y.; Penhale, D. W. H. *J. Controlled Rel.* **1985**, *1*, 225.
6. Leong, K. W.; Brott, B. C.; Langer, R. *J. Biomed. Mater. Res.* **1985**, *19*, 941.
7. Sharma, K.; Knutson, K.; Kim, S. W. *J. Controlled Rel.* **1988**, *7*, 197.
8. Yui, N.; Kataoka, K.; Yamada, A.; Sakurai, Y. *J. Controlled Rel.* **1987**, *6*, 329.
9. Song, C. X.; Sun, H. F.; Feng, X. D. *Polym. J.* **1987**, *19*, 485.
10. Zhu, K. J.; Lin, X.; Yang, S. *J. Appl. Polym. Sci.* **1990**, *39*, 1.
11. Cho, C. S.; Kim, S. W.; Komoto, T. *Makromol. Chem.* **1990**, *191*, 981.
12. Cho, C. S.; Kim, S. U. *J. Controlled Rel.* **1988**, *7*, 283.
13. Lee, K. C.; Chun, C. J.; Cho, C. S.; Kim, Y. H.; Sung, Y. K.; Kwon, J. K. *Proceed. Intern. Symp. Control. Rel. Bioact. Mater.* **1991**, *18*, 666.
14. Hayashi, T.; Ikada, Y. *Biomaterials* **1990**, *11*, 409.
15. Pytela, J.; Saudek, V.; Drobnik, J.; Rypacek, F. *J. Controlled Rel.* **1989**, *10*, 17.
16. Fuller, W. D.; Verlander, M. S.; Goodman, M. Biopolymers **1976**, 15, 1869.
17. Younes, H.; Cohn, D. *J. Biomed. Mater. Res.* **1987**, *21*, 1301.
18. Spurr, A. R. *J. Ultramicrostructure Research* **1969**, *26*, 31.
19. Mckinnon, A.; Tobolsky, A. *J. Phys. Chem.* **1968**, *72*, 1157.
20. Hagen, T. A.; Flynn, G. L. *J. Pharm. Sci.* **1983**, *72*, 409.
21. Chien, Y. W.; Mares, S. E.; Berg, J.; Huber, S.; Lambert, H. J.; King, K.F. *J. Pharm. Sci.* **1975,** *64*, 1776.
22. Graham, N. B.; McNeil, M. E. *Biomaterials* **1984**, *5*, 27.
23. Lee, E. S.; Kim, S. W.; Kim, S. H.; Cardinal, J. R.; Jacobs, H. *J. Membrane Sci.* **1980**, *7*, 293.
24. Bae, Y. H.; Okano, T.; Ebert, C.; Heiber, S.; Dave, S.; Kim, S. W. *J. Controlled Rel.* **1991**, *16*, 189.

RECEIVED October 5, 1992

Chapter 21

Release of a Calcium Channel Antagonist from Radiation-Copolymerized Acrylic Beads

A. B. Majali[1], Y. K. Bhardwaj[1], S. Sabharwal[1], H. L. Bhalla[2], and Piyush Raj[3]

[1]Bhabha Atomic Research Centre, Trombay, Bombay—400 085, India
[2]PERD Centre, Ahmedabad—380 054, India
[3]Bombay College of Pharmacy, Bombay—400 049, India

A series of 2-HEMA based polymer matrices having varying degree of microporous structure, crosslinking and hydrophilicity has been prepared by low temperature radiation induced polymerization. Diltiazem HCl (DTZ. HCl) a calcium channel antagonist was incorporated in the pre-polymerized hydrogels by swelling them in an ethanol-water system (70:30 w/w). The effect of polymer characteristics such as microporosity, crosslinking and hydrophilicity on the drug incorporation as well as release-rate profiles has been investigated. The results indicate that crosslinking within the polymer matrix, rather than the microporosity can effectively control the release. These results have been compared with the chemically polymerized 2-HEMA hydrogels.

Controlled release of a drug from polymer matrices is an effective way to minimize the side effects associated with the use of the drug thereby improving the safety of the drug in use (1-2). The polymer matrices can be prepared either by using chemical initiators at high temperature or by using ionizing radiation. Gamma irradiation offers two distinct advantages over conventional polymerization method viz. (i) it leaves no residual initiator in the gel and (ii) no undesired groups are added on to the hydrogel network. Kaetsu and coworkers have carried out entrapment of various drugs into the polymer matrix by means of radiation-induced polymerization of glass-forming synthetic monomers at low temperature (3-5). The characteristics of this method are (i) a polymer matrix with a microporous structure and large surface area is obtained by using crystallizable solvents, usually water and (ii) the drug release can be controlled by varying the hydrophilicity of the matrix and/or microporous structure (6-7). In all these studies the drug has been incorporated in the polymer matrix during the polymerization process to obtain implantable drug delivery systems. However, two disadvantages are encountered by incorporation of drug in this manner (i) any residual toxic monomer cannot be removed and (ii) radiation degradable drugs cannot be incorporated in the matrices. These problems can be

0097–6156/93/0520–0288$06.00/0
© 1993 American Chemical Society

eliminated if the polymerised matrices are first purified by swelling them in a proper solvent and the drug is subsequently incorporated by swelling the polymer in a saturated solution of drug in another suitable solvent.

In the present paper we report the controlled release of the drug Diltiazem hydrochloride (DTZ.HCl; d-cis-3-acetyloxy-5-[2-(dimethylamino)-ethyl]-2,3-dihydro-2-(4-methoxyphenyl)-1,5-benzothiazepin -4(5H)one), a calcium channel antagonist which has gained increasing acceptance in the treatment of various types of cardiovascular diseases and hypertension (8), from radiation polymerized polymer matrices. The drug has been incorporated by swelling the pre-polymerized polymer matrices in saturated solution of the drug in an ethanol-water (70:30) mixture. The desired release rates have been achieved by varying the crosslinking degree of the polymer network. These results have been compared with chemically polymerized polymer matrices containing DTZ.HCl.

EXPERIMENTAL SECTION

Materials. 2-Hydroxyethyl methacrylate (2-HEMA) and methyl methacrylate (MMA) from Aldrich Chemicals were used after vacuum distillation. Other monomers, N-Vinyl pyrrolidone (NVP) and ethylene glycol dimethacrylate (EGDMA), were 97% pure and used as received. Diltiazem hydrochloride sample was from M/s. Torrent Pharmaceuticals Ltd. and M/s. Cadilla Pharmaceuticals. Diltiazem assay was done spectrophotometrically by measuring its absorption at 234 nm.

Preparation of Radiation Polymerized Beads. The drops of synthetic mixtures of monomers were allowed to fall into a cold precipitation medium (-78°C). The solid globules obtained were irradiated at -78°C with Cobalt -60 gamma rays at a dose rate of 1 KGy hr^{-1}. After polymerization the polymerized beads were thawed gradually in ice water, separated and then dried. These beads were swelled in acetone to remove the unreacted monomer and then dried to constant weight. Chemically polymerized beads were prepared using 1% benzoylperoxide (BPO) as the initiator.

Dynamic Swelling. The dynamic swelling of the polymer matrices was carried out at room temperature (approx. 25°C) allowing the dry beads to swell in ethanol-water mixtures of various compositions. The bead samples were withdrawn at fixed time intervals and weighed with an analytical balance.

Drug Loading in Polymer Beads. A saturated solution of diltiazem hydrochloride was prepared in an ethanol water (70:30 w/w) mixture. About 5 gm. accurately weighed polymer beads were swelled in this saturated drug solution. On attainment of equilibrium swelling after about 48 hrs, the beads were filtered, vacuum dried and rinsed with methanol to wash off the surface drug. The beads were then dried to constant weight.

Drug Release Profiles. The dissolution profile of the beads was determined on USP XXI rotating basket dissolution apparatus. The basket was immersed in 900 ml. of distilled water maintained at 37°C ± 1°C and rotated at 50 rpm. Aliquots of 5 ml were withdrawn at regular intervals for a period of 10 hrs. Each withdrawal was replaced by 5 ml of fresh medium. The amount of drug released was estimated spectrophotometrically at 234 nm using a Shimadzu Spectrophotometer.

RESULTS AND DISCUSSION

Effect of Structural Modifications on Release Rates.

The structural modifications of the polymer matrix can be achieved in two ways viz. (i) by changing the micropore structure of the polymer and (ii) by changing the degree of crosslinking of the polymer network. It has been reported recently that the micropore structures within the polymer matrix can effectively control the rate of drug release (9-10). The radiation polymerization method at low temperature, consisting of glass forming monomers and water as additive, results in a microporous structure because of the formation of ice, which does not polymerize on irradiation, thus leaving a porous polymer matrix. We have also formed such micropore structures by radiation-induced polymerization of HEMA in presence of 30-50 wt.% water, with or without the addition of a crosslinking agent.

Figure 1(A,B) shows the time-dependent release of DTZ.HCl from polymer matrices with varying amounts of micropore structure. The results indicate that unlike previously reported studies (9-10), there is no appreciable change in the release rate profile of the drug although the two matrices have different degrees of swelling (TableI). This difference may be due to the different ways of incorporating the drug in the two systems. When the drug is added during the formation of the polymer network, it is homogeneously entrapped inside the polymer and is dissolved out by "Partition Mechanism" (11). This requires the thermodynamic dissolution of the drug from the polymer phase. Increasing the microporosity of the matrix leads to formation of more channels through which water can diffuse inside the matrix, effectively increasing the surface area of the drug-polymer interface with water, thus influencing the release rates.

In contrast, when the drug is incorporated by a swelling method, there is an inhomogeneous distribution of the drug as it is mainly entrapped in the solvent-filled regions of the gel, i.e. in the channels formed. When such a matrix is swelled in water drug leaching is only from these channels by "Pore Mechanism" which is much faster than from inside the polymer network (partition mechanism) (11). This explains the ineffectiveness of microporosity change in controlling the release rates here.

To control the rate of diltiazem release, 2-4 wt% of the crosslinking agent EGDMA was used to form crosslinks in these polymer matrices. Figure 1 (C,D) shows the variation of cummulative % drug release into solution as a function of time for different amounts of EGDMA. It is apparent from the results that the release of DTZ is markedly affected by changing the crosslinking of the network. The rate of Diltiazem release decreases with increasing EGDMA content indicating that increasing the crosslinking of the poly (HEMA) matrix decreases the mobility of diltiazem significantly as evident from the first-order release rate constant value (K_r) shown in Table I.

Effect of Hydrophilicity on Drug Uptake

Hydrogel properties can also be adjusted through an appropriate mix of

Table I : Parameters for Release of Diltiazem HCl from the Polymer Drug Matrices

S.No.	Monomer (%)				WATER	Dose (M Rad)	EWC* (%)	Drug Loading	K_r (hr^{-1})
	HEMA	NVP	MMA	EGDMA					
1.	70	-	-	-	30	0.6	56.81	145.84	1.4377
2.	50	-	-	-	50	0.6	66.75	167.12	1.5467
3.	68	-	-	2	30	0.6	41.32	82.53	0.8379
4.	66	-	-	4	30	0.6	39.99	80.41	0.6009
5.	56	14	-	-	30	0.6	59.13	94.48	0.6650
6.	50	18	-	2	30	0.6	-	-	0.5018
7.	63.5	-	2.5	4	30	0.6	51.65	49.52	0.7056
8.	61	-	5.0	4	30	0.6	52.22	38.22	0.5237

* Equilibrium water content

hydrophobic and hydrophilic monomers. N-Vinyl-2-Pyrrolidone (NVP) interacts strongly with water and can be useful for increasing the swelling of a gel, while methylmethacrylate (MMA) can be used for reducing the swelling of the gel. Various copolymers of HEMA with NVP or MMA were prepared and their swelling behaviour and drug loading characteristics were studied. These results are indicated in Table I. The swelling behaviour data shows that as expected, the addition of NVP marginally increases the water uptake at equilibrium as compared to poly(HEMA) alone. However, the drug loading in the presence of NVP was found to be lower from the ethanol-water solvent mixture. This can be explained on the basis of the effect of addition of ethanol on the swelling behaviour of poly(HEMA). Organic solvents like ethyl alcohol, which are much less polar than water, can break the hydrophobic bonds in poly(HEMA) thereby inducing a marked increase in the swelling of the gel above its equilibrium swelling in water (12). Addition of NVP makes the matrix more hydrophilic, reducing the hydrophobic content. Hence there is less swelling of the matrix in an ethanol-water mixture, leading to lower drug loading in spite of higher hydrophilicity of the matrix.

Effect of Hydrophilicity on Release Rates

Addition of NVP. Figure 2 shows the effect of addition of NVP to the polymer matrix on the release rate profile of DTZ.HCl. The results show that addition of NVP in radiation polymerization decreases the release rate of the drug as compared to poly(HEMA) alone. This is contrary to the expectation that increasing the hydrophilicity should increase the release rate, as observed in the chemically polymerized beads (Figure 3). Although a definite explanation is not available presently, a plausible explanation can be that since in radiation polymerization the beads are formed in a hydrophobic solvent at low temperature, NVP, being a more hydrophilic monomer, may try to form core of the bead. On swelling, the drug may concentrate inside the polymer matrix and there may be a lower concentration of the drug on the surface, resulting in lower release rate. Further work to confirm this is in progress.

The addition of crosslinking agent (EGDMA) to HEMA-NVP matrix further decreases the release rate as expected. These results are shown in Figure 2(c).

Addition of MMA. The results of addition of MMA on the release rate are shown in Figure 4. As expected the addition of MMA makes the gel more hydrophobic, and decreases the release rate. Further addition of crosslinking agent lowers the release rate of the drug. The results of chemically polymerised HEMA-MMA copolymer are presented in Figure 5 for comparision.

The results of % drug release vs square root of the immersion time are shown in Figure 6. It is evident from the results that a linear relationship is obtained only for matrices which are crosslinked whereas for others a non-linear curve is obtained. It can thus be concluded that rate of drug release is controlled by diffusion of the drug through the swollen polymer matrix to the solid-liquid interface only for the crosslinked matrices.

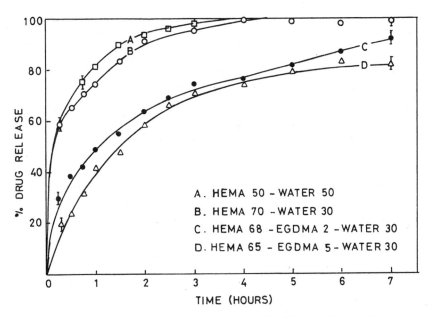

Figure 1 : Effect of microporosity and addition of crosslinking agent (EGDMA) on the release profiles of DTZ.HCl from radiation polymerized copolymer beads.

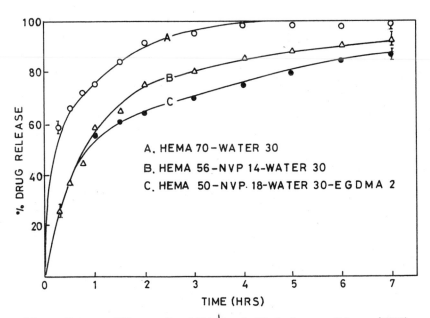

Figure 2 : Effect of addition of N-vinyl pyrrolidone (NVP) and crosslinking agent (EGDMA) on the release profiles of DTZ.HCl from radiation polymerized copolymer beads.

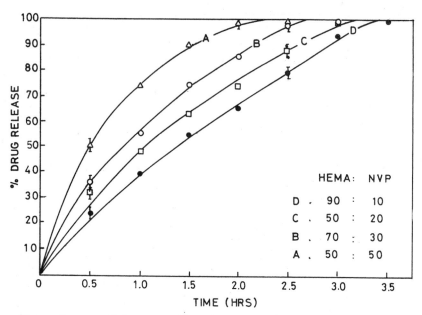

Figure 3 :   Release profiles of Diltiazem.HCl from chemically polymerized poly (HEMA:NVP) beads.

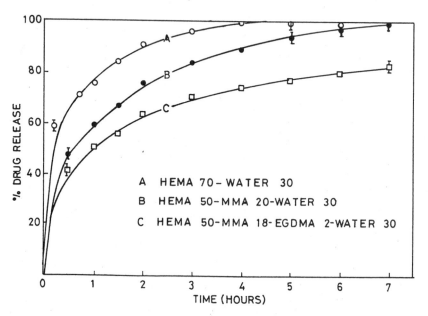

Figure 4 :   Effect of addition of methyl methacrylate (MMA) and crosslinking agent (EGDMA) on the release profiles of DTZ.HCl from radiation polymerized copolymer beads.

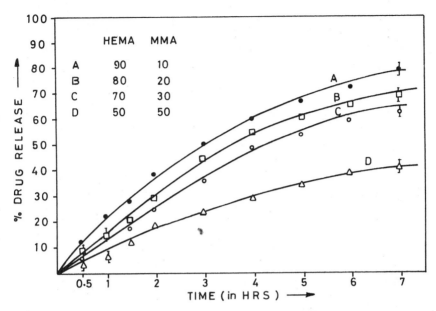

Figure 5 :   Release profiles of Diltiazem. HCl from chemically polymerized beads.

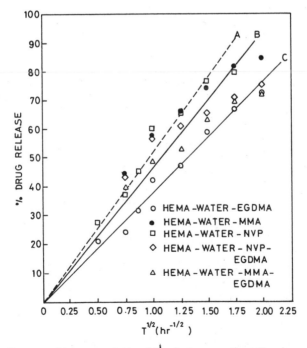

Figure 6 :   Linear relationship between the % drug release from polymer matrices and the square root of time.

SUMMARY

Following conclusions can be drawn from this work:

(i) When the drug is incorporated in radiation polymerised beads by swelling, changing of microporosity does not effectively control the release rate.
(ii) Addition of small amounts of crosslinking agent markedly effects the release rates however, increasing the crosslinking also increases the drug retention.
(iii) Addition of 20% (w/w) MMA decreases the hydrophilicity of the matrix and lowers the release rate but increases the drug retention.
(iv) The absence of micropore structures in the chemically-polymerised beads results in lower drug loading in the matrix as compared to the radiation-polymerised beads. This also leads to better release rate profiles in the case of chemically polymerised beads.

Based on the results, HEMA-WATER-EGDMA (68 : 30 : 2 w/w %) appears to be the most suitable polymer system for the controlled release of Diltiazem HCl as this matrix allows a large % drug loading along with a convenient release rate in which 90 mg. of drug can be released in about 8-10 hours. There is a problem of large % release in the first hour, further work is in progress to overcome this.

ACKNOWLEDGEMENTS

We gratefully acknowledge the free gift of Diltiazem HCl from M/s. Cadilla laboratories Ltd. Ahmedabad and M/s. Torrent Pharmaceuticals Ltd. Ahmedabad.

LITERATURE CITED

1. Veronese F.M.; Ceriotti,G.; Keller, G.; Lora, S. and Carenza, M. Rad. Phys. Chem. 1990, 35, 88.
2. Kaetsu, I. Rad. Phys. Chem. 1985, 25, 517.
3. Kaetsu, I. Yoshida, M. and Yamada, A.J. Biomed. Mater. Res. 1980, 14, 185.
4. Yoshida, M. Asano, M. and Kaetsu, I. Biomaterials, 1983, 4, 33.
5. Yoshida, M. Asano, M. Kaetsu, I. Nakaik; Yuasa, H. and Shida, K. J. Biomed. Mater. Res. 1985, 19, 615.
6. Yoshida M, Asano M. Morita, Y. and Kaetsum, I. Colloid. Polym. Sci. 1987, 265, 916.
7. Ximing, Li. Shen, W. Liu, C. Nishimoto Sei-Ichi and Kagiya, T. Rad. Phys. Chem. 1991, 38, 377.
8. Feld, G. ; Singh, B.N. ; Hosp. Formul. 1985, 20, 814.
9. Yoshida, M. ; Kumakura, M. ; and Kaetsu, I. Polymer 1978,19,1379.
10. Kaetsu, I. ; and Naka, Y. Proc. 4 Japan-China Bilateral symp. on Radiat. Chem. 1989, pp 165.
11. Gehrke, Stevin H. and Lee, Ping I. In "Special Drug Delivery System" (Ed.) Praveen Tyle, Drugs and Pharmaceutical Sciences Series/41, Marcel Dekker Inc. New York (1990) pp 346.
12. Refojo, M.F. J. Poly. Sci. 1967, 5, 3103.

RECEIVED September 24, 1992

Chapter 22

Degradable Polyphosphazene Derivatives
Synthesis and Evaluation

J. Crommen, J. Vandorpe, S. Vansteenkiste, and E. Schacht[1]

Laboratory of Organic Chemistry, University of Ghent,
Ghent B–9000, Belgium

Polyphosphazene derivatives having amino acid ester side groups were prepared by reaction of poly[(dichloro)phosphazene] with ethyl esters of glycine and small amounts of a depsipeptide ester co-substituent. The rate of hydrolytic degradation of the polyphosphazene materials could be controlled by the content of the depsipeptide ester side groups or by blending poly[(amino acid ester)phosphazenes] with poly[(amino acid ester)-co-(depsipeptide ester)phosphazenes]. A more detailed study of the degradation process indicated that in the initial stage of polymer degradation the amino acid ethyl ester is released, with subsequent formation of amino acid, ethanol and polymer backbone cleavage.

Polyphosphazenes (I) are polymers with an inorganic backbone consisting of alternating nitrogen and phosphorous atoms linked by alternating single and double bonds. Starting from the poly[(dichloro)phosphazene] (II) a variety of polymers with variable properties can be prepared by nucleophilic displacement reactions (1-4):

$$\begin{matrix} \text{Cl} & & \text{R} \\ | & & | \\ \text{-[P=N]}^-_n & \longrightarrow & \text{-[P=N]}^-_n \\ | & & | \\ \text{Cl} & & \text{R} \\ \text{(II)} & & \text{(I)} \end{matrix}$$

[1]Corresponding author

0097–6156/93/0520–0297$06.00/0
© 1993 American Chemical Society

Allcock and coworkers, who extensively explored the field of polyphosphazene synthesis, reported that amino-substituted polyphosphazenes (e.g. R, R' =
ethyl esters of amino acids, imidazol, methylamine) are susceptible towards
hydrolytic degradation (5-7) and hold promise as biodegradable materials. Poly[(a-
mino acid ester)phosphazenes] have been used for the preparation of macromolecular prodrugs (8,9) as well as for the production of drug loaded microspheres (10).
Imidazolide-substituted polyphosphazenes were prepared for the production of
bioerodible matrix systems (11).

Searching for biodegradable polyphosphazenes with an adjustable rate of
degradation, we recently prepared a series of polyphosphazenes substituted with
different amino acid ethyl esters. In addition, polyphosphazenes substituted with
ethyl glycinate (R) and variable amounts (0-10%) of a depsipeptide (-NH-CH$_2$-
COO-CH(CH$_3$)-COOEt, gly-lac-OEt) as co-substituent were prepared (12,13). In
the depsipeptide, the ester linkage between the amino acid and the lactide moiety is
sensitive towards hydrolysis. Hence, in aqueous media depsipeptide substituted
polyphosphazenes will undergo initial hydrolysis at the site of this ester linkage
leading to the formation of polymers with pendant acid groups. The degradation of
the polyphosphazene backbone is assumed to be catalysed by carboxylic acids (9).
It was therefore anticipated that introduction of a controlled amount of depsipeptide
side groups could be an elegant way to control the rate of degradation of the amino
acid substituted polyphosphazenes.

The hydrolytic degradation of the poly[(amino acid ester)phosphazenes] and
the poly[(amino acid ester)-co-(depsipeptide ester)phosphazenes] was examined in
phosphate buffer as well as during storage of the bulk materials.

Materials and Methods

Poly[(dichloro)phosphazene] [NPCl$_2$]$_n$ ($\overline{M}_w$ = 8.4 10^5; molecular weight distribution ~1.6) was obtained as a cyclohexane solution (12 wt%) from ETHYL Corporation (USA) and used as received.

Ethyl glycinate hydrochloride was a kind gift by Tessenderlo Chemie
(Belgium). All other L-amino acid ethyl esters and dipeptides were obtained via
Sigma. They were dried before use in a vacuum cabinet over P$_2$O$_5$ at 60°C for 6
h.

All organic solvents were obtained from Janssen Chimica (Beerse, Belgium).
They were dried and distilled prior to use. Triethylamine (Janssen Chimica) was
purified over tosylchloride, subsequently over ninhydrin and finally distilled from
CaH$_2$ before use.

IR-spectra were recorded on a Beckmann IR 4230 apparatus or a Perkin
Elmer (1600 Series) FT-IR, using KBr pellets or polymer films.

Intrinsic viscosities were determined in THF or CHCl$_3$ at 25°C starting from
a 2-3 % solution (g/dl). The determination of the reduced viscosity values were
carried out using an Ubelhode-dilution-viscosimeter (Schott-Gerate, K = 0.0049)
and Schott-Gerate AVS3-P22 equipment.

^{1}H-NMR-spectra were recorded using a 200 MHz (Brucker WP-200), a 360
MHz (Brucker WH-360) or 500 MHz (Brucker WH-500) NMR apparatus. ^{31}P-

NMR-Spectra were recorded using a 206 MHz Brucker AN-500 or a 32.44 MHz Brucker WP 80 SY NMR-apparatus.

Thermal analysis of the polymers was carried out using a Perkin-Elmer Differential Scanning Calorimeter (DSC-2C).

Synthesis of Poly[(amino acid ester)phosphazene]. The poly[(amino acid ester) phosphazenes] (III - VI) were prepared following the procedures described by Allcock (8). Reactions were carried out under an atmosphere of dry nitrogen. Contact with the atmosphere during filtration was kept to a minimum.

The preparation of poly[bis(ethylglycinate)phosphazene] (III) is given as an example. Dried ethyl glycinate hydrochloride (30.7 g; 0.22 mole) was transferred into a 1000 ml flask containing dry benzene (450 ml) and dry purified triethylamine (TEA) (30.5 ml; 0.22 mole). The suspension was stirred at reflux for 3.5 hours and a small amount of benzene was distilled off to remove traces of water. The mixture was then cooled and filtered under dry N_2 atmosphere into a 2000 ml three neck flask. The filtrate was cooled with ice and 11.9 ml (0.086 mole) TEA was added. To this mixture a solution of 5 g (0.043 mole) poly[(dichloro)phosphazene] (42 ml of a 12 % solution in cyclohexane, Ethyl Corporation USA) in 500 ml dry benzene was added dropwise. Stirring was continued first at 0°C for 6 h and then at 25°C for 16 h. After removal of the insoluble hydrochloride salts by filtration, the viscous polymer solution was concentrated by vacuum evaporation of solvent at 30-35°C on a rotary evaporator. Dropwise addition of this concentrate into 1l dry n-heptane (intense stirring) yielded a solid white elastomer, which was reprecipitated from benzene into 1l n-heptane. The obtained polymer was soluble in benzene, THF, $CHCl_3$, CH_2Cl_2 and insoluble in water. The polymer was stored at -25°C under nitrogen or argon atmosphere.

Synthesis of Depsipeptide-containing Poly[(amino acid ester)phosphazenes].

Synthesis of Ethyl N-(benzyloxycarbonyl)-amino-2-(O-glycyl)glycolate. 25.2 g 2-bromoethylacetate (0.15 mole, d=1.5 g/ml) was added dropwise to a solution of 20.1 g N-(benzyloxycarbonyl)-glycine (0.1 mole) and 10.1 g triethylamine ((0.1 mole, d=0.729 g/ml) in 150 ml ethylacetate. The reaction mixture was stirred at reflux during 2 hours. After removal of the insoluble hydrobromide salts, the mixture was washed with diluted HCl, Na_2CO_3 solution and finally with water. After drying over Na_2SO_4, the organic layer was concentrated to a viscous oil. Ethyl N-(benzyloxycarbonyl)-amino-2-(O-glycyl)glycolate(18.4 g, 0.062 mole, yield = 62 %) can be isolated by crystallization from n-pentane. TLC (EtOAc/hexane (1/1)): R_f=0.89. ^{1}H-NMR ($CDCl_3$): δ=7.2 (s, phenyl, 5H); δ=5.1 (s, benzyl, 2H); δ=4.1 (s, methylene, 2H); δ=4.7 (s, 2H); δ=4.2 (q, 2H(Et)); δ=1.2 (t, 2H(Et)). IR (KBr): cm^{-1} = 3400 (s, NH); 3000 (s, CH); 1740 (s, C=O); 1650 (s, C=C, phenyl).

Synthesis of Ethyl N-(benzyloxycarbonyl)-amino-2-(O-glycyl)lactate. 19.9 g ethyl-2-bromo propionate (0.11 mole, d=1.394 g/ml) was added dropwise to a solution of 20.1 g N-(benzyloxycarbonyl) glycine (0.1 mole) and 10.1 g triethylamine (0.1 mole, d=0.729 g/ml) in 150 ml ethyl acetate. The reaction mixture was stirred at reflux for 48 h. After removal of the insoluble hydrobromide salt by filtration the mixture was washed with diluted HCl, $NaCO_3$ solution and finally with water. After drying over Na_2SO_4, the organic layer was concentrated to a colorless viscous oil. Ethyl N-(benzyloxycarbonyl)-amino-2-(O-glycyl)lactate (28.5 g, 0.092 mole, yield = 92 % was isolated by adsorption chromatography on silica gel (EtOAc/hexane 1:11(v/v)).TLC (EtOAc/hexane (1/1)): R_f=0.57. ^{1}H-NMR (CDCl$_3$) : δ=7.3 (s, phenyl, 5H); δ=5.05 (s, benzyl, 2H); δ=3.95 (s, methylene, 2H); δ=5.05 (q, methyn, 1H); δ=1.4 (d, methyl, 3H); δ=4.1 (q, 2H(Et)); δ=1.2 (t, 2H(Et)). IR (KBr): cm^{-1} = 3400 (s, NH); 3000 (s, CH); 1740 (s, C=O); 1650 (s, C=C, phenyl).

Removal of the N-protective Group. A solution of 5 g ethyl N-(benzyloxy-carbonyl)-amino-2-(O-glycyl)glycolate and 1.07g oxalic acid in 500 ml absolute ethanol was hydrogenated in a Parr Reductor apparatus in presence of 1 g Pd/C catalyst (hydrogen pressure: $\approx$30 psi, 30-35°C). When hydrogen pressure remained constant (1-2 h), the catalyst was removed from the reaction mixture by filtration. The filtrate was allowed to cool to 0°C overnight. This resulted in the formation of ethyl 2-(O-glycyl)glycolate (gly-glycOEt) ammonium oxalate crystals which could be isolated by filtration and drying under vacuum. TLC (acetone/CHCl$_3$ (85/15)): R_f = 0.19), (yield = 90 %). ^{1}H-NMR (CD$_3$OD): δ=3.9 (s, methylene, 2H); δ=4.8 (s, 2H); δ=4.2 (q, 2H^{Et}); δ=1.2 (t, 3H^{Et}). IR (KBr): cm^{-1} = 3400 (s, NH); 3100-3400 (sh, NH$^+$); 3000 (s, CH); 2200-2600 (w, NH$_3$); 1740 (s, C=O), 1600 (w, NH bending); 1230 (s, ester).

The same procedure was used for removing the protective group from the other depsipeptide derivatives.

Synthesis of Poly[(ethyl 2-(O-glycyl)lactate)-co-(ethyl glycinate)phospha-zene] (VII - XIII). Changes in reaction conditions caused by different handling of analogous products can cause diversity in molecular weight of the resulting poly-mers. For this reason, polymers of different degrees of ethyl 2-(O-glycyl)lactate substitution were prepared simultaneously. Reaction times and temperatures were limited to a minimum to prevent intensive chain cleavage during synthesis.

To a suspension of dried ethyl 2-(O-glycyl)lactate ammonium oxalate (3.4 g; 0.0145 mole) in 75 ml anhydrous acetonitrile was added 2.02 ml (0.0145 mole) dry triethylamine (TEA). The mixture was then stirred for a few minutes until a clear solution was obtained. This solution was then added dropwise to an excess of dry THF (200 ml). After removal of insoluble oxalate salt by filtration, calculated volumes of the filtrate (containing the desired moles of ethyl 2-(O-glycyl)lactate) were transferred into separate 500 ml flasks, and diluted with dry THF to a total volume of 165 ml. These mixtures were then cooled to 0°C as dry triethylamine was added. To each of these reaction mixtures a solution of 1 g [NPCl$_2$]$_n$ (8.75 ml 11.4 % solution in cyclohexane) in 80 ml dry THF was added dropwise. Stirring was continued for 19 hours. Meanwhile glycine ethyl ester hydrochloride (19.25 g) was transferred into a 500 ml flask containing dry THF (240 ml) and TEA (19.1 ml). This mixture was stirred and refluxed for 3.5 hours and was then cooled to room temperature and filtered. When the reaction of ethyl 2-(O-glycyl)lactate with [NPCl$_2$]$_n$ was completed, the glycine ethyl ester solution is divided into four equal parts (60 ml) and each part is added to one of the reaction mixtures at 0°C together with dry purified TEA. The reaction mixtures were then stirred for an additional 19 hours. After removal of the insoluble hydrochloride salts by filtration, the polymer solution was concentrated by vacuum evaporation at 30-35°C. Addition of the resulting viscous polymer solutions to 300 ml dry n-heptane while stirring yielded a polymer gel. Reprecipitation from dry chloroform into 300 ml dry ether yielded the poly[(ethyl 2-(O-glycyl)lactate)-co-(ethyl glycinate)phosphazene] as a white solid polymer free of most HCl-salts (Beilstein test). Residual solvent was removed by drying under vacuum to give a slightly yellow-colored product.The degrees of substitution (x and y) for the different products were respectively: x = 0.45, y = 0.55; x = 0.25, y = 0.75; x = 0.10, y = 0.90. Upscaling of the reactions to 10-15 g quantities yielded no major experimental problems. Variation in the degree of ethyl 2-(O-glycyl)lactate substitution is possible by variation of the reactant to chloropolymer ratio. Addition of small amounts (0.5-3 equivalents relative to P-Cl) of ethyl 2-(O-glycyl)lactate in this procedure results in polymers containing the corresponding amount of side group.

In the ^{1}H-NMR (CDCl$_3$)of [NP(NH^{1}CH$_2$COO^{2}CH(^{3}CH$_3$)COOEt)$_{2x}$(NH^4-CH$_2$COOEt)$_{2y}$]$_n$ all signals were broadened and unresolved: δ=5.1 (b, 1H^2); δ=4.1 (b, 2H^{Et}+2H$^{Et'}$); δ=3.8 (b, 2H^1+2H^5); δ=1.5 (b; 3H^3); δ=1.25 (b, 3H^{Et}+-3H$^{Et'}$). ^{31}P-NMR (CDCl$_3$): polymer δ=3-4 ppm (broadened singlet, relative to H$_3$PO$_4$). IR (film on KBr): cm^{-1} = 3600-2800 (s, NH); 3000 (vs, CH), 1740 (vs, C=O ester); 1200-1400 (vs, P=N).

Preparation of Polymer Pellets (d=1 cm, h=1.5-2 mm) by Heat Compression.
For the preparation of the pellets, a matrix was used which allowed processing of four devices at the same time. 4 times 200 mg of polymer was placed into the cavities (d= 1cm) of the preheated matrix. The pressure is generated by means of a heated hydraulic press (Fontijne apparatus, Vlaardingen, Holland).

In Vitro Degradation. For each type of polymer, a number of pellets (ca 200 mg, d = 1.00 cm, h = 1.5-2 mm) were placed in 20 ml phosphate buffer solution (SØrensen buffer, pH = 7.40, containing 0.01 % (wt/v) NaN$_3$) at 37°C. The flasks were placed in a bench shaking apparatus. Mass loss during hydrolysis was determined by removal of pellets from the degradation medium at regular time intervals, drying in vacuum over P$_2$O$_5$, and subsequent weighing. Swelling of the pellets during degradation was monitored by removal of surface water with filter-paper and mass determination of the pellet prior to drying. Release of aminoacids and their ethyl esters from the degrading material can be observed by means of FmocCl-HPLC analysis as discussed below. Duplicate experiments, showed an acceptable reproducibility of the experimental method used in these hydrolysis experiments. Identical studies with model polymers were performed in pure water at 37°C in order to allow elemental analysis of the devices or to permit deter-mination of phosphates formed upon hydrolysis of the polymer in this medium.

Analytical Techniques.

Analysis of Amino Acids or Amino Acid Esters. The method of Goede-moed and Mense (*14*) was used for the determination of amines by means of High Performance Liquid Chromatography (HPLC). This procedure comprises reversed phase HPLC with gradient elution and precolumn derivatization with 9-fluorenyl-methyloxycarbonylchloride (Fmoc-Cl).

The following chromatographic system was used: a KONTRON 420 chromatographic system (two pumps, M800-high pressure eluent mixing cell) connected to a KONTRON SFM-25 fluorescence detector both controlled by a multi-tasking personal computer system (KONTRON-software). The excitation wavelength was set at 270 nm, the emission wavelength at 310 nm (High Voltage = 325 V, Response = 2.0 s). A Rheodyne Model 7125 injection valve with a 20 μl sample loop was used. Chromatographic separation was performed by the use of a Dupont ZORBAX ODS 5μm reversed phase column (4.6 * 150 mm) and with following gradient elution: solvent A contained 40 % CH_3CN and 60 % acetic acid buffer (pH = 4.1), solvent B contained 70 % CH_3CN and 30 % acetic acid buffer. The acetic acid buffer consisted of 3 ml 100 % acetic acid per liter water, adjusted with sodium hydroxide to pH 4.1.

The following derivatization procedure was applied: standard solutions were sampled as prepared. Samples taken from in vitro degradation experiments were centrifuged and supernatant was taken for further analysis. Reaction medium samples in benzene/THF were diluted first to the appropriate concentrations with 2 % CH_3OH in THF. 25 μl of sample was mixed with 50 μl of 1M borate buffer (pH = 5.0) and 425 ml water and with 500 ml of Fmoc-Cl reagent (5 mg into 20 ml CH_3CN) solution. The mixture was allowed to stand for exactly 10 min. at ambient temperature and 20 μl of this solution was analyzed as described above.

A linear calibration curve for amino acids and their esters was obtained in the 1-1000 nmole/ml range (peak hight and peak surface mode). The detection limit was far below 0.1 nmole/ml (0.01 ppm for glycine).

Determination of Phosphates in Aqueous Media. Pellets (d = 1 cm, h = 1.5-2 mm, m = ca 200 mg) of the polyphosphazene derivative $[NP(GlyLacOEt)_{0.2^-}(GlyOEt)_{1.8}]_n$ were prepared by means of heat compression and were placed in 20 ml water containing 0.01 % NaN_3 at 37°C. At certain time intervals the degradation medium was centrifuged and the precipitate was removed. Quantitative determination of phosphates formed upon degradation of the polymeric material can be carried out by a variant procedure of the Molybdic Blue UV-method of Jones for the determination of phosphates. The acidity of the derivatization mixture in the Molybdic-Blue method was reduced to a minimum to prevent acid catalyzed degradation of polymer(fragments). The principle of the method is based on the reduction of molybdo-phosphate to molybdenum blue. The reduction is carried out by ascorbic acid at 37°C, and the colour formed was compared spectrophotometrically using standard solutions: a) ammonium molybdate 1.25 wt % $(NH_4)_6Mo_7O_{24}.4H_2O$ in 3N H_2SO_4; b) 10 wt % ascorbic acid in Millipore water; c) CHEN-Reagent: 2

volumes a) plus 1 volume b) diluted with 2 volumes of Millipore water; d) standard PO_4^{3-}-solutions in Millipore water: 50 and 100 ppm KH_2PO_4 solutions.

Technique used: 100 μl of sample into a teflon test tube was diluted with 7 ml of CHEN-Reagent. The reaction mixture was allowed to stand for 30-35 minutes at 37°C. The absorbance of the coloured solution was measured at 822 nm wavelength. An identical procedure carried out for the standard solutions treated allows exact calibration of the method used: 50 ppm standard: A = 1.268; 100 ppm Standard: A = 2.529. As a blank, a test was carried out with pure in-vitro medium, which was allowed to stand for the same period during which the polymer was hydrolysed.

Results and Discussion

Poly[(amino acid ester)phosphazenes]. The preparation of polyphosphazenes containing amino acid ester sidegroups (III - VI), by reaction of an excess of the amino acid esters with the poly[(dichloro)phosphazene], has been described before by Allcock (*8*). It was reported that quantitative chlorine displacement could be obtained with ethyl glycinate and ethyl alanate. For amino acid esters with a more bulky side group (e.g. phenyl alanine, leucine) only partial displacement was achieved. The residual P-Cl moieties could be quantitatively substituted by reaction with methylamine.

In an analogous way, we have prepared a series of polyphosphazenes containing ethyl esters of amino acids (Table I). The consumption of amino acid esters during the reaction with the dichloropolymer was followed by means of HPLC, using the method developed by Goedemoed (*14*). For most esters, except glycine and L-alanine, the chlorine displacement was not quantitative. Residual chlorines were replaced by subsequent reaction with ethyl glycinate.

Table I. Characterization of the Poly[(amino acid ester)phosphazenes]
$[NP(AAOEt)_{2x}(glyOEt)_{2y}]$

#	AA	x^a	y^a	T_g (°C)[b]	T_m (°C)[b]	[η] (dl/g)[c]
III	gly	0.00	1.00	-18	34	1.07
IV	ala	1.00	0.00	-5	-	1.28
V	phe	0.70	0.30	23	-	1.24
VI	leu	0.45	0.55	-5	-	0.36

[a] $x+y=1$; substituent ratios and compositions (x,y) were determined by means of 360 MHz ^{1}H NMR spectroscopy.
[b] Determined by DSC.
[c] Intrinsic viscosity values measured in THF at 25°C.

Poly[(amino acid ester)-co-(depsipeptide ester)phosphazenes].

Synthesis of Depsipeptide-substituted Polymers (VII - XIII). Depsipeptides are hybrid dimers composed of an amino acid and a glycolic or lactic acid ester:

$$
\begin{array}{c}
\text{O} \\
\| \\
\text{H}_2\text{N-CH-C-O-CH}_2\text{-C-O-R}^{\prime} \\
| \qquad\qquad \| \\
\text{R} \qquad\qquad\; \text{O}
\end{array}
\qquad\qquad
\begin{array}{c}
\text{O} \quad\; \text{CH}_3 \\
\| \qquad | \\
\text{H}_2\text{N-CH-C-O-CH-C-O-R}^{\prime} \\
| \qquad\qquad\quad \| \\
\text{R} \qquad\qquad\; \text{O}
\end{array}
$$

amino acid glycolate   amino acid lactate

Depsipeptides have been proposed as suitable prodrugs for amino acids (*15*). Poly(depsipeptides) were prepared recently as an alternative class of biodegradable polymers suitable for biomedical applications (*16,17*).

The ester bond between the amino acid and the lactate or glycolate moiety is hydrolytically labile. Hence, depsipeptides are attractive candidates as hydrolysis-sensitive acid precursors. Therefore, they were selected for the preparation of polyphosphazenes with controllable rates of degradation. Glycine and alanine esters of ethyl glycolate and ethyl lactate were prepared in analogy to the methods of Schwyzer (*18*) and Wermuth (*15*). A methanol/water mixture was selected as a suitable medium for initial degradation studies since all tested depsipeptides are readily soluble in methanol. The half-life times for hydrolysis given in Table II demonstrate a faster hydrolysis for the amino acid esters of glycolate. This phenomenon can be explained by the more hydrophobic nature of the lactate residue (an additional methyl group) compared to the glycolate moiety.

Table II. Hydrolytic Stability[a] of Selected Depsipeptides
$H_2N\text{-}CH(R_1)\text{-}COO\text{-}CH(R_2)\text{-}COOEt$

code	R_1	R_2	$T\frac{1}{2}$ (h)	k_{obs} (s^{-1})[b]
gly-glyOEt	H	H	3.9	0.178
gly-lacOEt	H	CH$_3$	12.0	0.058
ala-glycOEt	CH$_3$	H	4.4	0.156
ala-lacOEt	CH$_3$	CH$_3$	15.2	0.045

[a]Hydrolysis of depsipeptides in methanol containing 10 equivalents of water (resp. to depsipeptide) at 25°C.
[b]Rate constant calculated assuming pseudo first order hydrolysis kinetics.

Polyphosphazenes containing varying amounts of gly-lacOEt or ala-lacOEt were prepared by reaction of poly[(dichloro)phosphazene] with a selected amount of the depsipeptide (0-0.10 equivalents with respect to P-Cl). Remaining chlorides were substituted by subsequent reaction with ethyl glycinate, as illustrated below:

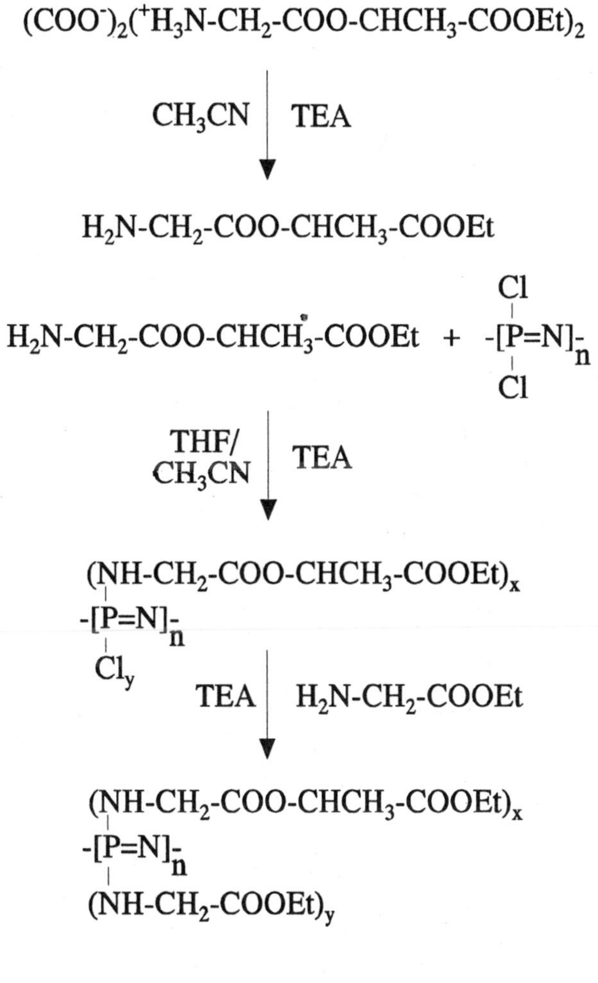

$$(COO^-)_2(^+H_3N-CH_2-COO-CHCH_3-COOEt)_2$$

$$\downarrow \quad CH_3CN \mid TEA$$

$$H_2N-CH_2-COO-CHCH_3-COOEt$$

$$H_2N-CH_2-COO-CHCH_3^{\bullet}-COOEt \quad + \quad -[P=N]_n^- \overset{Cl}{\underset{Cl}{|}}$$

$$\downarrow \quad \overset{THF/}{CH_3CN} \mid TEA$$

$$(NH-CH_2-COO-CHCH_3-COOEt)_x$$
$$-[P=N]_n^-$$
$$Cl_y$$

$$\downarrow \quad TEA \mid H_2N-CH_2-COOEt$$

$$(NH-CH_2-COO-CHCH_3-COOEt)_x$$
$$-[P=N]_n^-$$
$$(NH-CH_2-COOEt)_y$$

The consumption of the depsipeptides in the first stage of the substitution was followed by HPLC analysis of the reaction medium. It was observed that the alanyl esters did react much slower than the glycine esters. This is in agreement with the work of Allcock (*17*) and our own experience with the preparation of amino acid ester substituted polymers, indicating that the P-Cl substitution is more difficult to achieve with more bulky amines.

For the subsequent degradation studies, polyphosphazenes substituted with ethyl glycinate and co-substituted with varying amounts of gly-lacOEt (D.S.: 0.5 - 10%) were prepared. The composition and some properties of the polymers are summarized in Table III. It is clear that introduction of increasing amounts of the hydrolytic labile side groups results in a decreased intrinsic viscosity of the reaction product. Apparently, hydrolysis does occur to some degree during the preparation procedure.

Table III. Characterization of poly[(ethyl 2-(O-glycyl)lactate)-co-(ethyl glycinate) phosphazenes] [NP(gly-lacOEt)$_{2x}$(glyOEt)$_{2y}$]

#	x^a	y^a	T_g $(°C)^b$	T_m $(°C)^b$	$[\eta]$ $(dl/g)^c$
VII	0.45	0.55	23	-	0.19
VIII	0.23	0.77	20	-	0.20
IX	0.10	0.90	3	-	0.20
X	0.05	0.95	-6	-	0.75
XI	0.03	0.97	-6	-	0.78
XII	0.01	0.99	-15	-	0.72
XIII	0.005	0.995	-16	35	0.89

[a]$x+y=1$; substituent ratios and compositions (x,y) were determined by means of 360 MHz ^{1}H NMR spectroscopy.
[b]Determined by DSC.
[c]Intrinsic viscosity values measured in CHCl$_3$ at 25°C.

Introduction of gly-lacOEt side groups has a significant effect on the phase transition temperatures. The glass transition temperature of the polymers increases with increasing content of the gly-lacOEt cosubstituent. Whereas poly[bis(ethyl glycinate)phosphazene] is partially crystalline, the introduction of small amounts of gly-lacOEt side groups (> 1%) results in a loss of crystallinity of the material.

In Vitro Degradation Studies. Pellets of polyphosphazenes (VII - XIII) with varying content of gly-lacOEt side groups (d: 10 mm; h: 1.5 mm), prepared by heat compression (40°C; 8 kN/cm^2; 4 min) were immersed in phosphate buffer pH 7.4 at 37°C. At regular time intervals samples were withdrawn, dried and weighted. The mass loss as a function of incubation time is given for a series of polymers in Figure 1. The arrows indicate the time at which the pellets start to disintegrate. It is clear from this figure that the introduction of small amounts of depsipeptide side groups results in a remarkable increase of the sensitivity of the polymers towards hydrolytic degradation. Whereas the poly[bis(ethyl glycinate)phosphazene] (III) degrades slowly and a 20% mass loss is attained only after 100 days, the introduction of as little as 1% of the gly-lacOEt side groups results in a disintegration of the pellet after about 2 weeks. The half-life and the observed rate constant (k_{obs}) for the hydrolysis of ethyl 2-(O-glycyl) lactate and ethyl glycinate in the same medium at 37°C are given in Table IV.

Table IV. Influence of Polymer Side Groups on Hydrolysis Rates

side group	T½ (h)	k_{obs} (h^{-1})
ethyl 2-(O-glycyl lactate)	1.41	0.49
ethyl glycinate	15.84	0.045

The increased rate of hydrolysis of the depsipeptide substituted polymers is obviously due to the hydrolytic susceptibility of the depsipeptide.

The release of glycine and ethyl glycinate as well as the formation of phosphate ions during the in vitro degradation of product (IX) (10% depsipeptide side groups) is shown in Figure 2. It demonstrates that both glycine and ethyl glycinate are present in the degradation medium. Moreover, it was found that the mass loss is larger than that calculated from the amount of amino acid released. NMR analysis of the release medium indicated the presence of water soluble polymer fragments.

Polymer Blends. The above data demonstrate that the biodegradability of polyphosphazene materials can be controlled by the content of depsipeptide side groups. An alternative approach is the blending of a hydrolytic sensitive depsipeptide containing polymer with poly[bis(ethyl glycinate)phosphazene]. The feasibility of this approach is illustrated in Figure 3. The advantage of this method is that a variety of materials with variable degrees of degradability can be obtained starting from two master polymers. From a practical viewpoint this is more attractive than having to prepare a polymer with a given composition of side groups for each desired product. The straightforward explanation for the degradability of the blends would be that the acid side groups generated in the depsipeptide-containing polymer do promote the degradation of both polymers in the blend. However, it was found that for a 50/50 (w/w) blend of poly[(ethyl glycinate)-co-(ethyl 2-(O-glycyl lactate) phosphazenes] with poly[(ethyl alanate)phosphazene] the presence of the depsipeptide substituted polymer does not affect the release of the alanine side groups. Only the release of glycine and ethyl glycinate is significantly enhanced (Figure 4). At present, there is no satisfactory explanation for this unexpected phenomenon.

Conclusion

Polyphosphazenes having amino acid ester side groups are an interesting class of hydrolytically degradable polymers. The properties of the polymers can be widely varied by a proper choice of the side groups. The rate of degradation can be adjusted by the introduction of a controlled amount of depsipeptide side groups and by blending hydrolytically sensitive polymers with more stable derivatives.

Given the versatility in chemical composition, mechanical properties and physico-chemical properties, amino acid substituted polyphosphazenes are promising candidate materials for various biomedical applications.

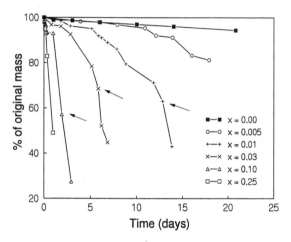

Figure 1.   Percent mass loss of poly[(ethyl 2-(O-glycyl)lactate)-co-(ethyl glycinate)phosphazenes] [NP(GlyLacOEt)$_{2x}$(GlyOEt)$_{2y}$] in PBS at 37°C.

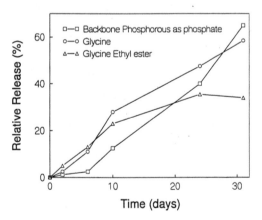

Figure 2.   Release of glycine, ethyl glycinate and phosphate during the degradation of polymer XII in water at 37°C.

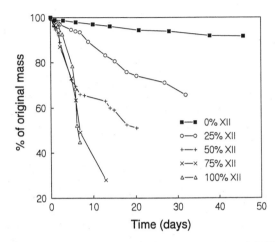

Figure 3. Percent mass loss of blends of products (III) and (XII) in PBS at 37°C (wt% XII indicated).

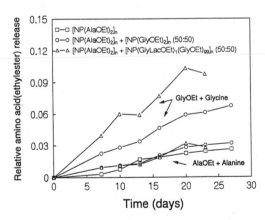

Figure 4. Release of amino acids and amino acid esters from blends (50/50, w/w) of product (IV) with product (III), resp. product (XII) in PBS at 37°C.

Acknowledgment

The authors thank the Belgian Institute for Encouragement of Research in Industry and Agriculture (I.W.O.N.L.) for providing a research fellowship to J.Crommen and J.Vandorpe. This work was also funded in part by the National Research Council (I.U.A.P. Programme).

Literature Cited

1. Allcock, H.R. *Chem.Rev.* **1972**, *72*, 315.
2. Allcock, H.R. *Contemporary Topics in Polym. Sci.* **1979**, *3*, 55.
3. Allcock, H.R. *Makromol.Chem., Macromol.Symp.* **1986**, *6*, 101.
4. Goedemoed, J.H.; Crommen J.H.; Schacht, E.H. In *Polyphosphazene drug delivery systems for antitumor treatment*; Goedemoed, J.H., Ed.; V.U. University press, Amsterdam, 1990, chapt.1.
5. Allcock, H.R.; Fuller, T.J.; Mack, D.P.; Matsumura, K.; Smeltz, K.M. *Macromolecules* **1977**, *10*, 824.
6. Allcock, H.R.; Fuller, T.J.; Matsumura, K. *Inorg.Chem.* **1982**, *21*, 515.
7. Allcock, H.R.; Scopelianis, A.G. *Macromolecules* **1983**, *16*, 715.
8. Grolleman, C.W.; Visser, A.C.; Wolcke, J.G.; Klein, C.P.; Van der Groot, H.; Timmerman, H. *J.Controlled release* **1986**, *4*, 119.
9. Allcock, H.R.; Hymer, W.C.; Austin, P.E. *Macromolecules* **1983**, *16*, 1401.
10. Goedemoed, J.H.; Mense, E.H.; De Groot, K.; Claessen, A.M.; Scheper, R.J.; *J.Controlled Release* **1991**, *17(3)*, 245.
11. Laurencin, T.J.; Koh, H.J.; Neenan, T.X.; Allcock, H.R.; Langer, R. *J.Biomed.Mater.Res.* **1987**, *21*, 1231.
12. Crommen, H.J.; Schacht, E.H.; Mense, E.H. *Biomaterials* **1992**, *13*, 511.
13. Crommen, H.J.; Schacht, E.H.; Mense, E.H. *Biomaterials* **1992**, *13*, 601.
14. Goedemoed, J.H.; Mense, E.H. In *Polyphosphazene drug delivery systems for antitumor treatment*; Goedemoed, J.H., Ed.; V.U. University press, Amsterdam, 1990, chapt.7.
15. Wermuth, C.G. *Chem.Industry* **1980**, *7*, 433.
16. Masaru, Y.; Masaharu, A.; Minoru, K. *Makromol.Chem.,Rapid commun.* **1990**, *11*, 337.
17. In 't Veld, P.J.; Dijkstra, P.J.; Van Lochem, J.H.; Feijen, J. *Makromol.Chem.* **1990**, *191*, 1831.
18. Schwyzer, R.; Iselin, B.; Feurer, M. *Helv.Chim.Acta* **1955**, *18*, 69.

RECEIVED October 1, 1992

Chapter 23

Mechanisms Governing Drug Release from Poly-α-Hydroxy Aliphatic Esters

Diltiazem Base Release from Poly-Lactide-*co*-Glycolide Delivery Systems

J. F. Fitzgerald and O. I. Corrigan

Department of Pharmaceutics, School of Pharmacy, Trinity College, 18 Shrewsbury Road, Dublin 4, Ireland

Diltiazem release from drug delivery systems, both microparticulate and pelleted, prepared from poly-lactide-co-glycolide co-evaporates is controlled by co-polymer decomposition. In contrast, drug release from systems prepared by direct compression of drug/co-polymer mixtures more closely approximates matrix type diffusion. Significant differences in water uptake and solid state form of the drug are observed depending on the method of preparation. Drug release from co-evaporate systems is fitted to a model which reflects the formation, spread and termination of activated nuclei of decomposition. Model parameters representing the induction and acceleration of decomposition, may prove useful in correlating release with the physicochemical properties of the drug and polymer, ultimately enabling a more quantitative approach to the design of biodegradable poly-α-hydroxy aliphatic ester based drug delivery systems.

The poly-α-hydroxy aliphatic esters are biodegradable polymers which have been extensively investigated as sustained drug delivery excipients *(1-3)* due to the apparent biocompatibility of the polymers and their degraded oligomeric subunits *(4-5)* . The most commonly researched polymers in this respect include co-polymers of glycolic and lactic acid, this being primarily as a result of their availability and versatility in polymer properties and performance characteristics *(6)* . By altering co-polymer molecular weight and ratio, changes in crystallinity, hydrophilicity and biodegradability are observed *(7)*. Drug release from these polymeric systems is not dependent on polymer characteristics alone, but also on physicochemical properties of the drug as well as morphology of the drug delivery system *(5,8)* . The release of small molecules from poly-lactic acid microspheres appears to follow a square root of time relationship *(1,9,10)* . Differences in release characteristics are also evident between bases and their acidic salts *(11)* . Release of acidic salts of basic drugs may occur by leaching or by a diffusion controlled process *(12)*, whereas release of basic drugs appears to be more dependent on degradation of the polymeric matrix, accelerated by an amine catalysed polymer hydrolysis *(11,13)*. Peptide release from biodegradable polymers appears to be bi-phasic, characterised by an initial burst followed by a slower release associated with polymer degradation *(8,14,15)*.

0097–6156/93/0520–0311$06.00/0
© 1993 American Chemical Society

The dependency of drug release on the physicochemical properties of the drug increases the difficulty of predicting drug release from biodegradable polymers. Furthermore, primary parameters which may govern drug release from biodegradable microspheres may not be readily evaluated due to the difficulty in examining the changing characteristics of microspheres during drug release. With this in mind, we produced poly-dl-lactide-co-glycolide (DL-PGLA) microspheres containing diltiazem base as well as diltiazem loaded DL-PGLA compressed discs, prepared from either a mechanical mix or solvent evaporate of the co-polymer and drug. Microsphere and disc characteristics were evaluated and compared, firstly to identify common parameters governing drug release, and ultimately with a view to developing a model to quantify drug release.

Experimental

Diltiazem base was produced by adding small aliquots of 1M sodium hydroxide (Riedel-de Haen 06205) to a solution of diltiazem hydrochloride (U.S.P. grade) in distilled water. The precipitate was collected under vacuum and dried over silica for a period of 7 days.

Poly-lactide-co-glycolide (Resomer RG 504, Boehringer Ingelheim, intrinsic viscosity approximately 0.5) microspheres containing diltiazem base were produced by the method of Beck et al. *(16)*. Microspheres were washed and collected, dried over silica and sieved into two size fractions, <125μm and 125 - 180μm. A scanning electron microscope (Hitachi 500) and X-ray diffractor (XRD-Siemens) were used to examine microsphere surface characteristics and differential scanning calorimetry (Mettler DSC20) was used to evaluate drug incorporation within the microsphere. Diltiazem base content was assayed using a UV spectrophotometer (HP8452A). Microsphere drug release was examined in isotonic phosphate buffer pH 7.4 at 37°C under sink conditions*(1)*. Co-polymer molecular weights (number average molecular weight (Mn), weight average molecular weight (Mw) and molecular weight distribution (Md)) were evaluated by gel permeation chromatography (GPC) against polystyrene standards (2-12K) before and after microsphere drug release.

Drug loaded discs of DL-PGLA were produced using the same solvents as those required for microsphere preparation. The solvents were evaporated from the drug/co-polymer solution over a hot plate to form a film which was dried over silica and under vacuum for 24 hours. The films were then cut into segments and heated in a die (13mm diameter) for 15 minutes. The die was allowed to cool and discs were compressed under 8000 kg/cm³ at room temperature for one minute. Discs were also prepared from mechanical mixtures of polymer (<180μm) and diltiazem base (<180μm). The drug and polymer were mixed in a mortar and pestle, and discs were compressed under the conditions previously detailed. The release of diltiazem base from discs was examined in phosphate buffer pH 7.4 at 37°C. and compared with diltiazem release from microspheres of equivalent drug loading. The changing dimensions of the disc were monitored during drug release.

Results and Discussion

The solvent evaporation procedure proved to be an effective method for preparing DL-PGLA microspheres containing diltiazem base. Microparticles produced were generally spherical and intact (Figure 1). All batches of microspheres contained between 50 - 60 % of the attempted drug loading. Microspheres in the smaller size range (<125μm) had a marginally lower drug content than microspheres in the larger size range (125-180μm). This result is consistent with a similar finding relating to the preparation of DL-PGLA microspheres containing procaine base *(17)* . Smaller

Figure 1: Scanning electron micrograph of microspheres containing 10.06% diltiazem base, size fraction <125µm.

microparticles have a greater surface area per unit weight from which a soluble drug will more readily leach during preparation.

X-ray diffraction patterns did not show the presence of crystalline diltiazem base on the surface of the microspheres (Figure 2). DSC showed the glass transition temperature of the amorphous co-polymer at 56.1°C (Figure 3). Diltiazem base had a crystalline melting point of 104°C. Endotherms corresponding to both the co-polymer and crystalline drug were evident in thermograms of physical mixtures. In contrast there was an apparent absence of crystalline drug within microspheres containing 20.22% diltiazem base. It is concluded that diltiazem base is not in crystalline form, either on the surface or within the co-polymer microspheres.

Diltiazem release from microspheres in phosphate buffer pH 7.4 was found to be dependent on drug loading (Figure 4). Drug release profiles were sigmoidal, characterised by an initial lag in release, the duration of which depended on drug loading. $T_{50\%}$ values ranged from 20.0 hours to 132.5 hours for 20.22 and 4.45% loaded microspheres respectively.

Photomicrographs of microspheres after 100% drug release showed that extensive degradation had occurred during drug release (Figure 5). The extent of co-polymer degradation during diltiazem release from DL-PGLA microspheres was quantified using GPC (Table I). Significant decreases were observed in microsphere co-polymer molecular weight values after 100% drug release, relative to values determined for freshly prepared drug free DL-PGLA microspheres. Furthermore, molecular weight decreases were similar for microspheres of both high and low drug content, irrespective of the time required for 100% drug release.

Table I: Co-polymer Molecular Weights after 100% Diltiazem Release from Microspheres

Drug Loading (%)	Time(hrs) for 100% Drug Release	Co-polymer Molecular Weight		
		Mn	Mw	Md
0.00[a]	-	14231	19912	46000-2000
4.45	310.10	5559	8419	31000-1200
10.06	167.40	5811	7810	29000-1300
20.22	75.25	5375	7543	27200-1600

[a] freshly prepared drug free microspheres

In an attempt to elucidate the drug release mechanism and determine the primary factors on which it depends, a series of co-polymer discs (13mm x 0.6mm) were produced containing diltiazem base. Drug release from discs prepared from a physical mixture of co-polymer and drug was compared with release from discs produced from drug co-polymer co-evaporate. Drug release from the two latter systems differed significantly. Percent drug release from discs formulated from mechanical mixtures was independent of diltiazem loading for the first 100 hours (Figure 6a). Subsequent drug release however appeared to be dependent on drug loading. In contrast, drug release from discs formulated by solvent evaporation was dependent on drug loading irrespective of percent drug released (Figure 6b).

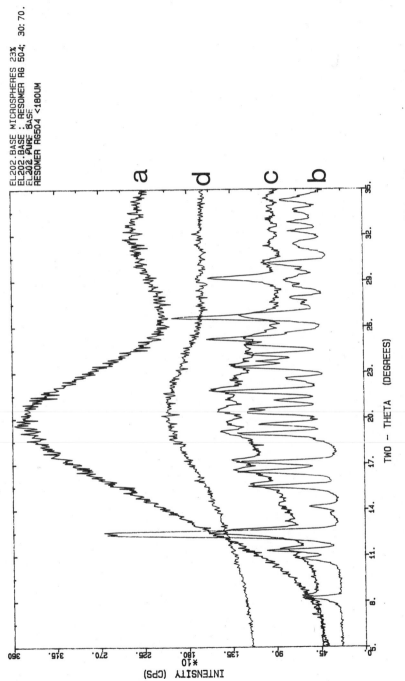

Figure 2: X Ray diffraction patterns of (a) Resomer RG 504, (b) diltiazem base, (c) mechanical mixture of diltiazem base and Resomer RG 504 30:70, and (d) microspheres containing 20.22% diltiazem base.

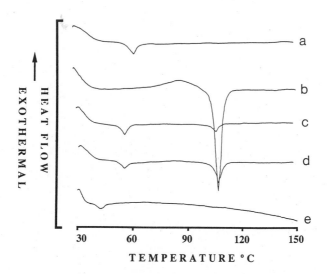

Figure 3: DSC thermograms of (a) Resomer RG 504, (b) diltiazem base, (c) and (d) mechanical mixtures of diltiazem base and Resomer RG 504, 10:90 and 20:80 respectively, and (e) microspheres containing 20.22% diltiazem base.

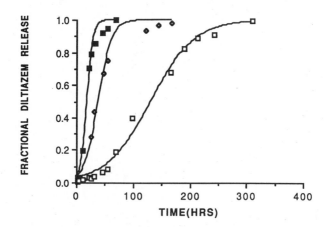

Figure 4: Diltiazem base release from DL-PGLA microspheres in phosphate buffer pH 7.4, <125μm size range, fitted to equation (13): □ - 4.45%, ◇ - 10.06%, ■ - 20.22% diltiazem base.

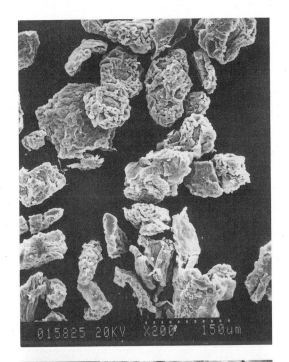

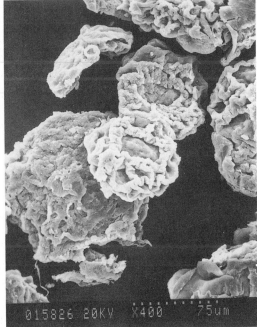

Figure 5: Microspheres containing (a) 20.22% and (b) 4.45% diltiazem base after 100% drug release in phosphate buffer.

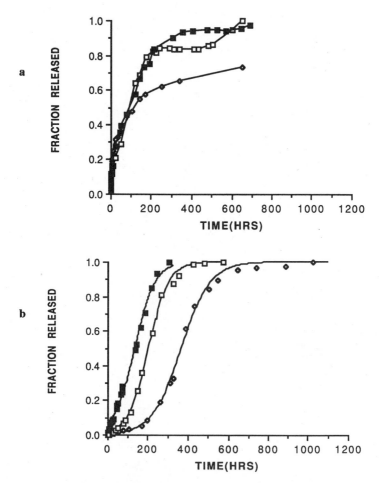

Figure 6: Drug release from discs in phosphate buffer pH 7.4; (a) MM- discs
compressed from a mechanical mixture; ◇ - 4.71%, ☐ - 10.00%, ■ - 20.00%
diltiazem base: (b) SE - discs compressed from a film prepared using the solvent
evaporation procedure; ◇ - 5.79%, ☐ - 10.03%, ■ - 20.62% diltiazem base,
fitted to equation 13.

The shape of the release profiles was also found to differ between the two formulations: diltiazem loaded discs produced by solvent evaporation gave rise to sigmoidal release profiles, characterised by a lag in initial drug release; in contrast, no lag period was evident for drug release from discs prepared from mechanical mixtures. It was further observed that drug release from the discs produced by the solvent evaporation procedure was similar in profile to drug release from microspheres. Although clear differences in lag times were evident, it was considered that the mechanism of drug release from diltiazem base loaded microspheres was similar to the mechanism of drug release from co-polymer/drug discs produced by solvent evaporation. This was substantiated by DSC analysis which showed the presence of crystalline diltiazem in discs prepared from a mechanical mixture of co-polymer and drug, but an absence of crystalline drug both in microspheres and discs produced by solvent evaporation.

During the early stages of drug release from discs prepared from the solvent evaporate, significant disc swelling was observed. To evaluate the ability of the co-polymer/drug discs to take up water, and examine the influence of water uptake on drug release, a series of drug loaded discs of varying height (0.6 to 1.6mm) was prepared. Disc diameter was measured during drug release, up to the point where disc fragmentation commenced. Drug released from discs containing 20.00% diltiazem base in a mechanical mixture with co-polymer, was dependent on disc height during the early stages (Figure7a). As drug release proceeded however this difference became insignificant. In contrast, the difference in drug released from diltiazem loaded discs of varying height prepared by the solvent evaporation method, was maintained up to 100% drug release (Figure 7b).

During drug release from the discs, polymer swelling was observed giving rise to changes in disc volume and surface area. If it is assumed that any change in disc volume prior to fragmentation is due solely to water uptake, then the mass of water taken up by a disc can be calculated given that the density of water at 37°C is 0.993333 *(18)*. As with the drug release profiles, significant differences in water uptake profiles between the two disc systems were observed (Figure 8). For discs prepared by solvent evaporation, initial water uptake was rapid leading to disc fragmentation within 50 - 125 hours, equivalent to 10 - 30% drug release. The rate of water uptake was dependent on disc diltiazem content and the total quantity of water taken up appeared to be dependent on initial disc volume/height. In contrast, there was insignificant water uptake during the first 200 hours of drug release from discs prepared from drug/co-polymer mechanical mixture. It should be noted that during this time period 80% of the total drug content was released from the discs. Disc fragmentation occurred after a sudden increase in water uptake, however not before 95% of total drug had been released.

The difference in drug release and water uptake profiles between the two disc systems may be explained in the following way. Preparation of diltiazem loaded DL-PGLA discs by solvent evaporation results in an amorphous co-polymer/drug matrix. Irrespective of whether the drug is dissolved or molecularly dispersed within the polymer, it may act as a polymer plasticizer, reducing the glass transition temperature T_g and enhancing polymer/water permeability *(19)*. As a result of water uptake, drug release is facilitated by degradation of the swollen polymer matrix. The presence of crystalline diltiazem within the co-polymer matrix does not, however, significantly alter co-polymer T_g (Figure 3). DL-PGLA may therefore retains its hydrophobic characteristics and poor permeability to water, and drug release may only occur by diffusion from a heterogeneous drug co-polymer matrix. Hutchinson has suggested that the sudden increase in water uptake prior to fragmentation may be due to random polymer degradation and the presence of more soluble oligomers within the polymer matrix, which enhance water permeability *(20)*.

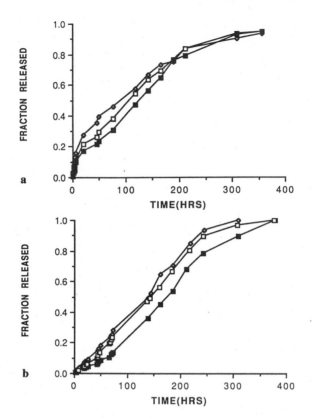

Figure 7: Diltiazem base release from discs of varying height: ◇ - 13mm x 0.6mm, ▫ - 13mm x 1.1mm, ■ - 13 x 1.6mm, prepared from (a) a mechanical mix of co-polymer and drug 20.00%, and (b) a solvent evaporate of co-polymer and drug 20.62%.

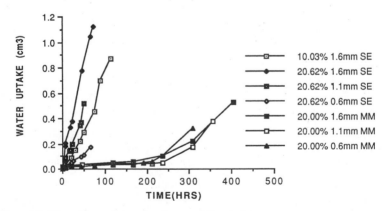

Figure 8: The effect of drug loading and disc height on water uptake into DL-PGLA/diltiazem base discs prepared from drug/co-polymer solvent evaporates (SE) and mechanical mixes (MM).

Co-polymer molecular weights were evaluated before and after drug release from each disc system (Table II). The starting material, DL-PGLA, had a number average molecular weight (Mn) of 14200 , a weight average molecular weight (Mw) of 19900 and a molecular weight distribution (Md) of 46000 - 2000. Significant reductions in co-polymer molecular weights were observed after drug release from all diltiazem loaded discs, prepared from either drug/co-polymer solvent evaporates or mechanical mixes. It should be noted, however, that discs prepared by solvent evaporation underwent a more extensive degradation over a shorter time period than did discs prepared from the mechanical mixtures.

Table II. Co-polymer Molecular Weights following Diltiazem Release from Discs

Disc Diltiazem Content %	Drug Released %	(hrs)	Mn	Mw	Md
5.79[a]	100	1026	3150	2900	30884 - 303
10.03[a]	100	574	3450	4980	29100 - 915
20.62[a]	100	228	3300	3400	31000 - 915
4.71[b]	80	653	3900	5839	33000-1400

[a] discs compressed from drug/co-polymer evaporates

[b] discs compressed from drug/co-polymer mechanical mixes

Model Development

Characterisation of drug release from diltiazem loaded microspheres and discs prepared by a solvent evaporation technique has suggested a release mechanism which is related to polymer decomposition. The following model describing a polymer decomposition dependent drug release is analogous to an original model evaluating the thermal decomposition of potassium permanganate crystals *(21)* .

The decomposition of a polymer may be considered to be dependent on the formation of activated nuclei on the surface of, and to a lesser extent, within the polymer matrix. Irrespective of location, these centres will exist along lines of strain within the matrix system. Polymer decomposition is initiated by polymer hydrolysis at active nuclei at or near the surface, resulting in the production of oligomeric subunits of varying molecular weight. Those oligomers below a specific threshold molecular weight display water solubility and permeability characteristics that differ significantly from the polymer itself *(20)* . As a result, there is a preferential water uptake at the centre of decomposition which promotes further polymer hydrolysis. The process of polymer hydrolysis is thus considered to progress outward along the surface of the polymer as well as penetrating into the polymer mass, generating a porous matrix. As degradation proceeds down the pore, new lines of strain are exposed perpendicular to the inner surface of the pore, resulting in the propagation of polymer degradation along many planes within the polymer matrix. The degradation of the polymer and generation of a porous structure facilitates the dissolution of low molecular weight oligomers and drug molecules from the bulk matrix.

The rate of increase in the production of oligomeric active sites which promote water uptake and further polymer hydrolysis, can be described by equation 1:

$$\frac{dn}{dt} = n_0 + \frac{\alpha.n}{\partial t} \tag{1}$$

where n is the number of active sites at any time t, n_0 is the initial number of sites, ∂t is the average time interval between production of oligomers and dissolution from the matrix, and α represents the probability of degradation branching from any given site. The process of hydrolytic propagation may proceed unhindered initially, but since there exists an increasing number of oligomeric active sites from which branching can originate as degradation proceeds, a degradation plane will eventually reach another internal or external plane resulting in the termination of decomposition along that plane. If b represents the probability of termination, equation 1 can be written:

$$\frac{dn}{dt} = n_0 + (\alpha - \beta).\frac{n}{\partial t} \tag{2}$$

If the relationship between α and β is described in a simple mathematical form, then

$$\beta = \alpha.\o \tag{3}$$

At t = 0, $\o$ is zero since planes of degradation cannot meet prior to the creation of those planes. Furthermore at some time t, the extent of branching arising from the exposure of new lateral strains must equal the extent of decomposition termination due to the meeting of degradation planes, for within any given matrix system polymer mass restrictions exist. This criteria is satisfied by the relationship

$$\o = \frac{x}{x_i} \tag{4}$$

where x is the extent of polymer decomposition and x_i is the extent of decomposition when the extent of branching equals the extent of decomposition termination. It is now noted that

$$(\alpha - \beta) = \alpha.(1 - \frac{x}{x_i}) \tag{5}$$

and that when $x > x_i$, the value of $(\alpha - \beta)$ reverses sign.
Where n_0 is considered small, then;

$$\frac{dn}{dt} = \alpha.(1 - \frac{x}{x_i}).\frac{n}{\partial t} = \frac{dn}{dx}.\frac{dx}{dt} \tag{6}$$

$$\frac{dn}{dx} = \alpha.(1 - \frac{x}{x_i}) \tag{7}$$

$$n = \alpha.(x - \frac{x^2}{2x_i}) \tag{8}$$

If it assumed that oligomer/drug dissolution occurs only as a function of polymer decomposition, then $x_i = c_i$, where c_i is the extent of drug release at x_i. Furthermore since the total extent of decomposition $x_f = 2.x_i$, then the total drug release $c_f = 2.c_i$. Therefore,

$$n = \alpha.c.(1 - \frac{c}{c_f}) \quad \text{and} \quad \frac{dc}{dt} = k.c.(1 - \frac{c}{c_f}) \tag{9},(10)$$

where $k = \frac{\alpha}{\partial t}$ and is a constant, provided that the time interval between oligomer generation and dissolution, and the probability of branching a are constant during this period. If t_{max} is the time when $c = c_i = c_f/2$ then,

$$\int_0^{\frac{c_f}{2}} \frac{c_f}{c.(c_f - c)} dc = k. \int_0^{t_{max}} dt \tag{11}$$

$$\ln \frac{c}{c_f - c} = k.(t - t_{max}) \tag{12}$$

Examining fractional drug release, $c_f = 1$ and

$$\ln \frac{c}{1 - c} = k.t + m, \text{ where } m = -k.t_{max} \tag{13}$$

Diltiazem release from microspheres and discs prepared by solvent evaporation techniques was fitted to equation 13 (Figure 4 & 6b). In all cases the fit was found to be good as measured by r^2 (Table III & IV), suggesting that diltiazem base release from these systems is dependent on polymer degradation and dissolution. Values of $k(hr^{-1})$ were found to be dependent on drug loading, as well as size and shape of the drug/co-polymer system. $t_{max}(hrs)$ was found to be similar to $T_{50\%}(hrs)$ and was dependent on both drug loading and size of the delivery system.

It is considered that size and shape of the drug/co-polymer system primarily determines the initial number of active sites at which polymer decomposition is initiated. For microspherical systems of relatively large surface area, the number of initial surface active sites from which branching may originate, is considered to be large. Furthermore because particle size is small, planes of decomposition may readily meet and so polymer decomposition termination is considered to take place early during the decomposition process. The combined effects of polymer decomposition branching and termination gives rise therefore to a sigmoidal release profile, the initial lag of which is dependent on the number of active sites initially present at or near the surface of the co-polymer drug system.

The acceleratory coefficient k is a measure of the relative ease of polymer decomposition branching within the polymer matrix. Where drug release is considered primarily dependent on polymer decomposition, any factor which alters the rate of polymer decomposition branching will alter the rate of drug release. As drug loading is increased a reduction in polymer T_g may increase water permeability, facilitating co-polymer degradation and the formation of secondary active sites from which branching can originate. Under such conditions, the rate of oligomer/drug dissolution from the polymer matrix is considered large. This satisfies the observation that the rate of water uptake and drug release from diltiazem loaded discs prepared by solvent evaporation, is dependent on diltiazem loading.

Table III: Values of k (hr^{-1}) and t$_{max}$(hrs) for Diltiazem Release from Microspheres

Microsphere Drug Loading %	k(hr^{-1})	t$_{max}$(hrs)	r^2
4.45	0.0261	124	0.99506
10.06	0.0846	32	0.99124
20.22	0.1807	20	0.99459

Table IV: Values of k (hr^{-1}) and t$_{max}$(hrs) for Diltiazem Release from Discs

Disc Drug Loading (%)	Disc Height (mm)	k (hr^{-1})	t$_{max}$(hrs)	r^2
5.79	0.6	0.01408	386.4	0.99866
10.03	0.6	0.01921	204.3	0.99970
20.62	0.6	0.01970	135.1	0.99504
20.62	1.1	0.02004	144.4	0.99700
20.62	1.6	0.01933	174.8	0.99817

This model is not exclusive to diltiazem release from DL-PGLA systems, but has been previously used to fit procaine *(17)* , and fluphenazine hydrochloride *(22)* release from other poly-a-hydroxy aliphatic esters. By evaluating the physicochemical properties of these basic drugs, it is hoped that a relationship between their properties and model parameters may be established to ultimately facilitate the prediction of drug release from biodegradable polymers.

Diltiazem release from discs prepared from mechanical mixtures was fitted to equation 13, however the fit was either poor or the data would not fit at all. The data was then fitted to an equation described by Cobby et al. *(23)*, based on the Higuchi diffusion principle characterising drug release from matrices having either a spherical, biconvex, or as in this particular case, cylindrical shape (Equation 14).

$$X = (q+2).k.t^{1/2} - (2.q+1).(k.t^{1/2})^2 + q.(k.t^{1/2})^3 \tag{14}$$

X is the fraction of drug released at time t, k is the rate constant and q is the shape parameter and is defined by the equation:

$$q = r_0 / h_0 \tag{15}$$

where r_0 is the radius of the disc and h_0 is half the initial disc height.
The rate constant k is defined by the Higuchi type equation

$$k = \frac{1}{A.r_0} \cdot \sqrt{D. \ C_S.(2.A - \text{ß}.C_S). \ \frac{\text{ß}}{\text{¥}}} \tag{16}$$

where A is the mass of drug initially present per unit initial disc volume, D is the diffusion coefficient and C_S is the solubility of the drug, ß is the porosity of the matrix and ¥ is the tortuosity factor of the matrix. q was evaluated at each sampling time interval, by measuring the dimensions of the discs using a verniers calipers. The data was fitted to equation 14, inputting q and t as independent variables, and the fit was considered relatively good, as measured by r^2 (Table V).

Table V: Values of k (hr⁻¹) for Diltiazem Release from Discs

Diltiazem Loading %	2.h_0 (mm)	k (hr⁻¹)	r^2
20.0	0.6	0.0017203	0.99542
20.0	1.1	0.0025573	0.98723
20.0	1.6	0.0031832	0.98916
10.0	0.6	0.0016362	0.99491
4.71	0.6	0.0016786	0.99101

Diltiazem release from compressed discs of drug and co-polymer, appeared to be primarily diffusion controlled. Values of k (hr⁻¹) were independent of drug loading at any given disc height, although k (hr⁻¹) did increase with disc height, probably as a result of changes in porosity and tortuosity.

Conclusion

The method of diltiazem incorporation into DL-PGLA delivery systems appears to be a primary determinant of drug release. Solvent evaporation gives rise to a homogeneous matrix which displays characteristics differing significantly from that of compressed mechanical mixtures of drug and co-polymer. It may be necessary for a drug substance to exist either in the form of a molecular dispersion or solid/solid solution within the polymer, in order for it to sufficiently alter polymer properties and induce a polymer decomposition dependent drug release.
The release profiles of diltiazem base from DL-PGLA microspheres and discs prepared from drug/co-polymer evaporates, were successfully fitted to a model describing a co-polymer decomposition controlled drug release, where decomposition is described as a hydrolytic polymer degradation followed by oligomeric/drug dissolution. Investigation

of the effect of drug/polymer physicochemical properties and process variables on the applicability of the model and model parameters, could lead to the optimal design of poly-a-hydroxy aliphatic ester based drug delivery systems.

References

1. Wakiyama, N., Kazuhiko, J. and Nakano, M., *Chem. Pharm. Bull.,* **1982,** *30, (7),* 2621 - 2628.
2. Phillips, M. and Gresser, J. D., *J. Pharm. Sci.,* **1984,** *73, (12),* 1718 - 1720 .
3. Suzuki, K. and Price, J. C., *J. Pharm. Sci.,* **1985,** *74, (1),* 21 - 24,
4. Spenlehauer, G., Vert, M., Benoit, J. P. and Boddaert, A., *Biomaterials,* **1989,** *10,* 557 - 563.
5. Visscher, G.E., Pearson, J.E., Fong, J.W., Argientieri, J.G., Robison, R.L. and Maulding, H.V., *J. Biomed. Mat. Res.,* **1988,** *22,* 733 - 746.
6. Lewis, D. H., *Biodegradable Polymers as Drug Delivery Systems* ; Chasin, M. and Langer, R., Eds.; Drugs and Pharmaceutical Sciences; J. Swarbrick: 1990, Vol. 45; 1 - 41.
7. Asano, M., Fukuzaki, H., Yoshida, M., Kumakura, M., Mashimo, T., Yuasa, H., Imai, K., Yamanaka, H. and Suzuki, K., *J. Cont. Rel.,* **1989,** *9,* 111 - 122.
8. Maulding, H. V., *J. Cont. Rel.,* **1987,** *6,* 167 - 176.
9. Jalil, R., and Nixon, J. R., *J. Microencap.,* **1990,** *7,* 53 - 66.
10. Ramtoola, Z., Corrigan, O. I. and Bourke, E., *Drug Devel. Ind. Pharm.,* **1991,** *17,* 695 - 708.
11. Maulding, H.V., Tice, T.R., Cowsar, D.R., Fong, J. W , Pearson, J. E. and Nazareno, J. P., *J. Cont. Rel.,* **1986,** *3,* 103 - 117.
12. Vidmar, V., Bubalo, A. and Jalsenjak, I., *J. Microencap.,* **1984,** *1,* 131 - 136.
13. Cha, Y. and Pitt, C. G., *J. Cont. Rel.,* **1989,** *8,* 259 - 265.
14. Sanders, L. M., Kent, J. S., McRae, G. I., Vickery, B. H., Tice, T. R. and Lewis, D. H.,*J. Pharm. Sci.,* **1984,** *73, (9),* 1294 - 1297.
15. Sanders, L. M., Kell, B. A., McRae, G. I. and Whitehead, G. W., *J. Pharm. Sci.,* **1986,** *75, (4),* 356 - 360.
16. Beck, L. R., Cowsar, D. R., Lewis, D. H., Cosgrove, R. J., Riddle, C. T., Lowry, S. L. and Epperly, T., *Fertility and Sterility,* **1979,** *31, (5),* 545 - 551.
17. Fitzgerald, J. F., Ramtoola, Z., Barrett, C. J. and Corrigan, O. I., *Proc. 10th Pharm. Tech. Conf., Bologna, Italy,* **1991,** *1,* 454 - 465.
18. *Documenta Geigy Scientific Tables* ; Diem, K. and Lantner, C., Eds.; J. R. Geigy S. A., Basle, Switzerland, 1970.
19. Pitt, C. G. and Schlinder, A., In *The Design of Controlled Drug Delivery Systems based on Biodegradable Polymers* ; Hafez, E. S. E. and van Os, W. A. A., Eds.; Progess in Contraceptive Delivery Systems; MTP Press: Lancaster, England, Vol. 1, 17 - 46.
20. Hutchinson, F. G., and Furr, B. J. A., *J. Ctrl. Rel.,* **1990,** *13,* 279 - 294.
21. Prout, E. and Tompkins, F., *Trans. Faraday Soc.,* **1944,** *40,* 448 - 458.
22. Ramtoola, Z., Corrigan, O. I. and Barrett, C. J., *J. Microencapsulation* , **1992,** *9, (4),* 415 - 423.
23. Cobby, J., *J. Biomed. Mater. Res.,* **1978,** *12,* 627 - 634.

RECEIVED October 1, 1992

Chapter 24

Synergistically Interacting Heterodisperse Polysaccharides

Function in Achieving Controllable Drug Delivery

John N. Staniforth[1] and Anand R. Baichwal[2]

[1]School of Pharmacy and Pharmacology, University of Bath,
Claverton Down, Bath BA2 7AY, United Kingdom
[2]Mendell Dynamics, Route 22, Patterson, NY 12563

A controlled release tableting excipient system is described which is composed of synergistic heterodisperse polysaccharides together with a saccharide component. The excipient has uniform packing characteristics over a range of different particle size distributions and is capable of processing into tablets using either direct compression, following addition of drug and lubricant powder or conventional wet granulation.

The synergism between the homo and heteropolysaccharide components of the excipient enables formulation manipulation of different rate- controlling mechanisms. The swelling and drug concentration gradient across the glassy and gelled regions of tablets prepared using drug/excipient blends were visualized using fluorescence imaging and quantified by digitized image analysis.

Drug release from tablets containing the polysaccharide excipient system was found to be capable of control using a variety of different formulations and process methods to provide a variety of different release modalities which were capable of matrix-dimension independence. Controllability of drug release was achieved by manipulation of the synergistic interactions of the heterodisperse polysaccharides.

A computer-aided pharmacokinetic model was used to predict likely *in vivo* drug blood levels from condition-independent in-vitro drug profiles. Phenylpropanolamine tablets containing the synergistic heterodisperse hydrogel excipient system used in the above study, were also used to provide in-vivo drug release data.

Oral Route

Oral delivery has been the most acceptable route for delivery of drugs to patients, whether this be in the form of a liquid preparation such as a suspension or an emulsion, or a solid dosage form such as a capsule or tablet. However, oral controlled release systems have potential problems associated with variable gastrointestinal transit times, variable gastrointestinal permeability along the gastrointestinal tract, variable lumenal contents which affect absorption and the effect of variable gastric emptying rate on first pass metabolism (*1*). Various

0097–6156/93/0520–0327$07.00/0
© 1993 American Chemical Society

mechanisms have been employed to deliver drug to the gastrointestinal tract in a controlled manner, including dissolution, diffusion, erosion and osmotic systems together with those having some muco-adhesive action.

Dissolution - Controlled Release.   One of the main release controlling mechanisms for sustained drug delivery is via alteration of dissolution properties. For this reason, if a drug has an inherently low solubility, then further reduction of the rate at which drug is released from the system may be unnecessary. However, as the solubility of the drug used increases, methods to retard the release are required. In the case of dissolution control, this may include different salt forms or conjugates of the drug, or materials added to the preparation that hinder the removal of drug by dissolution media and so reduce the rate at which drug is removed from the device. This type of dissolution-controlled system is dependent on the drug being removed from the surface of a device (whether this be a capsule, particulate system or a tablet) by the dissolution medium (ie the contents of the gastrointestinal tract). The process controlling the rate at which the drug is removed is the diffusion of drug across the static boundary layer that exists at the solid liquid interface into the bulk solution which is described by equation:

$$dm/dt = KA(C_s - C) \qquad \text{Equation (1)}$$

where dm/dt is the rate of dissolution, A is the surface area of the undissolved solid, C_s is the concentration of solute required to saturate the solvent at the experimental temperature, C is the solute concentration at time t and K is the intrinsic dissolution rate constant. The dissolution constant is a function of the diffusion coefficient of the solute in the dissolution media, the volume of dissolution medium and the thickness of the boundary layer.

A variant of equation (1) suggested by Danckwerts is sometimes more useful in quantifying the effect of a change in surface area on mass transfer. Release from microencapsulated particles can also be a dissolution-controlled process although equation (1) does not apply due to the fact that the change in surface area plays a major role in controlling the rate of drug release. Since individual particles are small, one dose of drug consists of a number of particles, and often release is controlled by coating particles with different thicknesses of polymeric material. The advantages of such systems are that they frequently exhibit more reproducible transport through the gastrointestinal tract, although the actual process of producing the encapsulated particles is subject to many process variables (2). Encapsulated particles have been produced by a variety of methods dependent on drug and excipient properties; for example, sustained release ibuprofen-wax microspheres were produced by a congealable disperse phase encapsulation method (3) whereas other formulations have been prepared using a coacervation process.

Diffusion - Controlled Release. The other main method used to control drug delivery is by altering the rate of diffusion, this being achieved either by a reservoir or monolithic matrix type device.

Reservoir Devices. The basic principle of reservoir devices is that a mass of drug is surrounded by a membrane or film that remains intact in the presence of the surrounding medium so that for the drug to be released, diffusion must occur across the membrane into the bulk solution. Such diffusive release is dependent on the difference in concentration between the bulk solution and that at the liquid film interface. This may be described using a modified form of Ficks first law:

$$dQ/dt \ = \ a_m D_i \ Pdc/l_c \qquad\qquad \text{Equation (2)}$$

where dQ/dt is the drug delivery rate, a_m is the membrane area, D_i is the diffusion coefficient of the species, P is the partition coefficient of the drug between the core and the membrane, d_c is the concentration gradient and l_c is the diffusional pathlength, l_c can be assumed to be the thickness of the membrane in ideal conditions.

A major factor determining the release rate of a particular drug is the permeability of the material used to form the film, to solvent and drug solution. A large amount of work has been carried out on diffusion of drugs across polymer films containing cellulose ethers such as hydroxypropyl cellulose. The major disadvantage of such devices is that if the film is damaged or imperfectly formed, there is a high risk that all the drug may be released in an uncontrolled manner which can lead to serious toxic effects from drugs with a low therapeutic index. Another performance related factor is that reservoir devices should yield zero order release but in reality they often do not because of core size changes which occur as the drug leaves the device and solute concentration levels fall.

Matrix systems.   Matrix systems can be subdivided into different categories, these being dispersed and porous systems where the matrix-forming material does not undergo dimensional changes in contact with the dissolution medium, and hydrogel systems where the matrix forming material undergoes dimensional changes in contact with the dissolution medium. The advantage of non-erodible dispersed matrix systems over reservoir and erodible systems is that they are relatively insensitive to changes in mixing and stirring conditions because diffusion is the rate controlling factor. Conventional dispersed systems suffer from non-linear concentration-time release, due to the longer distance that the drug in deeper layers of the matrix must travel to exit the delivery system. During both drug dissolution and diffusional processes, the boundary layer moves back into the matrix while its surface area is maintained. This is described by the Higuchi equation for drug release from dispersed matrices (*4*):-

$$Q \ = \ \{ \, D_m \ C_d (2A - C_d) \ t \, \}^{\,1/2} \qquad\qquad \text{Equation (3)}$$

where Q is the amount of drug released per unit area of matrix, D_m is the diffusion coefficient of the drug in the matrix, A is the initial amount of drug in unit volume of matrix, C_d is the solubility of the drug in the matrix and t is the time.

To overcome this problem of non-linear release and to facilitate zero order drug delivery, studies have been performed on disperse matrices that contain increasing concentrations of drug as the core is penetrated and have been shown to alleviate the problem of non-linear release (*5*).

Drug release from such systems is based upon the fact that the dissolution medium (eg water) surrounding the matrix device initially dissolves and leaches out drug from the surfaces of the device, but as this process continues with time, the dissolution medium travels further into the matrix and the drug then has to dissolve into the medium and then leave via diffusion along the porous water filled paths, created by the gradual ingress of the dissolution medium. Hence, before the tablet is placed in the dissolution medium, there are relatively few porous paths within the matrix. Drug release rates would therefore be expected to change with drug solubility and drug loading. Higuchi further modified the equation for dispersed matrix systems so that it was applicable to a porous system (*4*):

$$Q \ = \ \{ D \epsilon / \tau \ (2A - \epsilon C_s) \ C_s t \}^{\,1/2} \qquad\qquad \text{Equation (4)}$$

where τ is the tortuosity factor, ϵ is the porosity of the matrix after leaching of the drug, Q is the amount of drug released per unit area of matrix, D is the diffusion coefficient of the drug in the dissolution media, A is the initial amount of drug in unit volume of matrix, C_s is the solubility of the drug in the dissolution media, and t is the time.

Tortuosity is introduced into the equation to account for an increase in the path length of diffusion due to branching and bending of the pores, as compared to the 'straight through pores'. Tortuosity is defined as the ratio of the length of the average diffusional path to the thickness of the device, and tends to reduce the amount of drug released in a given time interval. A straight channel has a tortuosity of unity and a channel through a spherical bead has a tortuosity of 2 or 3. The Higuchi equation can be simplified to:-

$$Q = K_H t^{1/2}$$ Equation (5)

where K_H is the Higuchi constant.

Therefore a plot of drug release against square-root of time will be linear. However, the Higuchi equation assumes that the matrix device is completely wetted at time zero and that there is complete removal of air from the matrix; if this does not occur, then matrix tablets may not conform to the Higuchi model (7).

Hydrophilic matrices. Hydrophilic systems usually consist of a significant amount of drug dispersed in and compressed together with a hydrophilic hydrogel forming polymer and may be prepared together with either a soluble or insoluble filler. When these systems are placed in the dissolution medium, dissolution occurs by a process that is a composite of two phenomena : in the early stages of dissolution, polymer swelling occurs and tablet thickness increases. Soon thereafter, polymer (and drug) dissolution begins, the polymer dissolving due to chain disentanglement (6) or hydrogel formation as a result of cross-linking. The rate constant for drug release from a swellable matrix is a function of the diffusion coefficient of the drug matrix, which depends on the free volume of water (ie degree of swelling) and is shown by equation (4):

$$D = D_o e^{(-k'/q - 1)}$$ Equation (6)

where Do is the diffusion coefficient of drug in pure solvent, q is the degree of swelling, k' is a constant related to the characteristic solute volume and free volume of water (Zhang and Schwartz 1991).

Polysaccharide-based hydrophilic matrices. The principles of oral controlled release medication and the performance and mechanisms of action of hydrophilic controlled release matrices have been reviewed extensively elsewhere (8,9,10 for example). A wide variety of different hydrogel materials have been described for use in controlled release medicines, some of these are synthetic (11,12) but most are of semi-synthetic or natural origin (13,14,15,16) and relatively few contain both synthetic and non-synthetic material (17). However, some of the systems require special processes and production equipment; in addition, some are susceptible to variable drug release as the result of one or more of the following effects: pH dependency (18); Food effect variability (19); Ions and ionic strength dependency; Viscosity dependency; Corrasion/erosion variability; Content uniformity problems; Flow and weight uniformity problems; Carrying capacity and mechanical strength problems.

Additionally, formulation of a specific hydrophilic matrix-forming system to provide other than a first order drug delivery regime has often been problematical.

A hydrogel controlled release excipient system is described here which can be processed into tablets using direct compression without further formulation additives, other than drug and lubricant powder and using conventional processing equipment. Alternatively (20) the hydrogel excipient system may also be used in a conventional wet granulation process if desired (20). The excipient system contains a homo- and a hetero-polysaccharide together with a saccharide component (TimeR, Edward Mendell Co Inc Route 22, Patterson, NY 12563, USA). The excipient components exert an individual and synergistic action which aids tablet compaction and control of drug release with a variety of different modalities.

Experimental

(a)   Physical Characteristics of the Ungelled Excipient

Examples of the particle size distributions of some controlled release formulations developed for different drugs show that the modal diameters vary from approximately 195 μm to 215 μm. In some cases, the size distributions are unimodal, whilst in others, some bimodality is evident. Other differences include the degree of kurtosis and skewness of the distributions. These factors may be important in optimizing the mechanical properties, content uniformity and release-sustaining characteristics of a direct compression formulation for a given drug. Alternatively, in the case of a wet granulation system, the particle size distribution, as well as the agglomerating properties, may be important.

Notwithstanding the different particle size distributions available, it is interesting to note that the packing behaviour results in loose and consolidated bulk densities (and Hausner Ratios) which remain virtually constant.

This is a potentially valuable property in view of the dependency of powder flow and hence tablet weight and content uniformity on powder packing characteristics.

Figure 1 shows a scanning electron photomicrograph of a representative particle agglomerate designed for use in direct compression processing of verapamil hydrochloride. The surface morphology of the particles can be seen to provide a generally regular, near spherical particle shape, which helps confer good flow properties as shown by mass flow rate data. The surface texture is relatively rough with clefts and large pores capable of conferring segregation-resistance on the direct compression formulation.

Tablets compacted using an instrumented rotary tablet machine were found to possess strength profiles which were largely independent of the saccharide component (22). Scanning electron photomicrographs of ungelled tablet surfaces are shown in Figure 2a, and provide qualitative evidence of extensive plastic deformation on compaction, both at tablet surface (Figure 2a) and across the fracture surface (Figure 2b). Figure 2a also shows evidence of surface pores through which initial solvent ingress and solution egress may occur.

(b)   Physical Characteristics of the Hydrophilic Matrix

The properties and characteristics of a specific heterodisperse polysaccharide excipient system are dependent on the individual characteristics of polysaccharide constituents, in terms of polymer solubility, glass transition temperatures etc, as well as on the synergism both between different polysaccharides and between polysaccharides and saccharides in modifying dissolution fluid-excipient interactions.

For example, one heteropolysaccharide has a molecular weight of

Figure 1   Scanning electron photomicrograph of a single heterodisperse polysaccharide agglomerate containing verapamil hydrochloride.

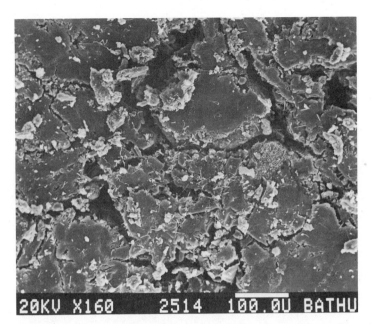

Figure 2a   Scanning electron photomicrograph of the compaction surface of a heterodisperse polysaccharide-containing tablet.   The SEM shows extent of ductile deformation and presence of surface pores.

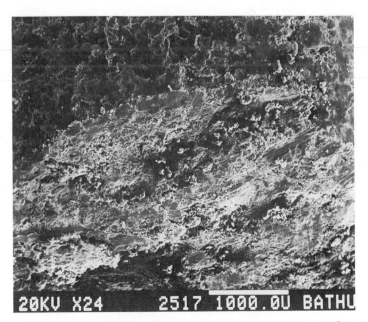

Figure 2b   Scanning electron photomicrograph of the fracture surface of a heterodisperse polysaccharide-containing caplet.   The SEM shows the extent of ductile deformation in the surface layer and within the caplet.

approximately 300,000 to 1,000,000, is readily soluble, which as a homodisperse system produces a highly ordered, helical or double helical molecular conformation which provides high viscosity without gel formation (Figure 3). In contrast, a homodisperse system of one homopolysaccharide is only slowly soluble and ungelled at low temperatures. Prolonged exposure to the dissolution fluid promotes solubilization which allows molecules to associate and undergo gelation as the result of intermacromolecular cross-linking in ribbon or helical "smooth" regions (Figure 4). The heterodisperse excipient contains both hetero- and homo-polysaccharides which exhibit synergism. The heteropolysaccharide component acts to produce a faster gelation of the homopolysaccharide component and the homopolysaccharide acts to cross-link the normally free heteropolysaccharide helices (Figure 5) the resultant gel is faster forming and more rigid. The viscosity and solubilization speed are further potentiated by the saccharide component and gel rigidity may also be further potentiated in the presence of some cations (Figure 6) and anions (22).

The mixed individual and synergistic viscosity modifying and gel-forming characteristics of the excipient systems produce a variety of possible release-controlling mechanisms in any given excipient system. Some of this behaviour can be described by an equation such as that reported for example by Frisch (21), Fan and Singh (8), and Korsmeyer and Peppas (9) amongst others:

$$Mt/M\infty = kt^n \hspace{3cm} \text{Equation (7)}$$

Where: $Mt/M\infty$ is the fractional solvent absorbed or drug released

t is the solvent absorption or drug release time; k and n are kinetic constants which depend on, and can be used to characterize, the mechanism of solvent sorption or drug release.

The mechanisms of solvent sorption and drug release vary from Case I or Fickian diffusion ($n = 0.5$), k = diffusion coefficient in initial half of solvent sorption drug release to Case II ($n = 1.0$), k α constant velocity of gel-glassy polymer interface, Super Case II ($n > 1$ probably with Fickian tailing ahead of slow Case II interface velocity causing accelerated solvent sorption and drug release when the solvent front meet at the three dimensional centroid, and Anomalous Behaviour ($0.5 < n < 1$), k is a characterisitic of solvent/drug solution diffusion and polymer relaxation.

In some cases, modulated drug release profiles are not well fitted to this model and require a more complex function, which takes into account a specific time at which the mechanism changes from one type to another, as the result of a time-dependent trigger for an excipient/excipient, excipient/fluid, drug/fluid, excipient/drug or mixed interaction. This may take the form:-

$$Mt/M\infty \hspace{1cm} = \hspace{1cm} k_1 + k_2 t^n \hspace{2.5cm} \text{Equation (8)}$$

alternatively, some other function may more accurately model this complex behaviour and this aspect is currently being investigated. Whichever mechanical or balance of mechanisms exists in a given excipient, the general functioning of the swelling controlled release tablet takes the form shown schematically in Figure 7. Figure 8 shows examples of changes in the degree of gel swelling and level of model drug in solution with time over a 24 hour period. The model drug used in this study was fluorescein sodium at a concentration of 10% w/w; magnesium stearate at a concentration of 1% w/w was used as a lubricant; the balance of the formulation comprised the heterodisperse hydrogel excipient, designed for use with Verapamil hydrochloride. The fluorescein sodium only fluoresces when dissolved

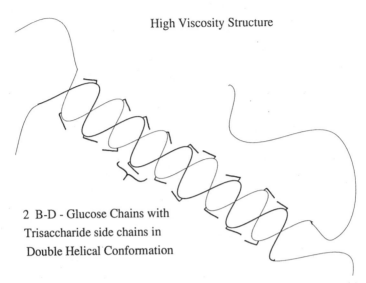

High Viscosity Structure

2 B-D - Glucose Chains with
Trisaccharide side chains in
Double Helical Conformation

Figure 3 Molecular conformation of a homodisperse heteropolysaccharide.

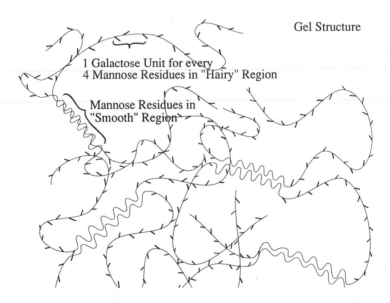

Gel Structure

1 Galactose Unit for every
4 Mannose Residues in "Hairy" Region

Mannose Residues in
"Smooth" Region

Figure 4 Molecular conformation of a homodisperse homopolysaccharide.

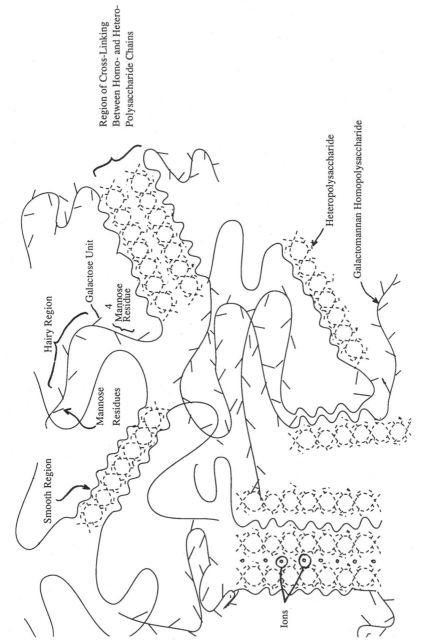

Figure 5   Molecular conformation of synergistically interacting heterodisperse polysaccharides.

Figure 6 Molecular conformation of a heterodisperse polysaccharide system in the presence of a saccharide component.

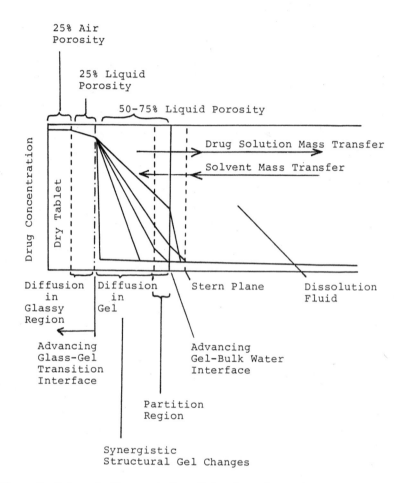

Figure 7   Schematic Representation of the General Functioning of Swelling Controlled Release Tablets using Heterodisperse Polysaccharides.

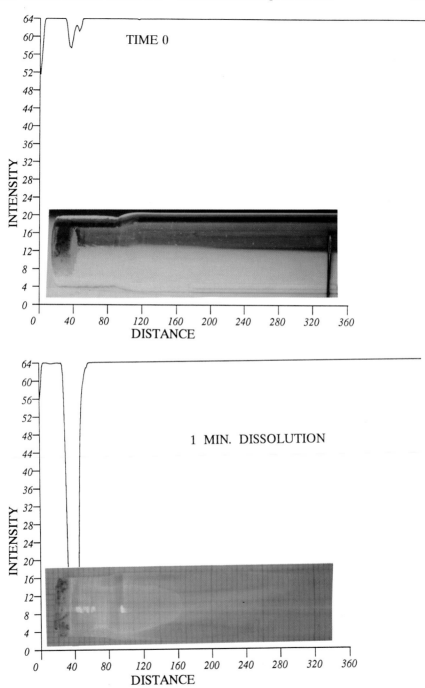

Figure 8 Semi-quantitative digital image analysis of gel formation and mass transfers over a 24 hour period in a heterodisperse polysaccharide tablet containing fluorescein sodium.   Note: Intensity (arbitrary units) in a measure of gray level from 0 = white to 64 = black.

Continued on next page

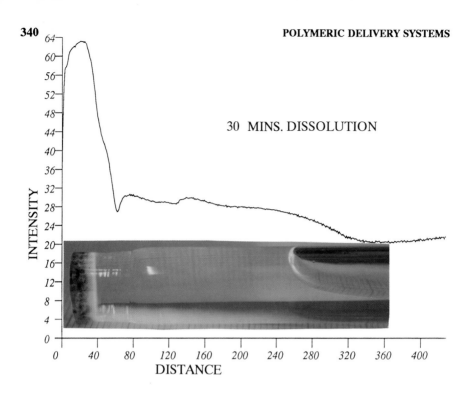

30 MINS. DISSOLUTION

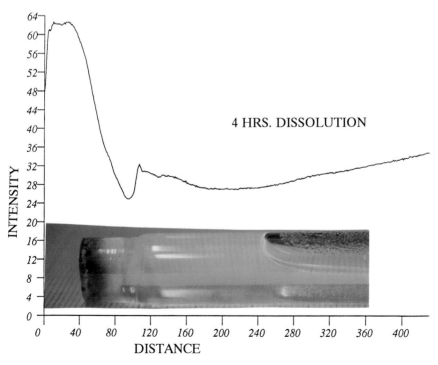

4 HRS. DISSOLUTION

Figure 8. Continued   *Continued on next page*

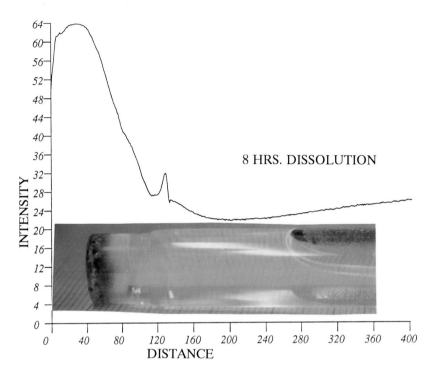

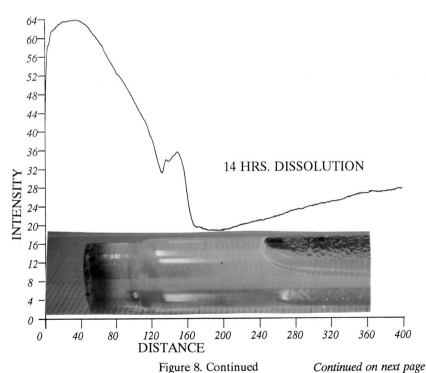

Figure 8. Continued   *Continued on next page*

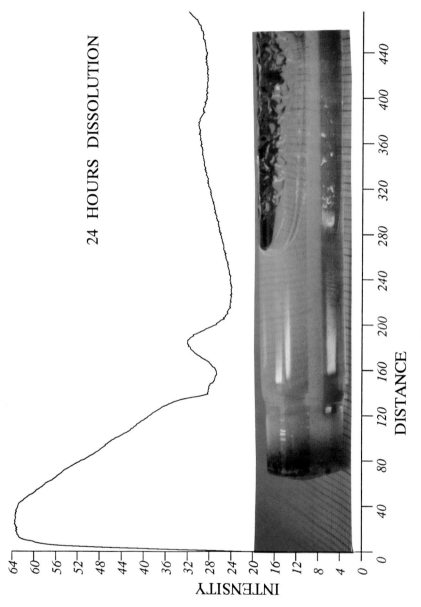

Figure 8. Continued

and it was this characteristic which allowed optical density differences in digitized images captured at different times to be used to produce a semi-quantitative interpretation of glass-gel transitions, extent and rate of gelation, any subsequent gel-sol transitions and appearance of drug in bulk solution over time. Figure 9 shows the relationship between gelled polymer dimensions and time. It can be seen that the increase in the dimensions of the gelled layer follows first order kinetics, comparable with those found for drug release from similar heterodisperse polysaccharide systems. This suggests that the type of gel formed by this heterodisperse polysaccharide system controlled drug release as a result of the increasing diffusion pathway through the expanding gel layer. In other heterodisperse polysaccharide systems, it has been found that the glass-gel transition at the solvent interface is the rate-controlling mechanism for drug release. In such cases, the rapid movement of drug in solution out of the gelled region means that in cases where erosion or corrasion occurs, the drug release profiles are not significantly affected.

The gels formed were found to be generally time dependent, thixotropic solids, which exhibit an open structure capable of allowing a very rapid movement of solvent and drug solution in the outer gelled regions, but which resists erosion and corrasion and is self-repairing following mechanical penetration. The rapid movement of drug in solution out of the gelled region means that in cases where erosion or corrasion occurs this does not influence drug release profiles. In tablets containing Verapamil hydrochloride, the presence or absence of the gel layer did not have a marked influence on drug release profiles. It appears therefore, that for this formulation, control of drug release is dependent on the glass-gel transformation rate, rather than prolonged diffusion through a tortuous swollen matrix.

(c)   **Biopharmaceutical Characteristics**

In-vitro **Characteristics.**   As a result of complex interactions between the component heterodisperse polysaccharides, it is possible to produce tablets having specific and reproducible dissolution profiles by manipulation of formulation components, excipient component ratios, drug-excipient ratios and process conditions. In addition, it has also been found possible to maintain such profiles constant whilst changing tablet dimensions.

Effect Of Formulation Components. Examples of *in-vitro* dissolution profiles for tablets containing metoprolol tartrate hydrochloride showed first order release kinetics. However, modification of the formulation components can be used to achieve zero order release kinetics. For example, Figure 10 shows examples of 2 formulations for propranolol hydrochloride, one of which was designed to have first order kinetics, and the other, zero order kinetics. The change in release kinetics results from changes in the rate of gel swelling, final gel strength and solvent and solution immobilization in core pores which result from changes in the heterodisperse polysaccharide component. In addition, the saccharide component may also be changed, although the effect of different saccharides on drug release is much smaller than that of different polysaccharides but may still be important in achieving fine control of drug release.

Effect of Modification of Heterodisperse Polysaccharide Component Ratios. The dissolution profiles of a tablet formulation containing an excipient system with a given heterodisperse polysaccharide component can be modified, firstly by alteration of the ratios of heteropolysaccharide to homopolysaccharide (Figure 11). In the example shown in Figure 11, change in the ratio of heterodisperse polysaccharides allowed the t_{50} for verapamil hydrochloride to be controlled over the range 30 minutes to approximately 5 hours.

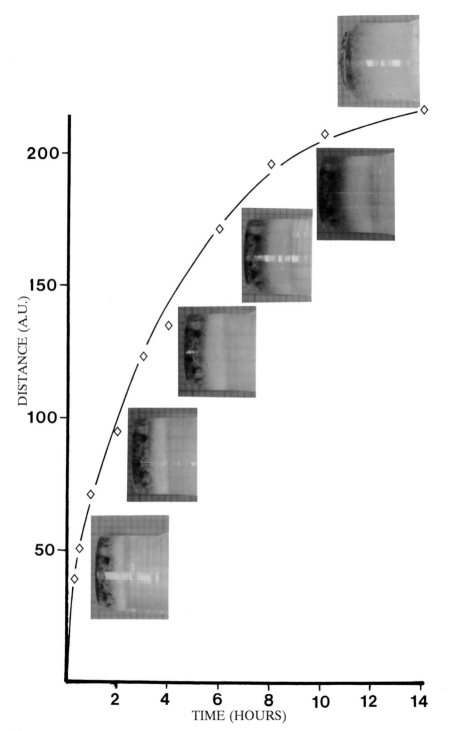

Figure 9 Relationship between gelled polymer dimensions and dissolution time.

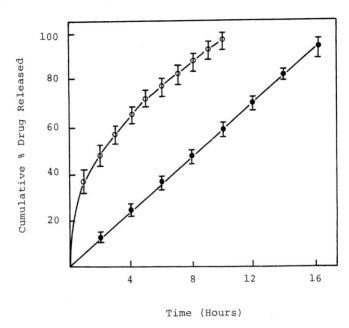

Figure 10   In vitro dissolution profiles for examples of 2 different heterodisperse polysaccharide formulations containing propranolol hydrochloride, showing first order (o) and zero order (•) release kinetics.

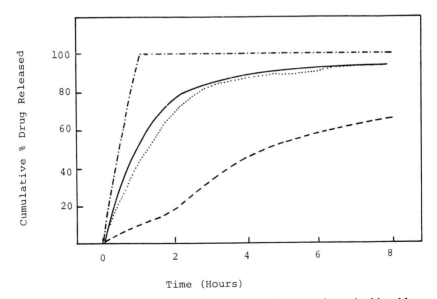

Figure 11 In vitro dissolution profiles for heterodisperse polysaccharide tablets containing verapamil hydrochloride and different ratios of homo- and hetero-(P) polysaccharides.
Key: — 55%G/45%P;   -·- 60%G/40%P;   ····· 50%G/50%P;
--- 25%G/75%P.

Utilization of Heterodisperse Polysaccharide Interactions to Produce Tablet Dimension-Independent Drug Release. Many hydrophilic tablets are known to show a direct relationship between release rate and tablet dimensions. A consequence of such behaviour is that an increase in tablet size causes a decrease in drug release rate. Another effect of release variability is evident when tablet shape is changed. For example, Figure 12 shows the effect of changing from 7/16" standard convex tablets to a caplet shaped core. It can be seen that a marked difference in drug release occurs, which may be due to differences in dry and wetted porosities between the caplet and biconvex tablets. However, other differences such as surface:volume ratios may also be important; for example, the dry surface of biconvex tablets was only approximately 60% of the equivalent caplet surface. Nevertheless, the influence of tablet shape and size on this type of formulation is significantly greater than that found using equivalent hydrophobic sustained release tablet formulations.This may be due to the relatively low importance of molecular transfer distances between the bulk solution and the ungelled surface in some heterodisperse polysaccharide systems revealed using the semi-quantitative fluorescence image processing method.

However, it was found that by careful control of the physical characteristics of the ungelled and gelled tablet cores such as those described above, drug release profiles could be closely matched for tablets of different sizes and shapes. Put another way, drug release from tablets produced using the heterodisperse polysaccharide excipient system were found to be tablet dimension independent, which may be an important formulation factor for achieving a desired release pattern within tablet weight and/or shape constraints. Comparable control of drug release from tablets of different dimensions was achieved by modifying the synergistic interactions between the homo and heteropolysaccharide and saccharide components of the heterodisperse polysaccharide excipient system. In this way, any weakening of the synergism might produce faster glass-gel transitions, but weaker gel strengths which might undergo eventual gel-sol transition producing a faster release rate. Conversely, enhancement of the level of synergism might produce a slower glass-gel transition, but result in a more highly cross-linked and stronger gel which would remain as a gel throughout the course of drug delivery.

Figure 13 shows an example of release profiles for 2 tablet formulations, each containing 240 mg Verapamil hydrochloride. In one formulation, 700 mg of heterodisperse polysaccharide excipient was found to control the release of Verapamil from tablets according to first order kinetics with a t_{50} of approximately 2.5 hours and a t_{90} of approximately 6 hours.

In the second formulation the excipient content was lowered to less than 250 mg. Figure 13 shows that Verapamil was released from these lighter tablets at a comparable rate to that obtained for tablets containing more than twice the controlled release excipient content.

In this example, the release profiles were maintained constant with different excipient levels by changing the level of interaction between polysaccharide chains. In the case of the lower tablet weight, the excipient controlled drug release as a result of the interaction between 1.5:1.5:7 parts of hetero-, homo-polysaccharide and saccharide. Whereas, for the higher tablet weight, the synergistic interaction between polymer chains of monomer or dimer conponent was produced using a 1:1:2 ratio of hetero-, homo-polysaccharide and saccharide. These components produced a larger tablet which was more porous in both ungelled and gelled states and in which rate of gelation was faster and gel strength lower. In the case of the smaller tablet with higher gel strength, the resistance to corrasion and erosion and other changes which might lead to differences in drug release rates was suggested from the co-incident *in-vitro* dissolution data produced using different agitation rates.

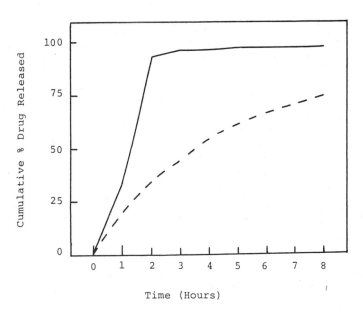

Figure 12   The effect of tablet shape on the release profile of verapamil hydrochloride from a heterodisperse polysaccharide matrix.
Key: --- 7/16" standard concave tablet; — 19 x 8 mm concave caplet

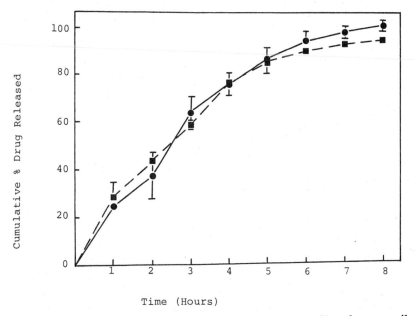

Figure 13   The effect of tablet mass on the release profile of verapamil hydrochloride from heterodisperse polysaccharide matrices.
Key to Tablet Masses: ● 1067 mg; ■ 492 mg

Effect of Modification of Drug-Excipient Component Ratios. Dissolution profiles show that as with other sustained release monolithic matrix systems, there is a direct (though non-linear) relationship between drug loading and drug release rates.

Effect of Process Conditions. Pre-wetting of the polysaccharide matrix appears to influence the time and rate of gel formation, which may result from modifications in hydrogen bonding analogous to those found in pre-wetting of other polysaccharides, notably cellulose (15).

In Vivo **Characteristics.** Excipient systems such as those described above were tested using a wide variety of different drugs. It was found that where required, formulations could be designed which were pH-, food effect- and agitation rate - independent. These properties may be especially important in achieving the desired level and reproducibility of controlled drug delivery from dosage forms *in vivo*.
   In-vitro dissolution characteristics obtained using formulations containing different drugs were used to produce an *in-vitro* prediction of *in-vivo* controlled release delivery system performance via drug-plasma profiles using a pharmacokinetic model processed via the STELLA computer software package. An example of predicted *in vivo* data is shown in Figure 14.
   The equivalent data from a 6 patient cross-over study of drug release *in vivo* is also shown in Figure 14 for comparison. Similar studies have been carried out for other drug-excipient systems.

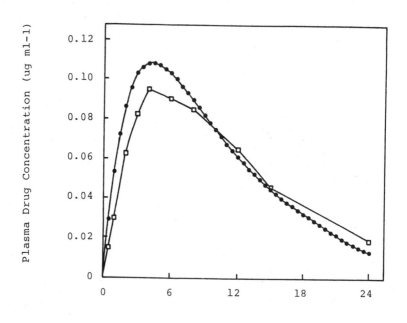

Time (Hours)

Figure 14   Relationship between plasma drug concentrations and time for heterodisperse polysaccharide tablets containing phenylpropanolamine.
Key: —•— Data predicted from in vitro dissolution studies using Stella pharmacokinetic computer model —□—. Data obtained in vitro from a 6 paitne study.

Literature Cited.
1. Amidon,G.L.; "Fasted-state variation of gastrointestinal variables: Implications for oral drug bioavailability", *Proc. Inter.Symp.Control.Rel. Bioact.Mater.*,Controlled Release Society, Inc., Illinois, USA,**1991**;*18.*
2. Eskilson,C.;"Controlled release microencapsulation", *Lejus Medical AB.*,Molndal, Sweden, **1985.**
3. Adeyeye,C.M. and Price,J.C.;"Development and evaluation of sustained-release ibuprofen-wax microspheres:effect of formulation variables on physical characteristics", *Pharmaceutical Research,* **1991,** *Vol 8,* 11.
4. Higuchi, T.; "Mechanism of sustained-action medication. Theoretical analysis of rate of release of solid drugs dispersed in solid matrices". *J.Pharm.Sci.,* **1988,** *5,* 622-624.
5. Scott,D.C. and Hollenbeck,R.G.; "Design and manufacture of a zero-order sustained-release pellet dosage form through non-uniform drug distribution in a diffusional matrix." *Pharmaceutical Research,***1991,** *8.*
6. Harland,R.S., Gazzaniga,A., Sangalli,M.E., Colombo, P. and Peppas,N.;"Drug/polymer matrix swelling and dissolution." *Pharmaceutical Research,* **1988,** *5,* 622-624.
7. Lockwood,P.J. and Staniforth, J.N.;"A novel technique for the determination of porous matrix tortuosity factors using measurement of gaseous diffusion coefficients". *Proc.Inter.Symp.Control. Rel.Bioact.Mater.*,**1991,***18.*
8. Fan,L.T. and Singh,S.K.;"*Controlled Release: A Quantitative Treatment*"; Springer Verlag, Berlin and London,**1989** p 111-156.
9. Korsmeyer,R.W., Gurny,R., Doelker,E., Buri,P. and Peppas,N.A.;"Mechanisms of Solute Release from Porous Hydrophilic Polymers" *Int.J. Pharm.,* **1983,** *28,* 25-35.
10. Graham,N.B. and McNeill,M.E.; "Hydrogels for controlled drug delivery", *Biomaterials,* **1984,** *5,* 27-36.
11. Roorda,W.E., Boddé,H.E., de Boer,A.G. and Junginger H.E.; "Synthetic Hydrogels as Drug Delivery Systems" *Pharmaceutisch Weekblad Sci.Ed.*,**1986,** *8,* 165-189.
12. Law,T.K., Whateley,T.L. and Florence,A.T.;"Cross-linked Poloxamer Hydrogels as Controlled Release Systems" *J.Pharm. Pharmacol.,* **1984,** *39,* 6P.
13. Alderman,D.A.; "A Review of Cellulose Ethers in Hydrophilic Matrices for Oral Controlled Release Systems" *Int.J.Pharm.Tech. and Prod.Mfr.,* **1984,** *5,* 1-9.
14. Van Aerde,P. and Remon,J.P.; "*In-vitro* Evaluation of Modified Starches as Matrices for Sustained Release Dosage Forms".*Int.J.Pharm.,* **1988,** *45,* 145-152.
15. Stockwell,A.F., Davis,S.S. and Walker,S.E. "In-vitro Evaluation of Alginate Gel Systems as Sustained Release Drug Delivery Systems".*J.Contr.Rel.,* **1986,** *3,* 167-175
16. Badwan,A.A., Abumalooh,A., Sallam,E., Abukalaf,A. and Jawan,O.; "A Sustained Release Drug Delivery System using Calcium Alginate Beads", *Drug Devel.and Ind. Pharm.,* **1985,** *11,* 239-256.
17. Shenouda,L.S., Adams,K.A. and Zoglio,M.A.;"A Controlled Release Delivery System using 2 Hydrophilic Polymers".*Int.J.Pharm.,* **1990,***61,*127-134.
18. Ford,J.L.,Rubinstein,M.H., Changala,A. and Hogan, J.E.; "Influence of pH on the Dissolution of Promethazine Hydrochloride from Hydroxypropylmethylcellulose Controlled Release Tablets". *J. Pharm.Pharmacol.,* **1984,** *39,* 115p.

19.   El Arini,S.K., Shiu,G.K. and Skelly,J.P.; "An In-vitro Study of Possible Food-Drug Interactions of the Controlled Release Propranolol Products ", *Int.J.Pharm.*, **1989,** *55,* 25-30.
20.   Baichwal,A.R. and Staniforth,J.N. "Directly Compressible Sustained Release Granulation", U.S.Patent No 4994276; **1991.**
21.   Frisch,H.L.; "Sorption and Trensport in Glassy Polymers - A Review".*Polymer Eng.Sci.*,**1980,***20,* 2-13.
22.   DeHaan,P. and Lerk,C.F.; "Oral Controlled Release Dosage Forms - A Review".*Pharmaceutisch Weekblad Sci.Ed.*, **1984,** *6,* 57-67.

RECEIVED October 1, 1992

Chapter 25

Drug Delivery System Using Biodegradable Carrier

S. Tokura[1], Y. Miura[1], Y. Kaneda[1], and Y. Uraki[2]

[1]Department of Polymer Science, Faculty of Science, and [2]Department of Forest Science, Faculty of Agriculture, Hokkaido University, Sapporo 060, Japan

6-O-Carboxymethyl-chitin (CM-chitin), one of the biodegradable chitin derivatives, shows several specificities such as chelating ability with calcium ion, specific adsorption of benzyl group following to the calcium chelation, and gel formation with trivalent iron ion. A prodrug was found to be released slowly into blood following the subcutaneous injection of polymeric drugs, in which the prodrug was either pendanted through a covalent bond to CM-chitin or entrapped within CM-chitin matrix in the presence of Fe^{3+}. The prodrug was then hydrolyzed, to become the active form, by enzymes in the blood.

Recently, a number of novel drug delivery approaches have been developed. These approaches include drug modification by chemical means, drug entrapment in small vesicles or within polymeric matrices. These arose from the fact that the conventional dosage forms of drug are not sufficiently effective. The potential use of macromolecules or macromolecular prodrugs as means to improve the efficiency of drugs has been studied, but a superior macromolecule for this is not yet developed because of the need for highly sophisticated materials. Chitin seems to have some advantageous properties to fulfill those requirements. Chitin is known to be hydrolyzed by lysozyme, a glycosidase distributed extensively in animals relating to bacteriolysis. Because it consists of naturally occurring substance, N-acetylglucosamine, chitin is non-toxic or has very low toxicity and immunogenicity.

6-O-Carboxymethyl-chitin (CM-chitin) is a chitin derivative which has high susceptibility to lysozyme, high water solubility, a reactive functional group, and chelating ability with calcium ions. CM-Chitin has known to chelate with calcium ion specifically (1) and form a gel-like matrix with ferric ion (2). Furthermore, it has been found that CM-chitin and its lysozymic hydrolysate showed little immunogenicity except slight mitogenic activity (3). These properties make it a very suitable material for a drug carrier. Several proposed types of CM-chitin drug carriers are shown in Figure 1.

Lysozyme Susceptibility of CM-chitin

Lysozyme hydrolyzes not only the β1,4-bond between N-acetylmuramic acid and N-acetyl-β-D-glucosamine (MurNAcβ1→4GlcNAc) but also the β 1, 4-bond of chitin.

0097–6156/93/0520–0351$06.00/0
© 1993 American Chemical Society

However, the latter hydrolysis proceeds very slowly owing to the high crystallinity, thus, chitin is degraded gradually in animals. The lysozyme susceptibility of chitin was found to be remarkably enhanced by the carboxymethylation of the C-6 hydroxyl group unless the C-3 hydroxyl group was substituted (3). A viscosity change, which was monitored using Ubbelohde viscometer, of CM-chitin aqueous solutions after the addition of lysozyme is shown in Figure 2. It is obvious that the further carboxymethylation of the C-3 position significantly reduced the lysozyme susceptibility of CM-chitin. The rate of hydrolysis seems to be influenced by the degree of substitution which depends on the reaction conditions such as alkaline concentration, temperature, etc.

Molecular Weight Distribution of Lysozymic Hydrolysate. CM-chitin was hydrolyzed with lysozyme to investigate the molecular weight distribution of lysozymic hydrolysate by gel permeation chromatography (GPC). A 500 ml of 0.1% (w/v) CM-chitin (D.S.= 0.8) phosphate buffer solution (pH 6.2, M/15) was hydrolyzed at 37°C by 2,500 unit/ml of egg white lysozyme which was a similar level as that found in human tear. Samples were taken from the reaction mixture for GPC analysis at 7, 24, 72 and 96 hr; these were treated by trichloroacetic acid-precipitation of the protein fraction. CM-chitin oligomers collected by centrifugation were redissolved in deionized water and analyzed by GPC using Sephadex G-50. The molecular weight of the CM-chitin oligomers were estimated both by the titration of reducing end groups to calculate the number of molecular chains (4) and by HCl-indole method to give the total N-acetylglucosamine (GlcNAc) residue (5). The molecular weight (M.W.) of the CM-chitin oligomer was calculated by applying following equation ;

$$\text{M.W.} = \frac{\text{Number of GlcNAc residue}}{\text{Number of molecular chain}} \times \text{R.W.} \qquad (1)$$

where R.W. is the residual molecular weight estimated from the degree of substitution of CM-chitin.

The time dependent molecular weight distribution of CM-chitin hydrolysate is shown in Figure 3. The molecular weight distribution of CM-chitin hydrolysates maintains sharpness under the experimental condition even in the low molecular weight region.

The Fate of CM-chitin in Mice

[125]I-Labeled CM-chitin was prepared using Bolton-Hunter reagent and ethylenediamine spacer to investigate the fate and body distribution of CM-chitin. When ^{125}I-labeled CM-chitin was injected subcutaneously into mice (Figure 4), the CM-chitin hydrolysate was found to be metabolized rather uniformly except the stomach/intestine system of rather higher accumulation than that of other organs. Thus, the carrier CM-chitin shows no specific retention in a certain organ and is available for the non-targeting delivery of drugs.

Sustained Release of Drug from CM-chitin-drug Conjugate

Covalently bonded drug-carrying CM-chitin conjugates were synthesized to examine the potential for use as a drug carrier for sustained release.

Study with Methamphetamine-bearing CM-chitin (6). A hypnotic drug, methamphetamine (MA), was coupled with CM-chitin through aminoethane spacer. A saline solution of the conjugate (MAEA–CM-chitin) or MAEA hydrochloride was injected subcutaneously into rabbits and the clearance from serum was detected by the

1. Pendant type

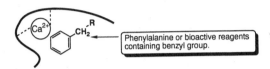

2. Adsorption type

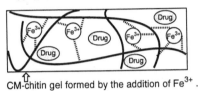

3. Entrapping type

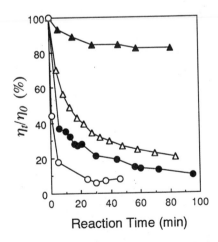

CM-chitin gel formed by the addition of Fe^{3+}.

4. Mixed type
Composed of adsorption type and entrapping type
using calcium and ferric ions.

Figure 1. Various applications of carboxymethyl-chitin for drug carrier.

Figure 2. Time course of lysozyme-catalyzed hydrolysis of CM-chitin. Relationship between the rate of hydrolysis and the degree of substitution of carboxymethyl group; D.S. = 1.2 ($\blacktriangle$), 0.46 ($\triangle$), 0.48 ($\bullet$), 0.78 ($\bigcirc$).

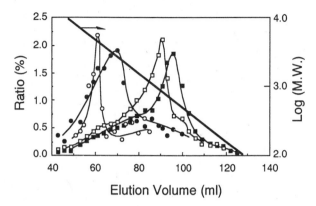

Figure 3. Gel permeation chromatograms of CM-chitin hydrolysate by lysozyme. O; 7 hr hydrolysis, ●; 24 hr hydrolysis, □; 72 hr hydrolysis, ■; 98 hr hydrolysis.

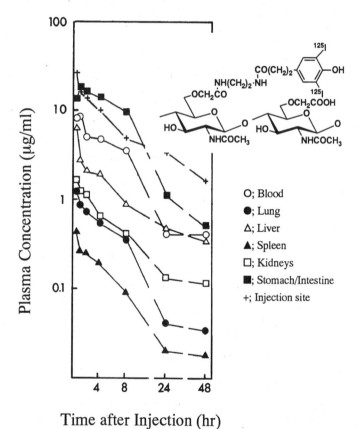

Figure 4. Body distribution of ^{125}I-labeled CM-chitin. 100 μg of ^{125}I-labeled CM-chitin was injected subcutaneously into mice.

ELISA method. Serum concentration of MAEA derivatives was maintained at a significant level for over 120 h after the administration (Figure 5). On the other hand, the serum concentration of MAEA was undetectable within 7 h after the injection, probably due to its rapid metabolism or secretion. This suggests that CM-chitin is a suitable drug carrier for sustained release. The MAEA or MAEA-bearing oligosaccharide was regarded to be an ineffective prodrug. The application of an enzyme susceptible peptidyl spacer is discussed below as a means to improve its effectiveness.

Study with Model Conjugates. Covalently bonded CM-chitin-drug conjugates were synthesized (7), in which drug analogue was coupled through an enzyme susceptible bond. In those conjugates, it is expected that the drug will be released through several hydrolysis processes. Oligomerization of the CM-chitin carrier by lysozymic hydrolysis is the first step and release of active drug by the cleavage of spacer-drug linkage with proteolytic hydrolysis is the second step. Thus, release of drug from the CM-chitin-drug conjugate will be achieved through the, at least, dual process proposed "Two Step Release of Drug".

To confirm the proposal, chromophore (p-nitroaniline, pNA)-terminating peptides were coupled to CM-chitin as a model for the CM-chitin-drug conjugate and the release profiles of drug analogue from these conjugates were investigated by lysozymic hydrolysis as a first factor and α-chymotryptic hydrolysis as a second factor of release.

Preparation of Polymeric Prodrugs. Polymeric prodrugs were prepared by the coupling of peptides bearing model drug to CM-chitin using morpho CDI [1-cyclohexyl-3-(2-morpholinoethyl)-carbodiimide-metho-p-toluenesulfonate] in deionized water as described elsewhere (7).

α-Chymotryptic Hydrolysis of Polymeric Prodrugs in the Presence of Lysozyme. As shown in Figure 6, a polymeric conjugate containing a spacer composed of single amino acid residue (CM-chitin-Phe-pNA) showed little release of model drug (pNA) within 6 h of reaction time when the conjugate was treated with α-chymotrypsin and lysozyme in 50 mM Tris-HCl buffer containing 50 mM calcium chloride, pH 7.4. After 6 h of the reaction, however, pNA started to release from the conjugate. It seems that the susceptibility of polymeric substrates to proteolytic hydrolysis is significantly related to the molecular weight of polymeric carrier, because the release of pNA observed for CM-chitin-Phe-pNA is considered to be the result of oligomerization of CM-chitin by lysozymic degradation during the first 6 h.

The smaller the molecular size of the conjugate, the larger the amount of pNA liberated. This results from a reduction of steric hindrance from the polymer backbone for the enzymatic hydrolysis. These results suggest the possibility of a two step release of drug and the usefulness of CM-chitin as a drug carrier. In conjugates with a spacer composed of two amino acid residues, such as CM-chitin-Gly-Phe-pNA, CM-chitin-Ala-Phe-pNA and CM-chitin-Abu-Phe-pNA (Abu; 4-aminobutyryl-), delay of drug release was not observed, though the release profile of model drug depended on the composition of spacer. The disappearance of the delay of drug release suggests the reduction of steric hindrance of proteolytic hydrolysis. As seen in Figure 6, the rate of drug release was increased with increasing length of the methylene chain of adjacent residue of Phe; -Abu-Phe-pNA> -Gly-Phe-pNA>> -Phe-pNA. This might be due to the increasing mobility of the side chain. The higher rate for chymotryptic hydrolysis of CM-chitin-Ala-Phe-pNA than those of other conjugates is regarded as a synergism of its higher susceptibility and the flexibility of the side chain for chymotrypsin-catalyzed hydrolysis.

α-Chymotrypsin Hydrolysis of Oligomeric Prodrugs. Figure 7 shows the dependence of the rate of α-chymotrypsin-catalyzed hydrolysis (v_0) on the number-

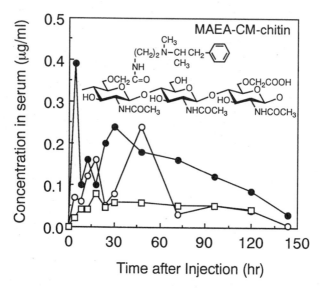

Figure 5. Time course of release of methamphetamine derivatives into serum after subcutaneous injection of MAEA-CM-chitin saline solution into rabbits; −●−: 4.2 mg of MA-containing polymer was injected; Both −O− and −□− : 2.8 mg of MA-containing polymer was injected.

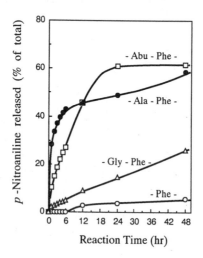

Figure 6. Chymotryptic hydrolysis of polymeric prodrugs in the presence of lysozyme (5,000 unit/ml). O; CM-chitin-Phe-pNA, Δ; CM-chitin-Gly-Phe-pNA, ●; CM-chitin-Ala-Phe-pNA, □; CM-chitin-Abu-Phe-pNA.

average molecular weight (MW) of oligomeric prodrugs, which might correspond to the steric hindrance of CM-chitin backbone. The release of chromophore from CM-chitin conjugate with a spacer of -Ala-Phe-(CM-chitin-Ala-Phe-pNA) was not clearly shown in the range of 4,000 to 50,000 of molecular weight, but a significant increase of the rate of α-chymotryptic hydrolysis was observed by the reduction of the MW to less than 4,000. This suggests that the susceptibility of polymeric substrates (MW= 4,000-50,000) for α-chymotrypsin is significantly affected by the molecular conformation of polymer chain, but the susceptibility of oligomeric prodrugs of molecular weight less than 4000 to α-chymotrypsin seems to be similar to that of low molecular weight prodrugs, because the spacer-chromophore bond might no longer be protected by the CM-chitin molecule. On the other hand, in case of CM-chitin-Abu-Phe-pNA, there is no significant change of v_0 in the MW range of 2,000-50,000. The independence of v_0 for CM-chitin-Abu-Phe-pNA seems to reflect the lower substrate specificity of the side chain for chymotryptic hydrolysis, *i.e.* the smaller the carrier chain, the higher the value of V_{max} or k_{cat}/K_m for CM-chitin-Ala-Phe-pNA but not for CM-chitin-Abu-Phe-pNA. It is likely that chymotrypsin recognizes the natural occurring Ala residue more readily than Abu residue. The results of CM-chitin-Ala-Phe-pNA and CM-chitin-Abu-Phe-pNA shows that both a difference in the structure of spacer and the molecular weight of drug carrier affects the α-chymotryptic hydrolysis. In other words, the structure of spacer is one of the factors that can be used to control the rate of drug release.

Entrapment of Drugs into CM-chitin Gel Formed by Ferric Ion

Since the addition of iron (III) chloride into CM-chitin solution induces gel formation, the anticancer agent doxorubicin or protein was incorporated into the CM-chitin gel during the gelation with iron (III) ion (2). In the present study, the peptidyl anticancer agent, neocarzinostatin (NCS), was incorporated into the gel and the release of NCS from the gel was investigated both *in vitro* and *in vivo*.

Preparation of CM-chitin Gel Containing NCS. CM-chitin gel was prepared by the method described previously (8). When calcium and ferric ion were added to the mixture of CM-chitin and NCS, more than 80% gel formation was observed. The use of calcium ion in addition to the ferric ion increased the amount of incorporation of NCS. Thus, a final concentration of 50 mM calcium chloride and 30 mM of iron (III) chloride was used for the preparation of NCS-containing gel with more than 50% recovery of NCS. The average size of the gel obtained was smaller than human platelets (2-3 µm in diameter) and showed a uniform distribution in gel size. The average size of a gel particle was estimated to be 0.70 µm by a Coulter Multisizer.

Release of NCS from the gel by lysozymic hydrolysis. CM-chitin gel containing NCS was incubated with 500 or 5,000 unit/ml of lysozyme (Figure 8). The release of NCS resulting from lysozymic digestion of the gel was observed in the presence of 500 or 5,000 unit/ml of lysozyme in a time- and dose-dependent manner. On day 7 after the incubation, about 80% of NCS incorporated was released by the exposure to 500 unit/ml of lysozyme, which is approximately equivalent to the physiological concentration in the human plasma. A spontaneous release of NCS from the gel, approximately 50% during four days, was also observed when the gel was incubated without lysozyme. However, when 500 unit/ml of lysozyme was added fresh on day 4, as indicated by an arrow in Figure 8, about 20% of NCS incorporated was additionally released from the gel during the last 3 days.

Release of NCS from the Gel *in vivo*. The NCS level in plasma after subcutaneous (s.c.) administration of CM-chitin gel containing NCS into mice was investigated to examine the potential of CM-chitin gel for sustained release of drugs.

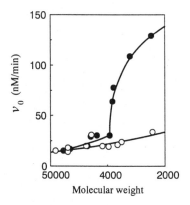

Figure 7. Dependence of the rate of α-chymotryptic hydrolysis (v_0) on the number-average molecular weight of polymeric prodrugs (MW); ●: CM-chitin-Ala-Phe-pNA, ○: CM-chitin-Abu-Phe-pNA.

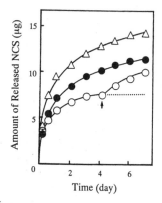

Figure 8. Release of ^{125}I-labeled NCS from CM-chitin gel by lysozyme digestion. CM-chitin gel containing ^{125}I-labeled NCS (1.5×10^8 cpm/15 μg NCS/45 μg gel) was incubated without (○) or with lysozyme of 500 unit/ml (●) or 5,000 unit/ml (△). Arrow; see text.

Figure 9 shows the plasma concentration of NCS and demonstrates that the clearance of NCS from the circulation was very rapid over a 6 h period after the s.c. injection of NCS alone. The NCS concentration in plasma was calculated from the standard curve by the growth inhibition of B16-BL6 melanoma cells *in vitro*. At 24 h after the injection of NCS alone, NCS in the plasma was undetectable (less than 1 ng/ml) as assessed by the growth inhibition of B16-BL6 melanoma cells by the plasma. Whereas, the NCS-containing gel gave a much smaller maximum plasma level at 6 h after the s.c. injection and NCS was gradually cleared from the circulation. NCS in the plasma was still detectable 48 h after the injection.

Stabilization of Peptide Drug by Adsorption to CM-chitin

Since a biodegradable CM-chitin has shown ability to adsorb neutral amino acids or peptides, especially phenylalanine (Phe), when calcium ion was chelated onto CM-chitin (9), stabilization of substances containing a hydrophobic moiety or Phe by CM-chitin would be expected.

Thus, the stabilization of a biologically active peptide to enzymatic degradation through the adsorption on CM-chitin was studied *in vitro* for the sustained release of biologically active or drug-bound peptides. The chromophore-bearing model peptide; enkephalin analogue (HCl·H-Tyr-Gly-Gly-Phe-pNA; [pNA]-Enk), was synthesized by the liquid phase peptide synthesis and the susceptibility of peptides adsorbed on CM-chitin to proteolytic hydrolysis was investigated to isolate the effect of CM-chitin on the stabilization of peptide from degradation.

α-Chymotrypsin-catalyzed Hydrolysis of Peptide Adsorbed on CM-chitin.
The peptide, [pNA]-Enk, was used to investigate the entrapping effect of CM-chitin on α-chymotryptic degradation. Profiles of α-chymotryptic hydrolysis of [pNA]-Enk (0.48 mM) preincubated with or without CM-chitin were shown in Figure 10 as time-dependent curves. *p*-Nitroaniline liberation was monitored at 410 nm. The rates of α-chymotryptic hydrolyses were influenced by CM-chitin concentration and the depression was maintained for a fairly long period in the absence of lysozyme probably due to the entrapping effect of CM-chitin because no marked difference in α-chymotrypsin-catalyzed hydrolysis of peptide was observed in the presence of lower concentration of CM-chitin. The CM-chitin-Ca^{2+} complex data showed any influence on the α-chymotryptic hydrolysis when the peptide was not pretreated with CM-chitin prior to an exposure to the α-chymotrypsin. Thus, the depression of α-chymotryptic hydrolysis seems to be due to the decrease of free substrate by the adsorption onto CM-chitin-Ca^{2+} complex. It seems that the peptides pretreated with CM-chitin were not bared to the proteolytic hydrolysis. The prevention of proteolytic hydrolysis through the adsorption on CM-chitin might have potential to stabilize the biologically active peptides and sustain the release of them followed by lysozymic hydrolysis of CM-chitin backbone.

Conclusion

Several types of drug delivery system was studied using CM-chitin as a carrier. In each case, the potential use of biodegradable and water soluble CM-chitin was indicated to be useful as the drug carrier of sustained release. Pendant and entrapping (including adsorption) type of drug delivery seems to fit for the sustained release of low molecular weight drugs. Although ferric ion entrapping seems to be favorable for the molecule of medium size, effective entrapping of drug of high molecular weight, such as proteins, was achieved by the use of ferric and calcium ions.

The biodegradability of CM-chitin would contribute to effective sustained release of drug in each case.

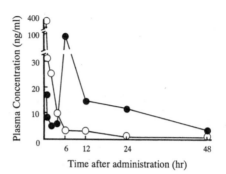

Figure 9. Plasma concentration of NCS after the s.c. administration of NCS alone or NCS-containing CM-chitin gel. Mice were administered s.c. with NCS (10μg) (O) or CM-chitin gel containing NCS (10μg NCS/330 μg gel) (●).

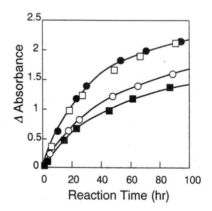

Figure 10. Time course of chymotryptic hydrolyses of [pNA]-Enk in the presence or absence of CM-chitin in 5% EtOH–50 mM Tris-HCl buffer, pH 7.4, containing 10 mM CaCl$_2$ at 30°C. The hydrolyses were carried out in the absence of CM-chitin (●) or in the presence of CM-chitin, (□); 0.29 mg/ml, (O); 1.03 mg/ml and (■); 1.87 mg/ ml.

Acknowledgments

This work was partially supported by a Grant-in-Aid from the Ministry of Education of Japan (No. 02555183).

Literature Cited

(1)   Tokura, S.; Nishimura, S.-i.; Nishi, N. *Polym. J.* **1983**, *15*, 597-602.
(2)   Watanabe, K.; Saiki, I.; Uraki, Y.; Tokura, S.; Azuma, I. *Chem. Pharm. Bull.* **1990**, *38(2)*, 506-509.
(3)   Nishimura, S.-I.; Nishi, N.; Tokura, S. *Carbohydr. Res.* **1986**, *146*, 251-258.
(4)   Imoto, T.; Yagishita, K. *Arg. Biol. Chem.* **1971**, *35*, 1154-1156.
(5)   Ohno, N.; Suzuki, I.; Yadomae, T. *Carbohydr. Res.* **1985**, *137*, 239-243.
(6)   Tokura, S.; Baba, S.; Uraki, Y.; Miura, Y.; Nishi, N.; Hasegawa, O. *Carbohydr. Polym.* **1990**, *13*, 273-281.
(7)   Tokura, S.; Kaneda, Y.; Miura, Y.; Uraki, Y. *Carbohydr. Polym.* in press.
(8)   Watanabe, K.; Saiki, I.; Matsumoto, Y.; Azuma, I.; Seo, H.; Okuyama, H.; Uraki, Y.; Miura, Y.; Tokura, S. *Carbohydr. Polym.* **1992**, *17*, 29-37.
(9)   Uraki, Y.; Tokura, S. *J. Macromol. Sci.-Chem.* **1988**, *A25*, 1427-1441.

RECEIVED September 11, 1992

Chapter 26

Blood Clearance and Body Distributions of Glycosylated Dextrans in Rats

S. Vansteenkiste[1], E. Schacht[1], L. Seymour[2], and R. Duncan[2]

[1]Laboratory of Organic Chemistry, University of Ghent,
Ghent B–9000, Belgium
[2]CRC Polymer-Controlled Drug Delivery Research Group, Department
of Biological Sciences, University of Keele, Keele, Staffordshire ST5 5BG,
United Kingdom

Dextrans with a narrow and well defined molecular weight distribution have potential as targetable carriers for pharmacological agents. Specific sugar moieties such as D-galactose, D-mannose and L-fucose were linked by a carbon C-6 spacer to the polymer backbone using the 4-nitrophenyl chloroformate activation method. Following ^{125}I-iodination and intravenous injection, the blood clearance and the body distribution of the glycosylated dextrans were assayed. The incorporation of the sugar moieties enhances the liver uptake from the bloodstream through receptor mediated endocytosis. The influence of a number of structural parameters on the in vivo fate of the macromolecular derivatives was investigated. Of great importance were the nature of the attached monosaccharide unit used as a ligand and the way those sugar residues are incorporated throughout the polymer chain: randomly or preclustered.

The discovery of the hepatic asialofetuin receptor in the rabbit liver by Morell et al. (1) has initiated extensive research on glycoside specific receptors. The resulting data have clearly demonstrated that various cell populations have glycoside specific receptors or lectins on their outer membranes. The occurrence of those natural recognition systems for pending sugars on glycoconjugates has been widely investigated and is well documented (2-8).

The recognition phenomena in the liver include several distinct pathways. The hepatocytes, the nonparenchymal cell types such as Kupffer cells (liver macrophages) and endothelial cells exhibit surface receptors for glycoproteins that differ qualitatively and quantitatively. Hepatocytes recognize and internalize mainly D-galactose- and N-acetyl-D-galactosamine-terminated glycoconjugates (8-13); Kupffer cells endocytose particulate material to which D-galactose groups are connected (9) and together with endothelial cells, also recognize L-fucose-, D-mannose-, or N-acetyl-D-glucosamine bearing glycoproteins (9,14-16). The role of these receptors in

0097–6156/93/0520–0362$06.00/0
© 1993 American Chemical Society

the cellular uptake of macromolecules by receptor-mediated endocytosis has been the subject of numerous investigations (*17-22*). The specificity of such interactions and the relatively simple structure of the sugar moieties makes these glycosylated macromolecular systems potentially useful for the enhanced delivery or targeting of drugs to specific cells (*23-26*).

In this paper the synthesis of L-tyrosinamide labelled dextrans substituted with either D-galactose, D-mannose or L-fucose (5-8 mol%) and their in vivo fate following intravenous administration to rats is described.

Dextrans have been proposed quite frequently as a carrier in the preparation of macromolecular prodrugs (*20-30*). Their in vivo fate is strongly dependent on their molecular size (*31*). Therefore, in this study only dextran derivatives with a narrow and well defined molecular weight were assayed. The influence of different chemical and structural parameters on the blood clearance and the body distribution is reported. This includes the nature of the sugar moiety facilitating the recognition of the polymers by cell lectins and the "preclustering" of some monosaccharide residues. A flexible C-6 spacer group was introduced in the structure of all ligands, enhancing possible receptor-ligand interactions following coupling to the polymer backbone (*32-33*).

Experimental Methods

Synthesis of L-tyrosinamide Labelled Glycosylated Dextrans.   All glycosylated dextrans were synthesized as described earlier (*34*) (Figure 1). Briefly, dextran fractions were activated by reacting a DMSO/Pyridine solution (1:1, v/v) of the polymer with 4-nitrophenyl chloroformate. The activated dextran was subsequently reacted with a calculated amount of L-tyrosinamide ($\pm$ 1 mol%) and the selected O-peracetylated glycoside, having an anomeric substituent with a terminal primary amino group (reviewed in Table I). After 18 hrs the reaction mixture was added to an excess of dry ethanol. In order to deacetylate the glycoside groups, the precipitate was redissolved in 0.1M NaOH and stirred for 5 minutes at room temperature. The reaction mixture was subsequently applied to a Sephadex G-25 column. The polymer fraction was collected and freeze dried. The content of the glycoside substituents was determined by ^{1}H NMR spectroscopy by comparison of the appropriate integration values. The characteristics of the various glycosylated dextrans are summarized in Table II.

Table I. List of the O-peracetylated Glycoside Derivatives

structure of NH_2-$Gly_n OAc_m$	code
$NH_2(CH_2)_6$-O-ß-D-gal$_1$(OAc)$_4$	ß-D-gal$_1$-OAc$_4$
$NH_2(CH_2)_5$-CONH-C-[CH_2O-ß-D-gal(OAc)$_4$]$_3$	ß-D-gal$_3$-OAc$_{12}$
$NH_2(CH_2)_6$-O-α-D-man$_1$(OAc)$_4$	α-D-man$_1$-OAc$_4$
$NH_2(CH_2)_6$-O-ß-L-fuc$_1$(OAc)$_4$	ß-L-fuc$_1$-OAc$_4$

Table II. Characteristics of L-tyrosinamide Labelled Dextran Conjugates

#	M_w	M_w/M_n	label	% label	ligand	% ligand
1	27,700	1.19	$TyrNH_2$	0.9	-	-
2	27,700	1.18	$TyrNH_2$	1.5	β-D-gal$_1$-OH$_4$	6.0
3	27,700	1.19	$TyrNH_2$	1.0	β-D-gal$_3$-OH$_{12}$	8.5
4	27,700	1.19	$TyrNH_2$	1.5	α-D-man$_1$-OH$_4$	5.0
5	27,700	1.19	$TyrNH_2$	1.5	β-L-fuc$_1$-OH$_4$	7.0

with: gal=galactoside, man=mannoside and fuc=fucoside

Radioiodination of the Dextrans Derivatives. Polymers were labelled by radioio-dination using the iodo-bead method (35). Following the addition of an excess of sodium metabisulphite, the dextrans were purified by dialysis against water (3-4 days). All preparations had a specific activity of approximately 50 μCi/mg polymer and were stored at -20°C. Immediately prior to use, all radioiodinated polymers were passed over prepacked Sephadex G-25 columns (PD-10 columns, Pharmacia, Sweden) to remove any free [^{125}I]iodide. This ensured that all radioactivity measured represented the macromolecular substrate.

Blood Clearance and Body Distribution of ^{125}I-glycosylated Dextrans. Female Wistar rats (160-200g, 3-4 months), under anaesthetic (Fluothane), were injected via the femoral vein with ^{125}I-labelled dextran derivatives, approximately 50μg in phosphate buffered saline (PBS) (100 μl). Blood samples (3x75μl) were taken after 2,5,10,15,20,25 and 30 minutes. At the end of the experiment, the rats were sacrifi-ced by cervical dislocation. The total radioactivity at t=0 was estimated by extrapo-lation, assuming 5.8 ml of blood per 100 g of rat body weight and based on the total amount of the recovered radioactivity.

In order to assess the body distribution of radioactivity, the major organs: lungs, spleen, liver and kidneys were removed and homogenized in a known volume of water. The intestines were dissolved (1h, 85°C) in a known volume of 1M NaOH. Finally the heart, muscle, skin and the residual carcass were dissolved (3hrs, 85°C) in a known volume of 10M NaOH. The urine and the faeces were not collected separately but were associated with the carcass content in all experiments. Bone, muscle and skin were scanned routinely and were found to contain only minor quantities of radioactivity.

Samples (3x 1ml) of each tissue solution were withdrawn and essayed for radioactivity. The amount in each organ was expressed as a percentage of the total recovered radioactivity, without any corrections being made for the blood content in the organs.

Blood Clearance of ^{125}I-Glycosylated Dextrans

In a first set of experiments (30 min.) the blood clearance of the glycosylated as well as the native dextrans after intravenous administration into the femoral vein of female Wistar rats was investigated. The results are represented in Figure 2.

The clearance from the blood circulation involves two distinct pathways: either through glomerular filtration by the kidneys or by cellular recognition and uptake. Due to the relatively low molecular weight of the injected dextran derivatives (M_w=27,700), excretion through the kidneys is still likely. Only dextran fractions exceeding the renal filtration threshold ($M_w \approx 45,000$) (*31*) will remain in circulation for a considerably longer time and eventually be captured by the cells of the reticulo-endothelial system (RES), in particular the Kupffer cells. Moreover, glycosylated dextrans can be recognized by specific receptors, present on the cell membranes, due to their exposure of sugar residues. The binding interactions between the polymers and the sugar-specific receptors or lectins can result in an increased disposition of the conjugates by a process of receptor-mediated endocytosis. It is well documented in the literature that galactose- bearing glycoconjugates are cleared rapidly from the blood stream by an asialoglycoprotein receptor present on the hepatocytes (*8-13*). On the other hand, Kupffer and endothelial cells may capture mannosylated or fucosylated macromolecules (*14-16*).

Since the number of receptors present on a cell surface is limited, quick saturation may occur. Therefore, a minimal dose of glycosylated dextran was injected in all experiments (ca. 50μg/rat) in order to avoid temporary blocking of the receptors, leading to an enhanced urinary excretion of the injected material. Both pathways, renal excretion or cellular uptake, were investigated further in body distribution experiments.

The data shown in figure 2 indicate quite clearly that all four glycosylated dextrans are quickly cleared from the plasma. They differ significantly from the native dextran. The half-life time is reduced from 12 min. (native dextran) to less than 2 min. for the glycosylated materials. This dramatic reduction in half-life time demonstrates the selective binding interactions between the glycosylated substrates and the receptors on the cell surface, resulting in an enhanced disposition in liver cells (see body distributions). This observation is in good agreement with the results already obtained with glycosylated N-(2-hydroxypropyl)methacrylamide (HPMA), as described extensively in the work of R.Duncan and J.Kopecek et al. (*36-39*).

The dextran conjugate bearing tri-galactose cluster moieties is cleared markedly faster from the blood circulation compared to the other polymers. The dependence of the rate of blood clearance on the nature of the D-galactose residues can be caused by the 3-fold increase of D-galactose moieties in a tri-galactose ligand or by the sterical arrangement of the glycosides in a tri-antennary side group. It was demonstrated by Lee et al. (*40-42*) that the interactions of synthetic galactose terminated oligosaccharides to the hepatic asialoglycoprotein receptor strongly depends on the fine structural features. Tri-antennary sugar clusters were found to exhibit a far more stronger affinity and binding capacity towards the hepatic lectin than mono-antennary oligosaccharides.

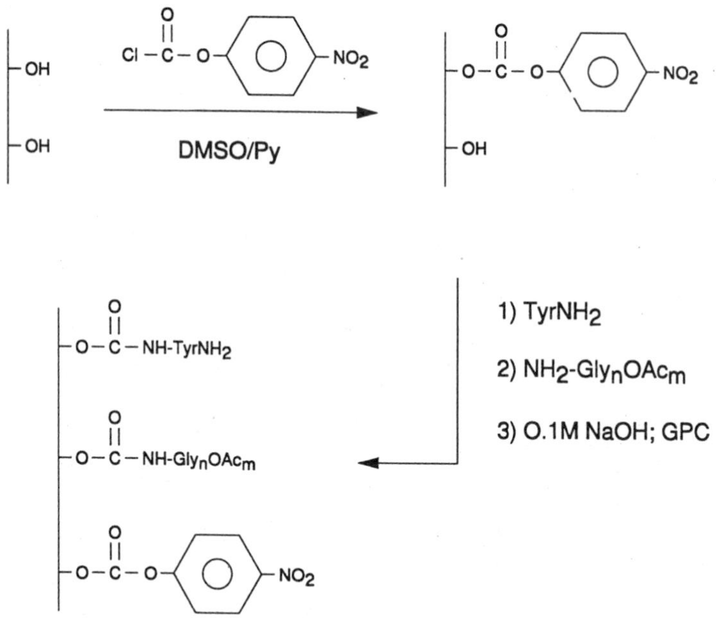

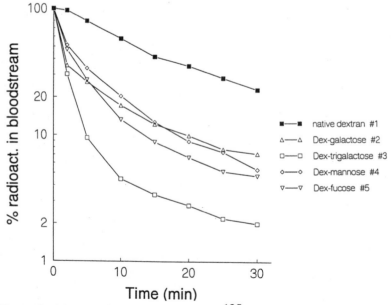

Figure 1. Schematic representation of the synthesis of L-Tyrosinamide labelled glycosylated dextrans. Tyr-NH$_2$ = L-Tyrosinamide; Gly = D-galactose, D-mannose or L-fucose; n=1, 3; m=4, 12.

Figure 2. Blood clearance of glycosylated ^{125}I-dextrans (#1-5) following femoral injection to rats.

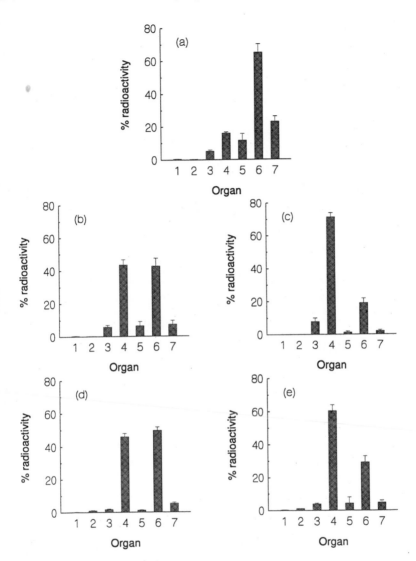

Figure 3. Body distribution of glycosylated ^{125}I-dextrans following intravenous injection (30 min.); (a) sample #1, (b) sample #2, (c) sample #3, (d) sample #4, (e) sample #5. The organ distribution of radioactivity is given according to the following key: 1-lung, 2-spleen, 3-intestine, 4-liver, 5-kidney, 6-carcass and 7-blood. Each column represents the mean value ± standard error of at least 3 determinations.

Body Distribution of ^{125}I-glycosylated Dextrans

The body distributions assayed 30 min. after the dextran derivatives ($50\mu g$/rat) were administered intravenously (femoral vein), are represented in Figure 3. All data are expressed as the percentage of the total recovered radioactivity per whole organ. No corrections were made for the blood content in the organs. In these 30 min. experiments, no attempt was made to isolate the urine or the bladder. As a consequence, the figures for the carcass content include the values for all excreted material. The injected native dextran is excreted by the kidney as could be expected from its molecular size (renal threshold: $M_w \approx 45,000$). Nevertheless, still 16% of the dose is captured by the liver in a non specific way. In contrast with this reference material, all glycosylated polymers do interact specifically with the receptors present on the cell surfaces, resulting in a major disposition of the materials in the liver cells (from 43% for the mono-galactose up to 71% for the tri-galactose dextran derivative). This significant difference between the tri-galactose substituted dextran (71%) and its mono-galactose analogue (43%), emphasizes the influence of the fine structural parameters on binding interactions between sugar-specific lectins and the synthetic glycosylated macromolecular carriers such as dextran. Preclustering of the incorporated sugar moieties strongly favours the affinity of polymers for the correspondent receptor. The faster rate of blood clearance of the same component, compared to the mono-galactose derivative, confirms this statement.

Unexpectedly, the fucosylated derivative expressed a pronounced higher affinity for the liver (60%) than the galactose (43%) and mannose (46%) substituted polymers.

Conclusion

In this paper we demonstrated that the incorporation of a number of selected monosaccharides such as D-galactose, D-mannose or L-fucose on dextran, had a dramatic effect on its in vivo fate following intravenous administration to rats. Compared to native dextran, the glycosylated polymers were cleared faster from the bloodstream. Due to the binding interactions with the cell surface receptors, the liver uptake was strongly enhanced. The significant difference in affinity for the galactose-specific receptor between the mono-galactose and the preclustered or tri-galactose substituted dextran stresses the importance of structural requirements to achieve an optimal binding interaction and site specific targeting of the dextran conjugate.

Acknowledgment

S.Vansteenkiste would like to thank the Belgium Institute for Encouragement of Research in Industry and Agriculture (I.W.O.N.L.) for providing a research fellowship.

Literature Cited

1   Morell,A.G.; Irvine,R.A.; Sternlieb,I.; Scheinberg,I.H.; Aswell,G. *J.Biol. Chem.* **1968**, *243*, 155.

2 Ashwell,G.; Harford,J. *Ann.Rev.Biochem.* **1982**, *51*, 531.
3 Monsigny,M.; Kieda,C.; Roche,A.C. *Biol.Cell* **1979**, *36*, 289.
4 Monsigny,M.; Kieda,C.; Roche,A.C. *Biol.Cell* **1983**, *47*, 95.
5 Gabius,M.J. *Angew.Chem.Int.Ed.Engl.* **1988**, *27*, 1267.
6 Neufeld,E.F.; Ashwell,G. In *The Biochemistry of Glycoproteins and Proteoglycans*; Lennarz,W.J.,Ed.; Plenum Press: New York, N.Y., 1980, 241-266.
7 Drickamer,K. *J.Biol.Chem.* **1988**, *263*, 9557.
8 Schwartz,A.L. *C.R.C. Critical Rev. Biochem.* **1984**, *16*, 207.
9 Kolb-Bachofen,V.; Schepper-Schafer,J.; Roos,P.; Hulsmann,D.; Kolb,H. *Biol.Cell* **1984**, *51*, 219.
10 Hudgin,R.L.; Pricer,W.E.; Ashwell,G.; Stockert,R.J.; Morell,A.G. *J.Biol.-Chem.* **1974**, *249*, 5536.
11 Stowell,C.P.; Lee,R.T.; Lee,Y.C. *Biochem.* **1980**, *19*, 4904.
12 Wall,D.A.; Hubbard,A.C. *J.Cell.Biol.* **1981**, *90*, 687.
13 Fallon,R.J.; Schwartz,A.L. *Adv.Drug Del.Rev.* **1989**, *4*, 49.
14 Ryan,U.S. *Adv.Drug Del.Rev.* **1989**, *4*, 65.
15 Stahl,P.D.; Schlesinger,P.H. *TIBS* **1980**, *5*, 194.
16 Largent,B.L.; Walton,K.M.; Hoppe,C.A.; Lee,Y.C.; Schnaar,R.L. *J.Biol.-Chem.* **1984**, *259*, 1764.
17 Janssen,R.W.; Molena,G.; Ching,T.L.; Oosting,R.; Harms,G.; Moolenaar,F.; Hardonk,M.J.; Meijer,D.K.F. *J.Biol.Chem.* **1991**, *266*, 3343.
18 Wall,D.A.; Hubbard,A.L. *J.Cell.Biol.* **1985**, *101*, 2104.
19 Lehrman,M.A.; Haltiwanger,R.S.; Hill,R.L. *J.Biol.Chem.* **1986**, *261*,7426.
20 Dautry-Varsat,A.; Lodish,H.F. *Scientific American* **1984**, *250*, 48.
21 Lloyd,J.B. In *Targeting of Drugs with Synthetic Systems*; Gregoriadis,G.; Senior,J.; Poste,G. Eds.; Plenum Press: New York and London, 1986, 57-63.
22 Rypacek,F.; Drobnik,J.; Kalal,J. *Ann.N.Y.Acad.Sci.U.S.A.* **1985**, 258.
23 Duncan,R. In *Sustained and Controlled Release Drug Delivery Systems*; Lee,-V.H.L.; Robinson J.R., Eds.; Marcel Dekker: New York, N.Y., 1987, 581-621.
24 Drobnik,J. *Adv.Drug Del.Rev.* **1989**, *3*, 229.
25 Meijer,D.K.F.; Van der Sluys,P. *Pharm.Res.* **1989**, *6*, 105.
26 Fiume,L.; Bassi,B.; Busi,C.; Mattioli,A.; Spinosa,G.; Faulstich,M. *Febs Letters* **1986**, *203*, 203.
27 Moltini,L. In *Drug Carriers in Biology and Medicine*; Gregoriadis,G. Ed.; Academic Press: London, U.K., 1979.
28 Larsen,C. *Adv.Drug Del.Rev.* **1989**, *3*, 103.
29 Yalpani,M. *C.R.C. Crit.Rev.Biotech.* **1986**, *3*, 375.
30 Schacht,E.H. In *Polymers in Controlled Drug Delivery*; Illum,L.; Davis S.S., Eds.; I.O.P. Plublishing Limited: Bristol, U.K., 1987.
31 Chang,R.L.S.; Ueki,I.F.; Troy,J.L.; Deen,W.M.; Robertson,C.R.; Brenner,-B.M. *Biophys.J.* **1975**, *15*, 887.
32 Wiegel,P.H.; Schnell,E.; Lee,Y.C.; Roseman,S. *J.Biol.Chem.* **1978**, *253*, 330.
33 Wiegel,P.H.; Schnaar,R.L.; Kuhlenschmidt,M.S.; Schnell,E.; Lee,R.T.; Lee,-Y.C.; Roseman,S. *J.Biol.Chem.* **1979**, *254*, 10830.

34 Vansteenkiste,S.; De Marre,A.; Schacht,E. *J.Bioact.Biocompat.Pol.* **1991**, *7(1)*, 4.
35 Markwell,M.A.K. *Anal.Biochem.* **1982**, *125*, 427.
36 Duncan,R.; Kopecek,J.; Rejmanova,P.; Lloyd,J.B. *Biochim.Biophys.Acta* **1983**, *755*, 518.
37 Chytry,V.; Kopecek,J.; Leibnitz,E.; O'Hare,K.; Scarlett,L.; Duncan,R. *New Polymeric Mat.* **1987**, *1*, 21.
38 Seymour,L.W.; Duncan,R.; Kopeckova,P.; Kopecek,J. *J.Bioact.Biocompat.Pol.* **1987**, *2*, 97.
39 Seymour,L.W.; Duncan,R.; Strohalm,J.; Kopecek,J. *J.Biomed.Mat.Res.* **1987**, *21*, 1341.
40 Kawaguchi,K.; Kuhlenschmidt,M.; Roseman,S.; Lee,Y.C. *Arch.Biochem.Biophys.* **1980**, *205*, 388.
41 Connolly,D.T.; Towsend,R.R.; Kawaguchi,K.; Bell,W.R.; Lee,Y.C. *J.Biol.Chem.* **1982**, *257*, 939.
42 Lee,Y.C.; Towsend,R.R.; Hardy,M.R.; Lonngren,J.; Arnarp,J.; Haraldsson,M. ; Lonn,M. *J.Biol.Chem.* **1983**, *258*, 199.

RECEIVED October 1, 1992

Chapter 27

Evaluation of Poly(*dl*-lactide) Encapsulated Radiopaque Microcapsules

David J. Yang, Li-Ren Kuang, Chun Li, Tony Tsai, Chun-Wei Liu, Walter J. Lin, Wayne Tansey, Sarah Nikiforow, Patricia McCuskey, Zuxing Kan, Kenneth C. Wright, and Sidney Wallace

Department of Diagnostic Radiology, M. D. Anderson Cancer Center, The University of Texas, Houston, TX 77030

Poly(d,l-lactide) (PLA) microcapsules encapsulated with ethyliopanoate (IOPA E) and ethyldiatrizoate (DZE) were evaluated for their potential use in computed tomography imaging. IOPA E was synthesized from iopanoic acid and thionyl chloride, yield 82.4%. DZE was synthesized from diatrizoic acid and ethyliodide, yield 90%. Both compounds were radioiodinated with $Na^{131}I$ and encapsulated in PLA microcapsules. In vivo tissue distribution of microcapsule groups and control groups (IOPA E and DZE) were evaluated. The percentage of injected dose per liver at 20 min, 2, 5 and 24 hours was 35.19, 20.90, 9.38 and 1.49 for the PLA ^{131}I-IOPA E group; and 46.25, 29.01, 13.19 and 1.27 for the PLA ^{131}I-DZE group. A significant difference ($P<0.05$) of liver uptake ratios between microcapsule groups and control groups within 5 hours was observed. In vivo microscopic studies demonstrated that both microcapsules showed good circulating properties in the liver and were actively phagocytized by Kupffer cells. These findings suggest that small microcapsules loaded with radiopaque materials have potential to produce a contrast density difference between normal liver tissue and abnormal hepatic focal lesions.

Because the liver of cancer patients is one of the most frequent sites for metastases, computed tomography (CT) scans of liver has become one of the most important examinations. The noncontrast CT examination is the simplest technique to perform. However, in noncontrast CT scans, less than 10% of focal hepatic lesions can go undetected (*1*). Application of iodinated material in CT has shown the possibility of detecting focal hepatic lesions by developing a contrast density difference between normal tissue and focal hepatic lesion (*2*). Conventional angiographic contrast media

[1]Current address: Department of Diagnostic Radiology, Box 59, University of Texas M. D. Anderson Cancer Center, 1515 Holcombe Boulevard, Houston, Texas 77030

0097–6156/93/0520–0371$06.00/0
© 1993 American Chemical Society

(CM) such as meglumine diatrizoate, iothalamate and metrizoate are water soluble materials. When CM is administered intravenously (iv), it distributes uniformly throughout the vasculature and the interstitial spaces of all tissues except the brain. The lesion detectability, which is dependent on the difference in attenuation between parenchyma and lesion, is often not significantly improved, despite the use of increasingly higher doses of CM (1).

CM is rapidly cleared from the liver, providing only short time enhancement sufficient for lesion detection. The greatest density difference is usually obtained within 2-3 mins after iv bolus of iodinated material. Therefore, this technique is limited because it requires 3-4 iv boluses of iodinated material to evaluate the entire liver. In order to improve the hepatic lesion detectability by enhancement CT imaging, a number of particulate agents have been investigated. These agents are: emulsion of ethiodol (EOE-13), liposomes carrying diatrizoate, iodipamide ethyl ester particulate suspensions (1,3,4). All these particulate contrast media are prime targets for removal by the reticuloendothelial system (RES). Since most tumor cells do not have an RES, they do not enhance by CT contrast media and appear as areas of decreased density (4-7). Therefore, these particulate agents selectively increase the density between liver and hepatic lesions. However, due to technical problems, such as toxicity and shelf-stability of these particular agents, it may limit their clinical use.

Poly (d,l-lactic acid) (PLA) is a non-toxic and widely used biodegradable polymer (8-13). When PLA microcapsules were loaded with anticancer drugs, the capsules produced sustained release properties (14). The administered routes of PLA capsules were intra-arterial (for chemoembolization), parenteral, subcutaneous, and oral (8). The capsular size employed was 7-350 μm. Microcapsules with particle size less than 3 μm are suitable for intravenous injection. This capsular size is crucial for intravenous administration, because if the particle size is above 5 μm, it can embolize in the pulmonary capillary bed (15). Radiopaque materials microencapsulated in 1 μm particles would provide an ideal size particle for bioimaging studies. Up to this point, little information is known about using PLA encapsulated radiopaques in CT enhancement imaging. This study evaluates biodistribution of PLA microcapsules loaded with ethyliopanoate and ethyldiatrizoate. The structures of both contrast agents are shown in Figure 1.

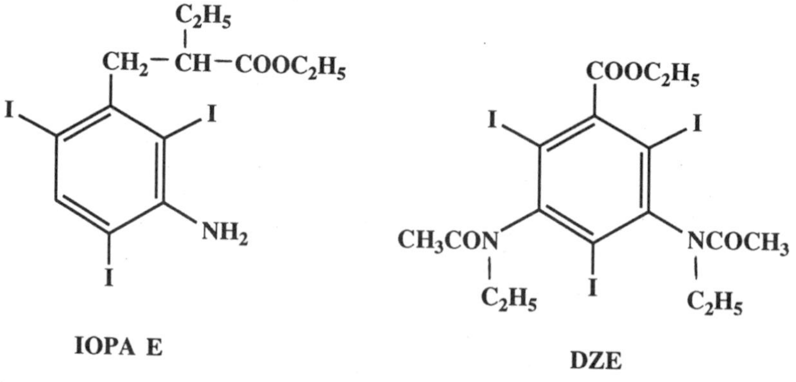

Figure 1. Structure of IOPA E and DZE

Methods

Materials. Poly(d,l-lactide) was obtained from Polysciences, Inc. (Warrington, PA). Polyvinylalcohol, iopanoic acid and diatrizoic acid were purchased from Sigma

Chemical Co. (St. Louis, MO). Na^{131}I (specific activity 7.75 Ci/mg, 680 mCi/ml) was obtained from Dupont New England Nuclear (Boston, MA). Female Sprague-Dawley rats (175-200 g) were purchased from Harlan, Inc. (Indianapolis, IN).

Synthesis of Ethyliopanoate (IOPA E). IOPA E was prepared by reacting iopanoic acid (2 g, 3.5 mmol) with thionyl chloride (0.4 ml, 5.25 mmol) in ethanol (50 ml) at 90°C for 3h. After evaporation of ethanol, the residue was dissolved in methylene chloride (100 ml) and washed with 5% NaOH. The methylene chloride layer was dried over MgSO$_4$ and evaporated to dryness. TLC (Silica; CHCl$_3$: MeOH; 9:1) indicated only one spot (R$_f$=0.8). The product yield was 1.73 g (82.4%), M$^+$=599 (shown in Figure 2), m.p. 64-67°C.

Radiolabeling of ^{131}I-IOPA E. Radiolabeling of IOPA E was achieved by solid-phase exchange technique (*15*). Briefly, IOPA E (5 mg) dissolved in THF (0.5 ml) was added with pivalic acid (5 mg) and Na^{131}I (1.6 mCi). The reaction was heated at 150°C for 3 h. The crude mixture was dissolved in CH$_2$Cl$_2$ (1 ml), loaded on a Sep-Pak column (Silica SPE Column, Whatman Lab., Clifton, New Jersey) and eluted with CHCl$_3$:MeOH (9:1). The product isolated was 1.17 mCi (73%).

Synthesis of Ethyldiatrizoate (DZE). Diatrizoic acid (6.14 g, 10 mmol), sodium ethoxide (3.4 g, 50 mmol) and ethyliodide (7.8 g, 62.5 mmol) were added to a stirred DMSO solution (50 ml). The reaction was stirred at room temperature for 3h. The solution was then poured in a beaker containing 1 liter of water. The white precipitate was collected. The solid was washed with ether and air dried, weighed 6.3 g (90% yield), (M+1)$^+$=699 (shown in Figure 3), m.p. 268-270°C. Anal. C$_{17}$H$_{21}$N$_2$O$_4$I$_3$ (C,H,N). Calc. C:29.25, H:3.03, N:4.01; Found. C:29.40, H:2.99, N:4.06.

Radiolabeling of ^{131}I-DZE. The radiolabeled DZE was prepared by the same technique as described in the preparation of ^{131}I-IOPA E. The radiochemical yield was 30%. This low yield could be due to the steric hindrance caused by the acetamide group neighboring the iodine atoms.

Preparation of PLA ^{131}I-IOPA E and PLA ^{131}I-DZE Microcapsules. PLA (0.6 g) and unlabeled IOPA E (0.4 g) or DZE (0.4 g) were dissolved in CH$_2$Cl$_2$ (15 ml). To this mixture ^{131}I-IOPA E or ^{131}I-DZE (250 µCi) was added. The organic phase was emulsified in 1% polyvinyl alcohol solution (220 ml). The mixture was stirred at 2200 rpm and warmed at 50°C for 4h to ensure complete evaporation of CH$_2$Cl$_2$. The suspension was filtered from a 5 µ nylon. The filtrate was then centrifuged (12,000 rpm, Damon/International Equipment Company, B-20A Centrifuge, Needham Heights, MA), washed with water, and centrifuged again. The resulting microcapsules were suspended in 10 ml of 25% human serum albumin (HSA) and stirred for 2h. This process has been used to prevent particle aggregation in vivo (*3*). The particles were cooled and allowed to sit overnight. The excess HSA was removed by centrifugation. The particles were resuspended in saline. The final concentration was 10 µCi/ml.

Particle Size Analysis. The size distribution of the microspheres was estimated with a Coulter Counter and Coulter Channelyzer (Coulter Electronics, Hialeah, FL). Microcapsules were suspended in saline solution containing Nonidet P40 (0.0001%) as dispersant. Measurements were made after sonication for 1 minute (Branson Cleaning Equipment Co., B-1200R-1, Shelton, CT) in order to cause deaggregation.

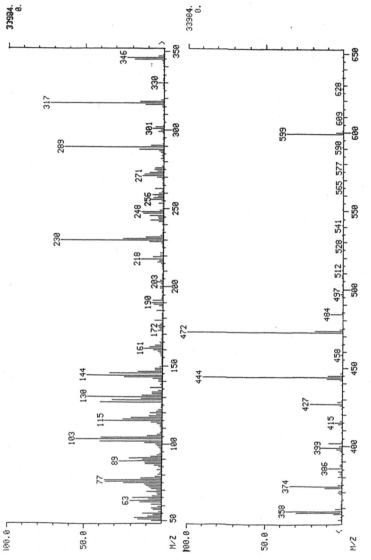

Figure 2. Mass Spectrum of Ethyliopanoate (IOPA E)

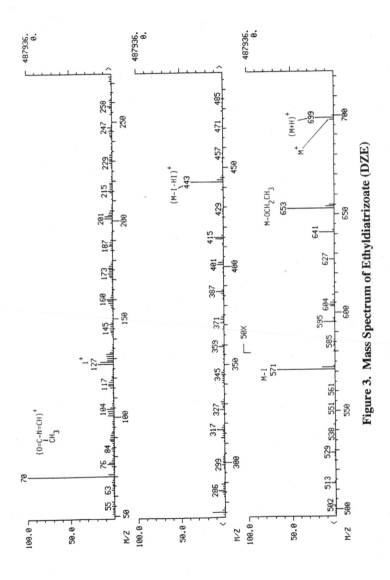

Figure 3. Mass Spectrum of Ethyldiatrizoate (DZE)

Electron Microscopic Examination. For scanning electron microscopic (SEM) examination, the microcapsules were diluted to different concentrations in water. Each sample (100 μl) was placed onto a 0.1 μm Nuclepore membrane and air dried. The dried filters were mounted onto stubs and sputter-coated with 200 Å gold-palladium, 80:20, in a Hummer VI (Technics, Springfield, VA) and examined in a Hitachi Model S520 scanning electron microscope.

Stability Test of PLA Microcapsules Loaded with IOPA E and DZE. Radiolabeled PLA-IOPA E and PLA-DZE microcapsules (20 μCi/mL) were incubated in 50% HSA at 37°C. At 20 min., 2, 5, and 24 hours, aliquots of serum were removed and centrifuged. The radioactivity of the supernatant was measured with a gamma counter (Packard, Model B5002, Downers Grove, IL).

In Vivo Biodistribution Studies. PLA microcapsules loaded with ^{131}I-IOPA E and ^{131}I-DZE were administered intravenously to rats (175-200 g, 5.0 μCi/rat, iodine content 75 mg I/kg body weight). The rats (N=3/time interval) were killed at 20 minutes, 2, 5 and 24 hours after injection. The percentage of injected dose in a given organ or weight of tissue was determined by a gamma counter.

In Vivo Microscopic Examination. The unlabeled PLA-IOPA E and PLA-DZE microcapsules were injected into rats (iv). The in vivo microscope, equipped with a video monitor (Axioplan, Zeiss Co, Germany), was then applied to observe the dynamic circulation of the microcapsules in liver.

Results and Discussion

Chemistry. Both IOPA E and DZE can be easily prepared with a high yield. Using standard solid phase exchange technique, both compounds were labeled in a reasonable yield.

Morphology of PLA Microcapsules. IOPA E and DZE capsules of 0.5-3 μm were prepared by the process described in the methods section. Figure 4 shows the average particle size of PLA encapsulated DZE. The particle sizes ranged from 0.5 to 3 μm in diameter. Figure 5 shows the SEM examination of PLA-IOPA E microcapsules. All the capsules prepared had smooth outer surfaces.

Stability Test of PLA-IOPA E and PLA-DZE Microcapsules. Less than 5% radioactivity was detected in serum after one hour incubation of both microcapsules. 20-25% of radioactivity was released in serum after 24 hours incubation of both microcapsules.

In Vivo Microscopic Observation. The in vivo microscopic observation revealed that (1) the microcapsules have uniform shape and little aggregation, (2) the Kupffer cells actively phagocytized PLA-IOPA E and PLA-DZE microcapsules and (3) the microcapsules demonstrated good circulating properties in the liver.

In Vivo Biodistribution Studies. The results of tissue distribution studies for ^{131}I-labeled IOPA E groups are shown in Tables I and II.

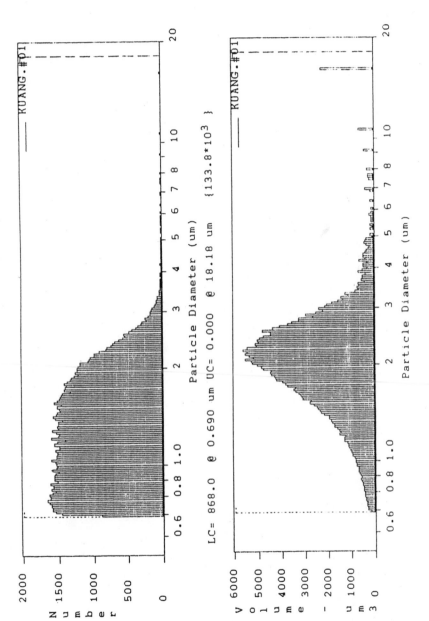

Figure 4. Particle Size Distribution of PLA-DZE Microcapsules

Figure 5. Scanning Electron Microscopic Observation of IOPA E
Microcapsules

Table I. Tissue Distribution of ^{131}I-IOPA E in Rats

Organ	20 minutes	2 Hours	5 Hours	24 Hours
Blood	1.51 ± 0.10[1]	1.41 ± 0.23	0.99 ± 0.04	0.28 ± 0.12
Lung	1.20 ± 0.16	0.64 ± 0.03	0.70 ± 0.31	0.13 ± 0.06
Liver	1.62 ± 0.18	0.97 ± 0.00	0.73 ± 0.06	0.39 ± 0.13
Spleen	1.89 ± 0.43	0.46 ± 0.01	0.21 ± 0.01	0.12 ± 0.05
Kidney	1.02 ± 0.10	0.61 ± 0.02	0.60 ± 0.33	0.24 ± 0.11
Intestine	2.79 ± 0.29	4.04 ± 0.53	3.66 ± 1.65	1.77 ± 0.80
Bone	0.47 ± 0.12	0.27 ± 0.03	0.32 ± 0.12	0.04 ± 0.01
Urine	0.34 ± 0.06	2.81 ± 0.40	2.68 ± 0.39	0.77 ± 0.35

[1]Data represents the percentage of injected dose per gram of tissue weight (% IND/G). Studies were conducted in three rats per time interval (N=3 / time interval).

Table II. Tissue Distribution of PLA Encapsulated ^{131}I-IOPA E in Rats

Organ	20 minutes	2 Hours	5 Hours	24 Hours
Blood	1.43 ± 0.71[1]	0.58 ± 0.18	0.87 ± 0.06	0.23 ± 0.10
Lung	1.56 ± 0.22	0.47 ± 0.20	0.48 ± 0.04	0.11 ± 0.04
Liver	4.13 ± 0.33	2.91 ± 0.75	1.61 ± 0.12	0.22 ± 0.05
Spleen	6.57 ± 0.43	6.53 ± 0.71	3.46 ± 0.90	0.28 ± 0.05
Kidney	0.54 ± 0.03	0.34 ± 0.09	0.38 ± 0.02	0.19 ± 0.07
Intestine	1.56 ± 0.46	1.56 ± 0.66	1.06 ± 0.03	0.36 ± 0.01
Bone	0.23 ± 0.03	0.20 ± 0.07	0.16 ± 0.01	0.05 ± 0.01
Urine	0.87 ± 0.44	8.75 ± 2.87	7.76 ± 1.56	1.33 ± 0.35

[1]Data represents the % IND / G (N=3 / time interval).

Tissue distribution for ^{131}I-labeled DZE groups are shown in Tables III and IV. The liver uptake in PLA-IOPA E microcapsule group was higher than IOPA E group. Similar results were shown in the PLA-DZE microcapsule group and the DZE group. A significant difference (P< 0.05) in percentage of injected dose per gram of liver between microcapsule groups and control groups within 5 hours was observed. The percentage of injected dose per liver at 20 min., 2, 5 and 24 hours for PLA-IOPA E and PLA-DZE microcapsules is shown in Figure 6. ^{131}I-IOPA E and ^{131}I-DZE were recovered in the intestine and urine at 20 minutes postinjection. The findings suggest that IOPA E and DZE released from PLA microcapsules were excreted through the gastrointestinal tract and kidneys.

Table III. Tissue Distribution of ^{131}I-DZE in Rats

Organ	20 minutes	2 Hours	5 Hours	24 Hours
Blood	0.00 ± 0.00[1]	0.00 ± 0.00	0.00 ± 0.00	0.00 ± 0.00
Lung	0.13 ± 0.09	0.02 ± 0.02	0.00 ± 0.00	0.00 ± 0.00
Liver	3.46 ± 0.79	1.86 ± 0.15	0.91 ± 0.17	0.00 ± 0.00
Spleen	2.30 ± 0.78	0.72 ± 0.27	0.08 ± 0.06	0.00 ± 0.00
Kidney	0.34 ± 0.04	0.23 ± 0.01	0.11 ± 0.04	0.00 ± 0.00
Intestine	0.69 ± 0.24	0.00 ± 0.00	0.00 ± 0.00	0.00 ± 0.00
Bone	0.00 ± 0.00	0.00 ± 0.00	0.00 ± 0.00	0.00 ± 0.00
Thyroid	0.00 ± 0.00	0.00 ± 0.00	0.00 ± 0.00	0.00 ± 0.00
Urine	1.39 ± 0.59	3.36 ± 0.11	2.97 ± 0.77	0.00 ± 0.00

[1]Data represents the % IND / G (N=3 / time interval).

Table IV.   Tissue Distribution of PLA Encapsulated ^{131}I-DZE in Rats

Organ	20 minutes	2 Hours	5 Hours	24 Hours
Blood	0.00 ± 0.00[1]	0.02 ± 0.00	0.05 ± 0.02	0.00 ± 0.00
Lung	4.33 ± 0.94	2.62 ± 0.35	1.29 ± 0.24	0.18 ± 0.03
Liver	5.40 ± 0.29	3.69 ± 0.18	1.52 ± 0.32	0.15 ± 0.02
Spleen	10.75 ± 2.62	7.66 ± 1.22	2.19 ± 1.36	0.12 ± 0.02
Kidney	0.45 ± 0.03	0.34 ± 0.03	0.17 ± 0.13	0.00 ± 0.00
Intestine	0.50 ± 0.08	0.37 ± 0.07	0.40 ± 0.07	0.00 ± 0.00
Bone	0.00 ± 0.00	0.00 ± 0.00	0.00 ± 0.00	0.00 ± 0.00
Thyroid	0.00 ± 0.00	0.01 ± 0.01	0.00 ± 0.00	0.00 ± 0.00
Urine	1.43 ± 1.12	4.61 ± 0.19	3.22 ± 2.90	0.47 ± 0.12

[1]Data represents the % IND / G (N=3 / time interval).

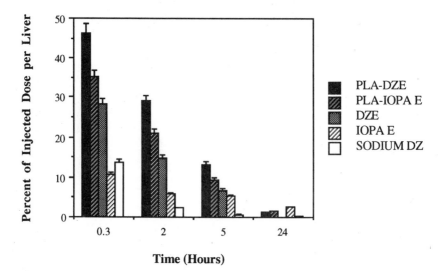

Time (Hours)

Figure 6.   Liver Uptake of PLA Encapsulated Radiopaque
Microcapsules

Summary

Diatrizoate is a water-based ionic contrast agent. Iopanoic acid is an oral radiopaque
medium for cholecystography and cholangiography. Approximately 50 percent of the
injected dose is excreted within 24 hours and completely in about 5 days (16). Both
contrast agents were formulated as microcapsules (0.5-3 μm) for intravenous
injection. Using particulate contrast media for CT enhancement of the liver was
reported (6,17,18). However, most studies show that the particles were quickly
degraded into water-soluble material and eliminated from the body. By using
microencapsulation techniques, (1) the microcapsules produced sustained release
properties in the liver, and (2) the administration route could be changed from oral to
iv. Both PLA encapsulated DZE and IOPA E microcapsules produced high uptake in
liver compared to the other organs. In vivo microscopic studies demonstrated that
both microcapsules were actively taken up by Kupffer cells. Because the Kupffer
cells are generally not present in lesions, it is possible to produce a contrast density
difference between normal liver tissue and abnormal hepatic focal lesions using the

microencapsulation technique. The information obtained from this study should provide a basis for using microcapsules to improve detectability of hepatic tumors by CT.

Acknowledgements

The authors wish to thank Dianne Perez-Onuogu for her excellent help in typing this manuscript. This study is supported by George and Cleo Cook Fund and the John S. Dunn Foundation. Scanning electron microscope facility is supported by NIH Core Grant P30-CA16672.

Literature Cited

1. Bernardino, M.E. In *Contrast Media: Biologic effect and Clinical Application*; Parver, Z., Ed.; CRC Press: Cleveland, Ohio, **1987**, pp 25-39.
2. Burgener, F.A; Hamlin, D.J. *Am J Roentgenol* **1983**, *140*, 291-296.
3. Lauteala, L.; Kormano, M.; Violante, M.R. *Invest Radiol.* **1984**, *19*, 133-141.
4. Seltzer, S.E.; Davis, M.A.; Adams, D.F.; Shulkin, P.M.; Landis, W.J.; Harvon, A. *Invest Radiol* **1984**, *19*, 142-151.
5. Violante, M.R.; Fischer, H.W. In *Contrast Media: Biologic effect and Clinical Application*; Parver, Z., Ed.; CRC Press: Cleveland, Ohio, **1987**, pp 89-103.
6. Violante, M.R.; Mare, K.; Fischer, H.W. *Invest Radiol.* **1981**, *16*, 40-45.
7. Sands, M.S.; Violante, M.R.; Gadeholt, G. *Invest Radiol.* **1987**, *22*, 408-416.
8. Conti, B.; Pavanetto, F.; Genta, I. *J Microencapsulation* **1992**, *9*, 153-166.
9. Bodmeier, R.; McGinity, J.W. *Int J Pharm* **1988**, *43*, 179-186.
10. Iwata, M.; McGinity, J.W. *J. Microencapsulation* **1992**, *9*, 201-214.
11. Jalil, R.; Nixon, J.R. *J. Microencapsulation* **1989**, *6*, 473-484.
12. Krause, H.J.; Schwarz, A.; Rohdewald, P. *Int J Pharm* **1985**, *27*, 145-155.
13. Ogawa, Y.; Yamamoto, M.; Okada, H.; Yashiki, T.; Shimamoto, T. *Chem. Pharm. Bull.* **1988**, *36*, 1095-1103.
14. Spenlehauer, G.; Vert, M.; Benoit, J.P; Chabot, F.; Veillard, M. *J Controlled Release* **1988**, *7*, 217-229.
15. Wieland, D.M.; Kilbourn, M.R.; Yang, D.J.; Laborde, E.; Gildersleeve, D.L.; Van Dort, M.E.; Pirat, J-L; Ciliax, B.J.; Young, A.B. *Int J Appl Radiat Isot* **1988**, *39*, 1219-1225.
16. Mcevoy, G.K. *AHES Drug Information*; American Society of Hospital Pharmacists, Inc.: Bethesda, MD, **1992**; pp 1440-1454.
17. Fischer, H.W. *Invest Radiol*, **1990**, *25*, 52-56.
18. Violante, M.R; Dean, P.B. *Radiology*, **1980**, *134*, 237-239.

RECEIVED October 5, 1992

Chapter 28

Design of Poly(α-malic acid)–Antitumor Drug–Saccharide Conjugate Exhibiting Cell-Specific Antitumor Activity

T. Ouchi, H. Kobayashi, K. Hirai, and Y. Ohya

Department of Applied Chemistry, Faculty of Engineering, Kansai University, Suita, Osaka 564, Japan

Poly(α-malic acid) is of interest as a biodegradable and bioadsorbable poly(lactide) type drug carrier, which is able to covalently attach both drug and targeting moiety. The macromolecular prodrug of antitumor drug will reduce the side-effects, have the ability to target tumor cells and exhibit high antitumor activity. The design of poly(α-malic acid)/ 5-fluorouracil(5FU)/saccharide and poly(α-malic acid)/ adriamycin(ADR) /saccharide conjugates were investigated. Monosaccharides such as galactosamine, glucosamine and mannosamine were used as targeting moieties. Growth inhibition against various tumor cells *in vitro* and the survival effect against mice bearing tumor cells *in vivo* were tested. Poly(α-malic acid)/5FU/galactosamine and poly(α-malic acid)/ADR /galactosamine conjugates showed stronger growth-inhibitory effects than poly(α-malic acid)/5FU and poly(α-malic acid)/ADR conjugates against *human hepatoma* cells *in vitro*. These results could be explained by galactose receptor-mediated specific uptake into *hepatocyte* cells. Moreover, the conjugates of poly(α-malic acid) with 5FU exhibited significant survival effects against p388 *lymphocytic leukemia* in mice by intraperitoneal(*ip*) transplantation/*ip* injection. The obtained conjugates did not display an acute toxicity in the high dose ranges.

Since poly(α-malic acid) has the application as a biodegradable and bioadsorbable polylactide type drug carrier having reactive carboxyl groups, which is able to attach drugs and targeting moieties (*1*). In comparison with low molecular weight prodrug, the macromolecule/drug conjugate is expected to overcome the problem of the side-effects of the parent anticancer agent. We have already reported that poly(α-malic acid) immobilizing 5-fluorouracil (5FU) exhibited an excellent survival effect against p388 *lymphocytic leukemia* cells in mice by intraperitoneal(*ip*) transplantation/*ip* injection (*2*). 5FU has a strong antitumor activity, which is accompanied, however, by undesirable side-effects (*3,4*). Doxorubicin (adriamycin; ADR) is one of the most prominent clinical antitumor agents, however its undesirable side-effects have also been cited. Additonally, some kinds of saccharides were recently reported to play important roles in biological recognition. So, in order to achieve the active targeting of macromolecule/drug conjugate, we have employed the saccharide residue as a targeting

0097–6156/93/0520–0382$06.00/0
© 1993 American Chemical Society

moiety. This has the advantage of reducing the side-effects while exhibiting high
antitumor activity. The present paper presents the design of poly(α-malic acid)/
5FU/saccharide and poly(α-malic acid)/ADR/saccharide conjugates, which are expected
to have affinity for tumor cells; the release behavior of parent drugs from the conjugates
in aqueous solution, the survival effect of the conjugates against p388 *lymphocytic
leukemia* in mice *ip/ip* and the growth-inhibitory effect of the conjugates against various
tumor cells *in vitro*.

Experimental

Synthesis of Poly(α-malic acid)/Saccharide Conjugate (1a-1c). Poly(α-
malic acid) (degree of polymerization: 11.3) was synthesized by ring-opening
polymerization of malide benzyl ester and consequent H2/Pd reduction according to the
method reported previously by our group (*1*). Three kinds of saccharide residues,
such as galactosamine, glucosamine, mannosamine, were attached to poly(α-malic acid)
via amide bonds using dicyclohexylcarbodiimide (DCC) to give poly(α-malic
acid)/saccharide conjugates 1a, 1b and 1c, respectively (Scheme I). 0.534 g (2.59
mmol) of DDC was added to an ice-cooled solution of 1.50 g (12.9 mmol) of
poly(α-malic acid) and 0.298 g (2.59 mmol) of 1-hydroxybenzotriazole (HOBt) in 30
ml of N,N'-dimethylformamide (DMF). The reaction mixture was stirred at 0°C for 1 h
and at room temperature for 24 h. A solution of 0.558 g (2.59 mmol) of galactosamine
hydroxychloride and 0.5 ml of triethylamine (TEA) in DMF (15 ml) was added to the
reaction mixture and stirred at room temperature for 24 h. The N,N'-dicyclohexylurea
(DCU) formed was removed by filtration and washed with DMF. The filtrate was
evaporated under reduced pressure to afford poly(α-malic acid)/galactosamine conjugate
1a.
   Poly(α-malic acid)/glucosamine 1b and poly(α-malic acid)/ mannosamine 1c were
prepared by the method described above. The degree of introduction of saccharide in
mol% per carboxyl group of conjugates 1a-1c was estimated by the 3,4-dinitrosalicylic
acid (DNS) reduction method (*5*).

**Synthesis of 1-[(Amino-n-pentyl)-ester]-methylene-5FU Hydrochloride
(4).**   The preparation of the 5FU amino derivative, 1-[(amino-n-pentyl)-ester]-
methylene-5FU hydrochloride 4, is shown in Scheme II (*6*). 14.4 ml (100 mmol) of 37
vol% formalin was added to 17.8 g (220 mmol) of 5FU and the reaction mixture was
stirred at 60°C for 45 min and then evaporated to give the oily 1,3-dimethylol-5FU (2).
12.3 g (120 mmol) of DCC, 15.6 g (120 mmol) of t-butoxycarbonyl- amino-n-caproic
acid and 0.7 g of 4-dimethylaminopyridine (DMAP) were added to the solution of 2 in
300 ml of acetonitrile and then stirred for 4 h at room temperature. The DCU formed
was filtered off and washed with dichloromethane. The resulting solution was washed
three times with 1 N HCl and aq. satd. NaCl solution. The oil layer, dried with
anhydrous sodium sulfate, was evaporated to give the oily residue. The crude product
was purified by recrystallization from diethylether to afford the white solid of
1-[(t-butoxycarbonyl-amino-n-pentyl)-ester]- methylene-5FU (3). Compound 3 was
dissolved in dioxane containing 4 N HCl, stirred for 1 h at room temperature. The
solution was evaporated to give the white powder of 1-[(amino-n-pentyl)-ester]-
methylene-5FU hydrochloride (4). M.p. 188-189° C; yield 30%. IR (KBr): 3050
($N^{+}H_3$), 3000, 2950 (CH2), 1720 (C=O of 5FU), 1740, (COO), 1270 (C-F), 1200
(COO), 1110cm^{-1}(C-N). ^{1}H-NMR((CD3)2SO), δ1.34 (m, 2H, CH2), 1.6 (m, 4H,
CH2), 2.35 (t, 2H, CH2), 2.73 (m, 2H, CH2), 5.58 (s, 2H, CH2), 8.05 (s, 3H,
$N^{+}H_3$), 8.25 (d, 1H, H-6). ^{13}C-NNMR((CD3)2SO)), δ23.9, 25.4, 26.8, 33.2, 40.3,
70.8(CH2), 129.7, 138.6, 140.9, 157.7(5FU), 172.8 (C=O). UV(MeOH): λmax
262nm, εmax 8800.

Synthesis of Poly(α-malic acid)/5FU Saccharide (5) and Poly(α-malic acid)/5FU/Saccharide Conjugate (5a-5c). The conjugate of 5FUs attached to poly(α-malic acid) *via* amide bonds through pentamethylene, monomethylene spacer groups and ester bonds (5) and the conjugate of poly(α-malic acid) fixing 5FU *via* amide bonds through pentamethylene, monomethylene spacer groups and ester bonds, and fixing saccharide units *via* amide or ester bonds (5a-5c) were prepared by the reaction steps shown in Scheme III. Conjugate 5 has no saccharide unit. Conjugate 5a, 5b and 5c have galactosamine, glucosamine, and mannosamine units as side saccharide residues, respectively.

1.05 g (5.09 mmol) of DDC was added to an ice-cooled solution of 0.590 g(5.09 mmol carboxyl unit) poly(α-malic acid) (*1*) and 1.58 g (5.09 mmol) of 4 ' in acetonitrile (70 ml). 1 ml of TEA was added dropwise to the above solution at 0°C for 3 h and then the reaction mixture was stirred at room temperature for 20 h. After the DCU formed was removed by filtration, the acetonitrile-soluble part was evaporated under reduced pressure to afford crude 5. In order to completely remove DCU, crude 5 was subjected to gel chromatography on Sephadex LH-20. Elution with THF and evaporation afforded a hygroscopic pale yellow powder. This powder was then reprecipitated using a THF/excess Et2O-n-hexane (1:1, v/v) system to give the desired poly(α-malic acid)/5FU conjugate (*5*). 0.664 g (3.22 mmol) of DCC was added to an ice-cooled solution of 1.2 g (8.06 mmol carboxyl unit) of 1a and 0.371 g (3.22 mmol) of HOSu in DMF (20 ml). The reaction mixture was stirred at 0°C for 1 h and then at room temperature for 24 h. A solution of 1.25 g (4.03 mmol) of 4 ' and 0.7 ml of TEA was added to the reaction mixture and stirred at room temperature for 24 h. The DCU formed was removed by filtration and washed with DMF. The filtrate was evaporated under reduced pressure. In order to completely remove DCU, the residue was subjected to gel chromatography on Sephadex LH-20 (Pharmacia Co.). Elution with DMF afforded poly(α-malic acid)/5FU/galactosamine conjugate 5a. Poly(α-malic acid)/5FU/glucosamine conjugate 5b and poly(α-malic acid)/5FU/mannosamine conjugate 5c were prepared by the method described above. The values of the degree of saccharide on the polymer in mol% saccharide based on the number of substituted carboxylic acid groups (DS) for 5a, 5b and 5c were determined to be 24, 16 and 13 mol%, respectively. Since only free 5FU was obtained by hydrolysis of the conjugates under refluxing in aq. 3 N NaOH solution for 24 h, the degree of substitution of 5FU in mol% per carboxyl group of conjugate (D5FU) of 5, 5a, 5b, 5c were determined to be 20, 26, 24 and 25 mol%, respectively. The D5FU value was obtained by GPC measurement of the released amount of 5FU in the hydrolyzed solution (column; Shodex OHpak KB-803: eluent; 0.013 M KH2PO4-Na2HPO4 buffer solution, pH 7.0: detector; UV 265 nm).

Synthesis of Poly(α-malic acid)/ADR Conjugate (6) and Poly(α-malic acid)/ ADR/Saccharide (6a-6c). The poly(α-malic acid)/amide/ADR conjugate 6 and poly(α-malic acid)/amide/ADR/saccharide conjugate 6a-6c were prepared by reacting ADR with poly(α-malic acid) or poly(α-malic acid)/saccharide conjugates 1a-1c, respectively (Scheme IV). The obtained conjugates were purified by gel filtration chromatography. The degree of substitution of ADR in mol% per carboxyl group of the conjugate(DADR) was estimated by UV measurement in water. The synthetic results of 6 and 6a-6c are summarized in Table I.

Determination of Extent of Release of 5FU and ADR from the Conjugates. The release behavior of 5FU from poly(α-malic acid) /5FU/saccharide

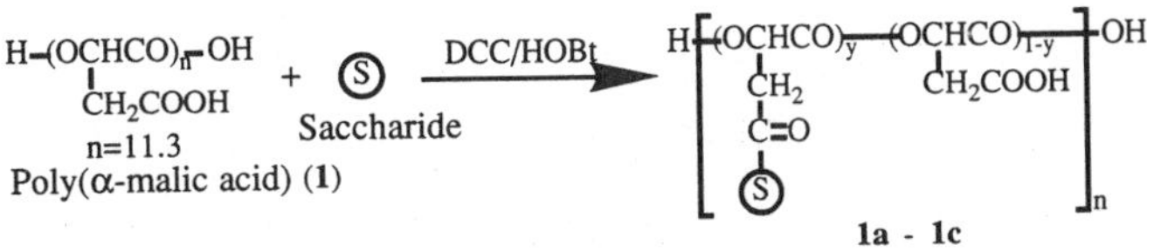

1a - 1c

Scheme I. Preparation route for poly(α-malic acid)/amide/saccharide conjugates **1a-1c**. **1a**(Ⓢ :galactosamine), **1b**(Ⓢ :glucosamine), **1c**(Ⓢ:mannosamine).

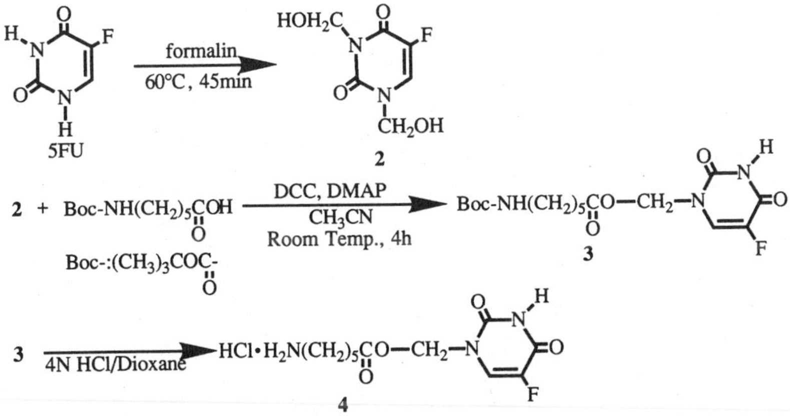

Scheme II. Preparation route for 1-[(amino-n-pentyl)-ester]-methylene-5FU hydrochloride **4**.

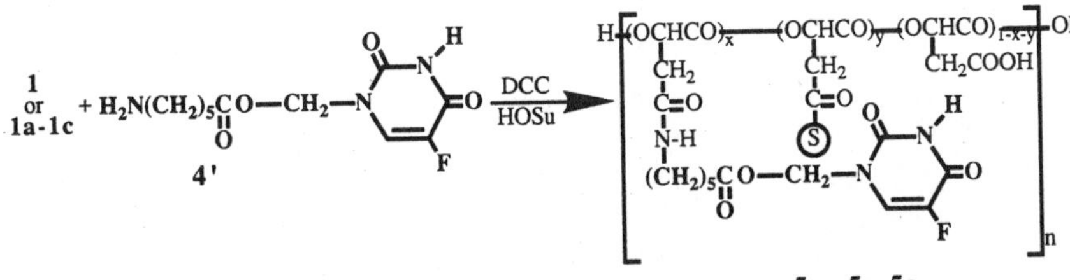

5 or 5a-5c

Scheme III. Preparation route for poly(α-malic acid)/5FU/saccharide conjugates **5**(y=0), **5a**(Ⓢ :galactosamine), **5b**(Ⓢ :glucosamine), **5c**(Ⓢ :mannosamine).

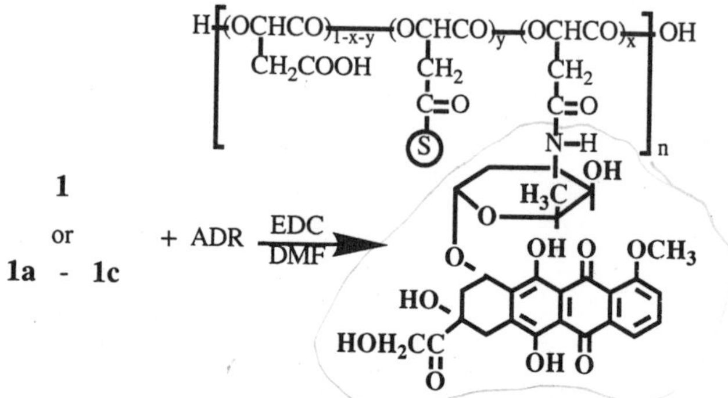

Scheme IV. Preparation route for poly(α-malic acid)/amide/ADR/saccharide conjugates **6**(y=0), **6a**(Ⓢ:galactosamine), **6b**(Ⓢ :glucosamine), **6c**(Ⓢ : mannosamine).

Table I. Synthesis of poly(α-malic acid)/amide/ADR conjugate **6** and poly(α-malic acid)/amide/ADR/saccharide conjugates **6a-6c**[a]

sample	**1** or **1a-1c** mg(mmol)	ADR · HCl mg(mmol)	WSC mg(mmol)	DS[b] mol%	DADR[c] mol%
6	100.0(0.860)	125.0(0.215)	61.5(0.323)		16.7
6a	14.2(0.104)	15.0(0.026)	10.0(0.052)	16.0	14.8
6b	13.9(0.104)	15.0(0.026)	10.0(0.052)	15.7	13.7
6c	16.0(0.104)	15.0(0.026)	10.0(0.052)	22.2	7.0

[a]1.3eq of TEA was added to the reaction mixture first for 2h at 0°C and overnight at room temperature.
[b]degree of introduction of saccharide was determined by DNS method.
[c]degree of introduction of ADR was determined by absorption at 495nm in H_2O.

conjugate **5a** was investigated *in vitro* at 37° C by shaking in 0.067 M KH2PO4-Na2HPO4 buffer solution (pH 7.4). The amount of 5FU released from these conjugates was estimated by GPC (column; Shodex OHpak B-805: eluent; 0.013 M KH2PO4-Na2HPO4 buffer solution, pH 7.0: detector; UV270nm). The release behavior of ADR from poly(α-malic acid)/amide/ADR/conjugate **6** *in vitro* was investigated in 0.067 M KH2PO4-Na2HPO4 buffer solution (pH=7.4) or 0.20 M-Na2PO4-0.1 M- citric acid buffer solution (pH=4.0). The amount of ADR released was estimated by HPLC method (column; TSK gel ODS-120T: eluent: acetonitrile/0.010 M NH4H2PO4 aqueous solution, vol. ratio 65:35: detector; fluorescence with excitation at 480nm and emission at 590nm).

Growth-inhibitory Effect of the Conjugates *in vitro.* The growth inhibition of conjugates **5a**, **5b** and **5c** was examined after exposure against HLE *human hepatoma* for 2 or 4 h. Aliquots of tumor cell suspension (100 µl) containing 4×10^4 cells in culture medium containing 10% fetal calf serum (FCS) were distributed in wells of a Corning 24-well plate and then incubated in a humidified atmosphere containing 5% CO2/95% air at 37° C for 2 days. The cells were distributed in aliquots of DMSO solution (10 µl) of conjugates **5a**, **5b** and **5c** and incubated at 4°C for 2 or 4 h. In order to remove the conjugate not bound to the cells, the cells were washed several times with culture medium. The washed cells were added to 1 ml of fresh culture medium, and then cultured in a CO2 incubator at 37°C for 2 days. The cells were harvested and viable cells were determined by 3-(4,5-dimethylthiazol-2-yl)-2,5-diphenyltetrazolium bromide (MTT) assay (7). The growth-inhibitory effect was calculated by equation (1):

Growth-inhibitory effect (%)

$$= 1 - \left(\frac{\log N_T - \log N_0}{\log N_C - \log N_0} \right) \times 100 \qquad (1)$$

N_T, N_C : Number of cells after incubation for 48 h
for test and control samples, respectively
N_0 : Initial number of cells

The blocking effect of the trimer of galactosamino-saccharide (GOS3) on the affinity of poly(α-malic acid)/5FU/galactosamine conjugate **5a** for HLE *human hepatoma* cells was examined. Aliquots of tumor cell suspension (100 µl) containing 4×10^4 cells in culture medium containing 10% FCS were distributed in wells of a Corning 24-well plate and incubated in a humidified atmosphere containing 5% CO2/95% air 37°C for 2 days. The cells were distributed into portions ($0-10^3$ µg) of GOS3 in culture medium and incubated at 4°C for 1 h, and then added to aliquots of DMSO solution (20 µl) of conjugate **5a** and 5FU and incubated at 4° C for 2 h. The cells, washed several times with culture medium, were added to 2 ml of fresh culture medium, and then cultured in a CO2 incubator at 37° C for 30 h. The cells were harvested and the viable cells were determined by the Trypan Blue dye exclusion test. Growth inhibition was calculated with the same equation described above. The growth-inhibitory effect of the conjugates **6, 6a-6c** was measured against HLE *human hepatoma* cells or Hela *utrecervical carcinoma* cells *in vitro*. Aliquots of tumor cell suspension (100 µl) containing 8.5×10^4 cells in culture medium containing 10% FCS were distributed in wells of 96-well multi-plate (Corning 25860MP) and incubated in a humidified atmosphere containing

5% CO2/95% air at 37° C for 48 h. The cells washed with culture medium were distributed in 100 μl of fresh culture medium containing 20 μl of PBS solution of the conjugate **6**, **6a** or free ADR and incubated in the same condition for 48 h. The number of viable cells was determined by means of MTT assay (*7*) using a microplate reader (MTP-120, Corona Electric Co.). The growth-inhibitory effect was calculated by equation (2) as follows:

$$\text{Growth-inhibitory effect (\%)} = (N_C - N_T) / N_C \times 100 \qquad (2)$$

N_C: Number of control cells
N_T: Number of treated cells

Galactose receptor-mediated growth-inhibitory effect against HLE *human hepatoma* cells *in vitro* was evaluated by the following two methods. One is the measurement of growth-inhibitory effect by poly(α-malic acid)/amide/ADR/galactosamine conjugate **6a** by the conjugates binding on HLE *human hepatoma* cell surface. Aliquots of tumor cell suspension (100 μl) containing 6.5×10^4 cells in culture medium containing 10% FCS were distributed in wells of 96-well multi-plate and incubated in a humidified atmosphere containing 5% CO2/95% air at 37°C for 48 h. The cells washed with culture medium were distributed in 100 μl of fresh culture medium containing 20 μl of PBS solution of the conjugates **6**, **6a**, **6b**, or **6c** (2.4×10^4 ADRmol/l) and incubated at 4° C for 0.25, 1.0 or 4.0 h. In order to remove the conjugates not bound to the cells, the cells were washed several times with culture medium. The washed cells were added to 100 μl of fresh culture medium and then cultured in a humidified atmosphere containing 5% CO2/95% air at 37° C for 48 h. The number of viable cells and growth inhibitory effect were estimated according to equation (2). The other is the test of blocking effect of the addition of inhibitor on uptake of the poly(α-malic acid)/ADR/galactosamine conjugate **6a** into HLE *human hepatoma* cells *in vitro*. Aliquots of tumor cell suspension (100 μl) containing 4.0×10^4 cells in culture medium containing 10% FCS were distributed in wells of 96-well multi-plate and incubated in a humidified atmosphere containing 5% CO2/95% air at 37°C for 48 h. The cells washed with culture medium were distributed in 100 μl of fresh culture medium containing 0.20, 40 or 60 μg of poly(α-malic acid)/galactosamine conjugate **1a** and incubated at 4° C for 1.0 h. After washing with culture medium, the cells were distributed in 20 μl of PBS solution of the conjugate **6** (2.4×10^4 ADRmol/l) or free ADR and incubated at 4°C for 2.0 h. In order to remove the conjugates or free ADR not bound to the cells, the cells were washed several times with culture medium. The washed cells were added to 100 μl of fresh culture medium and then cultured in a humidified atmosphere containing 5% CO2/95% air at 37° C for 48 h. The number of viable cells and growth-inhibitory effect were estimated by equation (2).

Measurement of Survival Effect of Mice Bearing p388 *leukemia in vivo.*

The survival effect was tested against p388 *lymphocytic leukemia* in female CDF1 mice (30 untreated mice/group and 6 treated mice/group) by *ip/ip* according to the protocol of the Japanese Foundation for Cancer Research (JFCR). 1×10^6 *leukemia* cells were injected by *ip* on day 0. The samples were sonicated in 0.05% sorbate 80 in sterile saline and administerd by *ip*. The mice received a dose of 200-800 mg/kg at day 1 and 5. The ratio of prolongation of life of the test mice, T/C(%), i.e., the ratio of the median survival of treated mice (T) to that of the control (C), was evaluated as antitumor activity *in vivo*. The average C value was generally 10 days. When the T/C value exceeds 120, the sample can be evaluated as active. These survival effect data are the results of the screening performed at the Cancer Chemotherapy Center of JFCR.

Results and Discussion

Release Behavior of 5FU or ADR from the Conjugates. In order to evaluate the release behavior of 5FU from poly(α-malic acid)/5FU/saccharide conjugates, the hydrolysis of poly(α-malic acid)/5FU/galactosamine conjugate **5a** was studied in 0.067 M phosphate buffer solution (pH = 7.4) at 37°C *in vitro*. It was confirmed by TLC that the **5a** conjugate was hydrolyzed to give only free 5FU itself but did not afford any 5FU derivative. From Figure 1, about 20% of 5FU was shown to be released within 24 h; the release rate of 5FU from the conjugate *in vitro* was found to be relatively slow. On the other hand, the hydrolytic rate for the main-chain ester bonds was found to be much slower than that for the pendant ester bonds. Therefore, after free 5FU was released from the conjugate, the main-chain ester bonds of the backbone polymer were thought to be degraded over a long period. The results of release behavior of ADR from poly(α-malic acid)/amide/ADR conjugate **6** is shown in Figure 2. In the region of neutral pH, the release rate of ADR from the conjugate **6** was very slow. On the other hand, the release rate of ADR from **6** increased under the acidic condition. Therefore, it can be expected that the amide bonds of the conjugate can be cleaved to release free ADR in the lysosomal acidic condition after uptake into tumor cells.

Growth-inhibitory Effect of the Conjugates against HLE *human hepatoma* Cells *in vitro*. The growth-inhibitory effects by conjugates **5a**, **5b** and **5c** after exposure to HLE *human hepatoma* for 2 or 4 h are shown in Figure 3. The exposure of conjugate **5a** to *hepatoma* tended also to have the highest growth-inhibitory effect among these conjugates. From the results shown in Figure 3, such an appearance of a cell specific growth-inhibitory effect based on the difference of the kind of saccharide residue can be explained by the active targeting of 5FU onto the surfaces of cells. As the uptaken mechanisms of 5FU into the cells, the saccharide-receptor mediated free 5FU diffusion route and the saccharide-receptor mediated poly(α-malic acid)/5FU/saccharide conjugate endocytosis route can be presumed, as shown in Scheme V (8). Since HLE *human hepatoma* cells have receptor specific affinities for galactosamine, poly(α-malic acid)/5FU/galactosamine conjugate **5a** seemed to achieve cell specific targeting to HLE *hepatoma* cells and to exhibit higher growth-inhibitory effect against *hepatoma* cells.

The results of the blocking effect of GOS3 on the appearance of the affinity of poly(α-malic acid)/5FU/galactosamine conjugate **5a** to HLE *human hepatoma* is shown in Figure 4. The growth-inhibitory effect of conjugate **5a** against *hepatoma* tended to decrease with an increase in the amount of added GOS3. On the other hand, the growth-inhibitory effect of 5FU against *hepatoma* was not affected by addition of GOS3. These results supported the suggestion that the appearance of a high level growth- inhibitory effect of poly(α-malic acid)/5FU/galactosamine against *hepatoma* was attributable to the cell specific targeting of 5FU to *hepatoma* via galactose-receptor mediated affinity.

The growth-inhibitory effect of the conjugate **6** and **6a** obtained was investigated against HLE *human hepatoma* cells or Hela *utrecervical carcinoma* cells *in vitro*. The IC50 values determined from the cytotoxic activity are summarized in Table II. The growth-inhibitory effects of conjugates **6** and **6a** were lower than of that of free ADR. Although the IC50 values of conjugates **6a** against Hela *utrecervical carcinoma* cells was the same value to that of conjugate **6**, the IC50 values of conjugates **6a** against HLE *human hepatoma* cells was smaller than that of conjugate **6**. These results mean that the conjugate **6a** having galactosamine residues showed higher growth-inhibitory effect

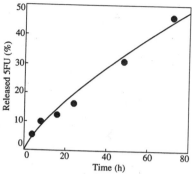

Figure 1. Release rate of 5FU from conjugate **5a**[a] in 0.067 M-KH$_2$PO$_4$- Na$_2$HPO$_4$ buffer solution (pH 7.4) at 37°C.

[a]DS = 24.1 mol%, D5FU = 25.6mol%.

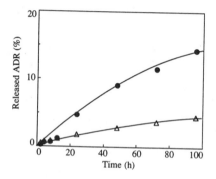

Figure 2. Release rate of ADR from conjugate **6** in aqueous solution at 37°C. △:0.067 M KH$_2$PO$_4$-Na$_2$HPO$_4$ buffer solution; pH = 7.4, ●:0.20 M Na$_2$HPO$_4$-0.10 M citric acid buffer; pH = 4.0.

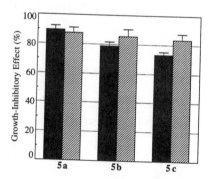

Figure 3. Growth-inhibitory effect by poly(α-malic acid)/5FU/saccharide conjugates **5a**, **5b** and **5c** against HLE *human hepatoma* cells *in vitro*.

The cells exposed with the conjugates for 2(■) or 4 h(▨) were cultured for 2

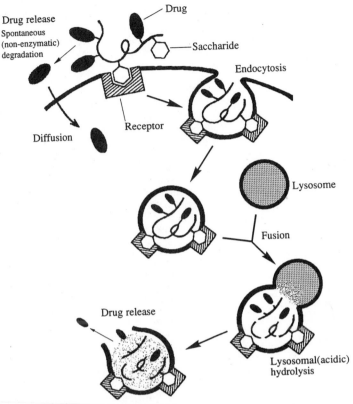

Scheme V. Routes of drug uptaken into the cells.

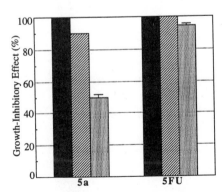

Figure 4. Blocking effect of GOS3 on growth inhibition by poly(α-malic acid)/5FU/ galactosamine conjugate **5a** and 5FU against HLE *human hepatoma* cells *in vitro*[a].

Dose of GOS3 (mg/well), ■ : 0 , ▨ : 625 , ▦ : 1000.

[a]The cells were treated with GOS3 for 1 h, exposed with conjugate **5a** and 5FU for 2 h, and cultured for 30 h.

than conjugate **6** having no saccharide residues against HLE *human hepatoma* cells having receptor specific affinities for galactose.

The results of growth-inhibitory effect of conjugates **6**, **6a**, **6b** and **6c** against HLE *human hepatoma* cells *in vitro* through short exposure at 4°C are shown in Figure 5. The growth-inhibitory effect of conjugate **6a** having galactosamine residues was larger than that of the other conjugates having the other kind saccharide residues or no saccharide residue. The results mean that the number of conjugate **6a** binding on the surface of HLE *human hepatoma* cells was larger than that of the other conjugates. The effect of poly(α-malic acid)/galactosamine conjugate **1a** as an inhibitor on the growth-inhibitory effect of conjugate **6a** or free ADR against HLE *human hepatoma* cells are shown in Figure 6. The growth-inhibitory effect of free ADR was not affected by the addition of conjugate **1a**, however, the growth-inhibitory effect of the conjugate **6a** was inhibited by the addition of conjugate **1a**. These results suggest that the high growth-inhibition effect of conjugate **6a** resulted from the receptor-mediated uptake of conjugate **6a** into HLE *human hepatoma* cells. Therefore, the galactosamine residue is expected to act as a targeting moiety to *hepatoma* cells. Thus, poly(α-malic acid)/ADR/galactosamine conjugate **6a** can be expected to be used as the biodegradable and water-soluble macromolecular prodrug of ADR having targetability to *hepatoma* cells, releasing free ADR in the lysosomal acidic condition after uptake into *hepatoma* cells.

Survival Effect of Mice Bearing p388 *leukemia in vivo*. The results of the survival effect of 5FU, 5FU amino derivative **4**, poly(α-malic acid)/5FU conjugate **5**, poly(α-malic acid)/5FU/saccharide conjugates **5a**, **5b** and **5c** against *leukemia* in female CDF1 mice by *ip/ip* are summarized in Figure 7. Although the survival effects per mol of 5FU in these conjugates were not higher than those of free 5FU and monomeric 5FU amino derivative **4** against *leukemia* mice, these conjugates exhibited significant antitumor activities. Moreover, mice treated with these conjugates did not exhibit a rapid decrease of body weight of the treated mice, even in the high dose ranges (the upper dose of conjugate: 400 g/kg) shown in Figure 7; they did not display an acute toxicity in these dose ranges. The dose capacity of 5FU was found to increase by the polymer/5FU conjugate technique. Thus, it was suggested that the side effects of monomeric antitumor agent was suppressed by such a macromolecule/antitumor drug conjugate technique.

Table II. IC_{50} values of poly(α-malic acid)/amide/ADR conjugate **6** and poly(α-malic acid)/amide/ ADR/saccharide conjugate **6a** against HLE *human hepatoma* cells and Hela *utrecervical carcinoma* cells

	tumor cells	
sample	HLE	Hela
free ADR	5.2×10^{-6}	6.2×10^{-6}
6	7.6×10^{-4}	4.9×10^{-4}
6a	1.8×10^{-4}	4.9×10^{-4}

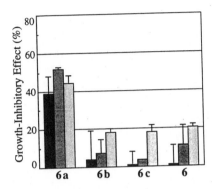

Figure 5. Growth-inhibitory effect by poly(α-malic acid)/amide/ADR/ saccharide conjugates **6a**, **6b**, **6c** and poly(α-malic acid)/amide/ADR conjugate **6** against HLE *human hepatoma* cells *in vitro*.
Incubation time (h), ■ : 0.25 , ▨ : 1.0 , ▦ : 4.0.

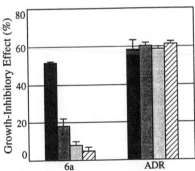

Figure 6. Effect of poly(α-malic acid)/galactosamine conjugate **1a** on cytotoxicity of poly(α-malic acid)/amide/ADR/galactosamine conjugate **6a** and ADR against HLE *human hepatom* cells *in vitro*.
Amount of inhibitor **1a** (mg), ■ : 0 , ▨ : 20 , ▦ : 40 , ▧: 60.

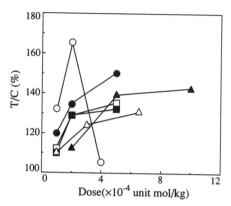

Figure 7. Survival effects per unit mol of 5FU in 5FU, **4**, **5**, **5a**, **5b** and **5c**
against p388 *leukemia* in mice by *ip/ip* : ○: 5FU, △: **4**,
●: **5**(D5FU = 20.0mol%),□: **5a**(DS = 24.1 mol%, D5FU = 25.6 mol%), ■:
5b(DS = 15.6 mol%, D5FU = 23.5 mol%), ▲: **5c**(DS = 12.9 mol%, D5FU =
25.2 mol%).

Acknowledgement

The authors wish to express their sincere appreciation to Dr. Tazuko Tashiro of the
Cancer Chemotherapy Center of the Japanese Foundation for Cancer Research for the
screening test for the survival effect against p-388 *lymphocytic leukemia* in mice by
ip/ip. The authors wish to give their thanks to Meiji Seika Co. Ltd. for providing ADR.

Literature Cited

1. Ouchi, T.; Fujino, A. *Makromol. Chem.* **1989**, 190, 1523.
2. Ouchi, T.; Kobayashi, H.; Banba, T. *Brit. Polym. J.* **1990**, 23, 221.
3. Bosch, L.; Harbers, E.; Heidelberger, C. *Cancer Res.* **1958**, 18, 335.
4. Bounous, G.; Pageau, R.; Regoli, D. *Int. J. Chem. Pharmacol. Biopharm.* **1978**,
   16, 519.
5. Hostetter, F.; Borel, E.; Denel, H. *Helv. Chim. Acta* **1951**, 34, 2132.
6. Ohya, Y.; Kobayashi, H.; Ouchi, T. *Reactive Polymers* **1991**, 15, 153.
7. Mosmann, T. *J. Immunol. Methods* **1983**, 65, 55.
8. Wong, T. C.; Townsend, R. R.; Lee, Y. C. *Carbohydr. Res.* **1987**, 170, 27.

RECEIVED September 11, 1992

INDEXES

Author Index

Affiliation Index

Subject Index

Production: Peggy D. Smith
Indexing: Deborah H. Steiner
Acquisition: Anne Wilson
Cover design: Sue Schafer

Printed and bound by Book Crafters

163699